South America's Natural Wonders

This book guides readers through the most iconic, geologically significant scenery in South America, points out features of interest, and describes how these features came to be. Starting in the glacial landscapes of southern Patagonia, this field trip guidebook examines the foothills of the Andes of western Argentina to understand its foreland deformation. Across the Andes, one observes deformation, volcanism, and mineral deposits associated with an onshore volcanic arc and uplift in the Atacama Desert of Chile. A transect across the Andes from Mendoza to Valparaíso follows in the footsteps of Darwin and, as an added bonus, explores the premier wine country around Mendoza, Argentina, and the Colchagua Valley, Chile.

Features:

- Clearly explains the geology of regions with an emphasis on landscape formation.
- Lavishly illustrated with numerous colorful maps, diagrams, and photos of breathtaking landscapes and their geological features.
- Describes the major geologic features of South America through the device of a geologic tour, making it an accessible read for those without any geologic training, as well as for professionals.
- Written in easy-to-understand language, the author brings his own experience to readers who want to explore and understand geologic sites first-hand.

South America's Natural Wonders is an inviting text that gives individuals with no background in geology the opportunity to understand key geologic aspects of local landscapes. It also serves as a guide to undergraduate and graduate-level students taking courses in earth science programs, such as geology, geophysics, geochemistry, mining engineering, and petroleum engineering. Teachers of these courses can also use this book to better understand their local geologic environment and geography.

Geologic Tours of the World
Gary L. Prost

Published Titles

North America's Natural Wonders
Appalachians, Colorado Rockies, Austin-Big Bend Country, Sierra Madre
Gary Prost

North America's Natural Wonders
Canadian Rockies, California, The Southwest, Great Basin, Tetons-Yellowstone Country
Gary Prost

The United Kingdom's Natural Wonders
Scotland and Northern Ireland, Lake District and Yorkshire Dales, Wales and West Midlands, England
Gary Prost

South America's Natural Wonders
Patagonia, Neuquén Basin, Atacama Desert, and Across the Andes
Gary Prost

South America's Natural Wonders

Patagonia, Neuquén Basin, Atacama Desert, and Across the Andes

Gary L. Prost

CRC Press
Taylor & Francis Group
Boca Raton London New York

CRC Press is an imprint of the
Taylor & Francis Group, an **Informa** business

Designed cover image: © Gary Prost. Front Cover: Road to El Chaltén, Argentina, and Mt. Fitz Roy. Back Cover: Sunrise at Torres del Paine, Chile.

First edition published 2024
by CRC Press
2385 NW Executive Center Drive, Suite 320, Boca Raton FL 33431

and by CRC Press
4 Park Square, Milton Park, Abingdon, Oxon, OX14 4RN

CRC Press is an imprint of Taylor & Francis Group, LLC

ISBN: 978-1-032-67006-5 (hbk)
ISBN: 978-0-8153-4804-7 (pbk)
ISBN: 978-1-351-16828-1 (ebk)

DOI: 10.1201/9781351168281

Typeset in Palatino
by Deanta Global Publishing Services, Chennai, India

To my wife Nancy, who lets me entertain my passion for geology, and

my good friend and traveling companion, Guy Peasley.

Contents

Preface

As a professional geologist, I have worked in many parts of the world. While on business trips I often wished I could take more time to just go and look at local geology. Sometimes a co-worker or partner would take me out into the field, or there would be a professional field trip I could take advantage of. Most of the time, however, the demands of business meant there was no time to investigate on my own.

When I retired, I was determined to remedy this situation and explore areas with spectacular geology. It was not easy to find guidebooks that explained even the better-known geologic wonders. So, I decided to pull the information together into a series of field guides that anyone could use.

The purpose of this field trip guidebook, then, is to inspire and inform. Inspire you to think, "I want to go there and see that." To inform you about what you are seeing, about the geologic history, and the human story. To answer the question, "What are those rocks, and how old are they? What is the story behind those towering volcanoes? Why is this the driest desert in the world? Why is this area so rich in oil or copper? Where can I best see giant condors? Why is the wine so wonderful?" Together we will unlock the secrets behind the beauty of the landscape. In terms anyone can understand, I explain what you are looking at, how it came to be, and why it is important. And yet, there should also be enough information to keep a geologist interested. Along the way, historical context is provided, a word or two is said about plants and animals, and you will learn where to find interesting minerals and fossils.

This geological guidebook is a collection of excursions across some of the best-known and least-known natural wonders in South America. Obviously, many important locations will have been left out because you just can't include everything. The transects and stops described in this volume were chosen for their classic beauty, abundance of geologically important sites, ease of access, and safety. Among other considerations, I do not want anyone to venture into areas that are not safe. For example, in early 2023 US State Department warnings kept me from adding a tour from Macchu Picchu to Lake Titicaca. I would like to include this and others in future volumes.

Acknowledgments

The assistance and good company of Guy Peasley was instrumental in allowing me to complete the fieldwork on these tours and is gratefully acknowledged. In addition to appreciating a good wine, he is a great navigator and lets me know when a stop is not safe.

About the Author

Gary L. Prost obtained his BSc in geology from Northern Arizona University in 1973 and an MSc (1975) and PhD (1986) in geology at Colorado School of Mines. Over the past 45 years, he has worked for Norandex (mineral exploration), Shell USA (petroleum exploration worldwide), the US Geological Survey (geologic mapping, coal), the Superior Oil Company (mineral and oil exploration), Amoco Production Company (worldwide oil exploration, remote sensing, and structural geology), Gulf Canada (international new ventures), and ConocoPhillips Canada (Canadian-Arctic exploration, gas field development, oil sands development, and reservoir characterization). He spent over 20 years working as a satellite image analyst in the search for hydrocarbons and minerals in more than 30 countries. During this time, he applied structural geology and remote sensing to exploration, development, and environmental projects. The second half of his career was spent working on regional studies, new ventures and frontier exploration, and oil and gas field development. Now retired, his most recent work has been in public outreach, leading field trips and educating the public on topics of geological interest. He is the principal geologist for GL Prost GeoConsulting of El Cerrito, California, and has been a registered professional geologist in Wyoming (United States) and in Alberta and the Northwest Territories (Canada). He has published six books, including *The United Kingdom's Natural Wonders* (Taylor and Francis, 2023), *North America's Natural Wonders (2 volumes;* Taylor and Francis, 2020), *The Geology Companion: Essentials for Understanding the Earth* (Taylor & Francis, 2018), *Remote Sensing for Geoscientists: Image Analysis and Integration* (third edition, Taylor & Francis, 2013), and *the English-Spanish and Spanish-English Glossary of Geoscience Terms* (Taylor & Francis, 1997). He is currently working on his next volume, *Geologic Tours of the World – Europe's Natural Wonders Volume 1, Western Europe.*

Introduction and a Few Words on Traversing South America

The objective of this travel guide is to explain to the curious explorer, the rock hound, the student, and the geologist what they are seeing when they look at the magnificent spires of the Torres del Paine, bathe in the geysers of El Tatio, experience the dry expanse of the Salar de Atacama, or gaze into the mile-deep pit at Chuquicamata. In simple terms, but not talking down, I explain what you are looking at, how it came to be, and why it is significant. In each area, I set the stage by providing geological and historical context. In wine country, we talk about how geology influences the fruit of the vine, and in each area, I say a word or two about the resources of the region.

Information has been provided about the mines in the areas we visit, even though most are not open to public tours at the time of writing (2023). It is unfortunate and, I feel, shortsighted that mining companies have not resumed public tours since the end of the COVID crisis. Despite some historical injustices, these mines have been and should be a source of pride for local communities and the mining companies themselves. They are impressive achievements of technology and labor. Perhaps in the future they will once again provide this service in order to inform the public about the important contributions of the mining industry.

These trips can be run in any direction, but most start and stop where there is easy access (a major town or airport). The transects as laid out can take up to five or six days, but the trips can easily be broken into smaller bits that can take as little time as a few hours. Times and distances are provided, but times are meant as a guide only and will change depending on traffic, weather, and road conditions. Each stop has, in addition to the distance from the previous stop, GPS coordinates in latitude and longitude. Although described stops are provided, there are many more stops that can be made along these routes: Do not feel constrained to use only those in the guidebook.

Please be careful when pulling off the road, especially when there are narrow shoulders or curves. Be aware of traffic at all times. Safety first!

When renting a vehicle, be sure to ask about road conditions. If the recommendation is "don't drive in this area," then DO NOT DRIVE in that area. There is a reason, whether it be bad roads, lack of water/gas stations, or even narcotics trafficking. Be aware that some routes may be closed in winter. They should all be accessible with a standard road car: Four-wheel or all-wheel drive is not necessary for these trips. Find out what import/export documents and insurance papers you will need and what restrictions apply if you are crossing a border in a rented vehicle. Expect to spend at least an hour going through paperwork, having luggage x-rayed, and your vehicle inspected at the border. At the time of writing, fresh food was not allowed across the border.

I would be remiss if I didn't say a few words about driving in and around large cities. Few if any rental cars have automatic transmissions, so know how to drive a manual. Standard driving rules are not followed consistently. I have encountered drivers weaving in and out of lanes at high speed, passing on the right, changing lanes abruptly, neglecting to use turn signals, and trying to be the last one through a red light. Tailgating is common. Drivers tend to ignore "bus and taxi only" lanes.

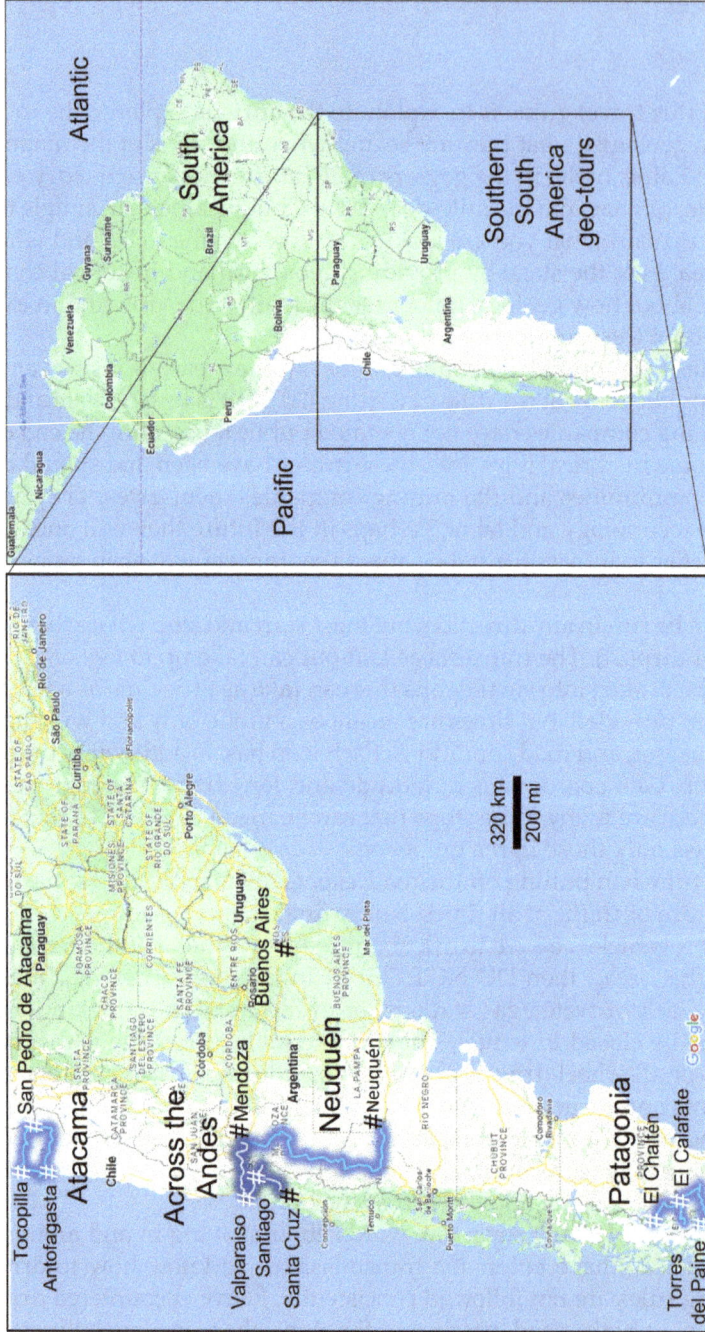

Overview of the geo-tours in this volume.

In some places you cannot turn right at a red light unless specifically permitted. Beware one-way streets: They may be marked with small arrows that are hard to see. Check which way cars are moving and the direction they are parked if you are not sure. Hotels may not have pull-outs for checking in or may not have parking at all. Parking areas are indicated by a large "E" (Estacionamiento = Parking).

Use a GPS or navigation system that adjusts to real-time traffic conditions if you can. I used a Tom Tom loaded with local maps, as well as Google Maps, and Apple Maps, but these last rely on cell phone coverage which can be spotty or absent in remote areas. I always travel with detailed paper maps as backup.

Toll roads are common in some areas. Some require cash only, while others use the automated TAG system that sends you a bill. Some toll booths let you drive through without stopping, while others have a bar that will be raised after you come to a stop and pay the toll.

Most roads are in good condition. Gas stations are well-spaced, and always have someone to pump gas for you.

Avoid driving in large cities if you can. Walk, use a taxi, or take the metro train system.

Some readers will be concerned about language issues. Whereas it is always convenient to speak the language, in this case Spanish, it is not necessary in most places. I try to converse in Spanish whenever possible, but my travel companion is not at all shy about asking questions in English, and usually gets a reply in decent English. Hotels and restaurants almost always have English speakers, and many South Americans are fluent in multiple languages.

Most towns and airports have Automatic Teller Machines that dispense local currency. If they don't, try a bank. Most hotels, restaurants, and gas stations take credit cards. Local markets may not.

Charts are provided to show what rock strata exist in each area, how old the layers are, and what they are made of. Each region is placed in a plate-tectonic context so that you know how and when mountains and basins formed. The geologic influences on our economy are discussed so that you know, for instance, why the wine is so good around Mendoza, and why the Atacama Desert has large deposits of copper, nitrates, and lithium.

Units are provided metric/si first, then their American English equivalent.

Abbreviations are explained, then used. For example, millions of years is abbreviated to Ma (for "mega-annum"), Ga (billions of years), Fm is used for Formation, Gp for Group, mbr for member, and billions of barrels of oil equivalent is BBOE.

Road descriptions, pullouts, travel times, mine and other special tours, contact names, websites, and emails are current as of the time of writing. Contact a park or museum website for hours and current entry fees.

A geologic time scale is provided here as a useful reference for those not familiar with the different geologic periods referred to in the trip descriptions. Age in millions of years is shown on the right.

EON	ERA	PERIOD	EPOCH		Ma
Phanerozoic	Cenozoic	Quaternary	Holocene		0.01
			Pleistocene	Late	0.8
				Early	1.8
		Tertiary (Neogene)	Pliocene	Late	3.6
				Early	5.3
			Miocene	Late	11.2
				Middle	16.4
				Early	23.7
		Tertiary (Paleogene)	Oligocene	Late	28.5
				Early	33.7
			Eocene	Late	41.3
				Middle	49.0
				Early	54.8
			Paleocene	Late	61.0
				Early	65.0
	Mesozoic	Cretaceous	Late		99.0
			Early		144
		Jurassic	Late		159
			Middle		180
			Early		206
		Triassic	Late		227
			Middle		242
			Early		248
	Paleozoic	Permian	Late		256
			Early		290
		Pennsylvanian			323
		Mississippian			354
		Devonian	Late		370
			Middle		391
			Early		417
		Silurian	Late		423
			Early		443
		Ordovician	Late		458
			Middle		470
			Early		490
		Cambrian	D		500
			C		512
			B		520
			A		543
Precambrian	Proterozoic	Late			900
		Middle			1600
		Early			2500
	Archean	Late			3000
		Middle			3400
		Early			3800?

Geologic time scale. Ma indicates millions of years ago. Mississippian and Pennsylvanian are often combined and referred to as the Carboniferous. From the U.S. Geological Survey.

1

Patagonia: Torres del Paine, Chile, to Fitz Roy Massif, Argentina and Chile

Sunrise at Torres del Paine is awe-inspiring.

Overview

This trip takes us to some of the most iconic geologic wonders of the world, including the towering spires and turquoise lakes of Torres del Paine, the sea of ice known as Perito Moreno, and the Fitz Roy Massif that symbolizes Patagonia. This guide provides an overview of the rocks and structure of the southern Andean Thrust Belt and foreland basin between Torres del Paine National Park in southern Chile and the Fitz Roy Massif in Los Glaciares National Park, Argentina. We examine massive intrusions, typical thrust-belt structures, and the sedimentation history of the World Heritage site at Torres del Paine National Park. Classic turbidite deposits are seen near Torres del Paine and Perito Moreno, and Perito Moreno is the vanishingly rare glacier that is not in retreat. The valley glaciers and moraines are extensions of the Andean ice field, the fourth largest after Antarctica, Greenland, and Iceland. We pass through landscapes carved by glaciers and inhabited by the descendants of Gondwana and examine how the Andean fold-thrust belt evolved along with the adjacent sedimentary basins.

DOI: 10.1201/9781351168281-1

Overview of the Patagonia geo-tour.

Itinerary

Begin either in Puerto Natales (Chile) or El Calafate (Argentina)

    Stop 1 Torres del Paine National Park
    Stop 1.1 Lago Nordenskjöld Overlook
    Stop 1.2 Cerro Toro Turbidite Sand Channel
    Stop 1.3 Mirador Salto Grande
    Stop 1.4 Mirador Cuernos
    Stop 1.5 Mirador Condor
    Stop 1.6 Lago Grey
    Stop 1.7 Mirador Río Serrano
    Stop 1.8 Mirador Lago del Toro
    Stop 2 Frontal Folds of Main Ranges
    Stop 3 Cerro Castillo
    Stop 3.1 Deformed Cerro Toro Formation, Cerro Castillo
    Stop 3.2 Cerro Castillo Overview
    Stop 4 Dorotea Formation Cuesta
    Stop 5 Cerro Calafate, Lago Argentino, and the Glaciarium

A Brief History of the Southern Andes

Tectonics

The Andean Cordillera is the result of horizontal shortening and magmatism since Late Cretaceous. From west to east, the cordillera can be divided into four zones:

1) A western coastal belt, where a highly deformed, accretionary prism complex of late Paleozoic metasedimentary rocks are intruded by the granitic Patagonian Batholith.

2) The main cordillera, where late Paleozoic sedimentary and metamorphic rocks, Jurassic volcanics, and Mesozoic to Miocene granitic rocks of the Patagonian Batholith are deformed by thick-skinned, high-angle reverse and mainly east-directed thrust deformation.

3) The Patagonian fold-thrust belt, where late Paleozoic sedimentary and metamorphic rocks, Jurassic volcanics, and Early Cretaceous to Tertiary sediments are thrust eastward.

4) The Magallanes (or Austral) foreland or back-arc basin, containing up to 8,000 m of gently deformed to undisturbed Early Cretaceous to Tertiary sediments (Gorring, 2008).

We will be touring Patagonia through the third and fourth of these zones.

The area that is now the southern Patagonian Andes was at one time on the southwest margin of Gondwana, the supercontinent that later broke up into South America, Africa, India, Australia, and Antarctica. While still part of Gondwana it was subject to the late Paleozoic Gondwanan Orogeny. Late Paleozoic (Permian?) structures include an early phase consisting of isoclinal (tight) folding and overturned west-northwest-trending folds and thrusts. A later deformation phase includes north-northeast-oriented open folds (Giacosa et al., 2012). Late Paleozoic basement rocks include the low-grade metasedimentary rocks of the Río Lacteo Formation, mainly quartzites, mica schists, and phyllites with minor greenstones and marbles, and the thick sedimentary sequence of the Bahía de La Lancha Formation, mainly graywacke, quartz sandstone, and shale up to 2,000 m (6,560 ft) thick. Deformation and metamorphism of these units are thought to be related to the collision of exotic terranes with the Pacific margin of Gondwana during the late Paleozoic Gondwanan Orogeny (Gorring, 2008).

The margin of Gondwana that would later become Patagonia experienced widespread Triassic to Middle Jurassic extension prior to the breakup of the supercontinent. Basins such as the north-northwest-oriented Rocas Verdes rift basin are filled with massive rhyolitic volcanics. These Middle-to-Upper Jurassic Chon Aike volcanics can be greater than 2,000 m (6,560 ft) thick. Locally they are known as the Tobifera Formation, the El Quemado Complex, and the Ibáñez Formation (Gorring, 2008).

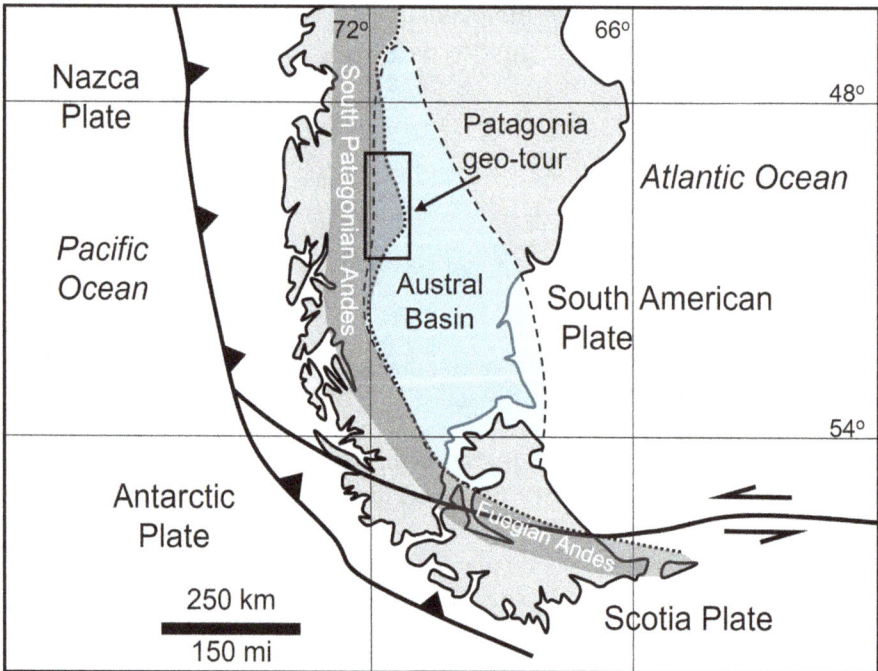

Plate tectonic setting of the Andean orogeny. Modified after Fosdick et al., 2011; Giacosa et al., 2012.

The Andean Orogeny began in Late Cretaceous time as a result of subduction along the western margin of South America. Intense compression began about 75 Ma, causing the Andean thrust front to migrate eastward. The Patagonian Cordillera was uplifted, most of the Patagonian Batholith was emplaced, and the Magallanes back-arc basin developed. Erosion of the Chon Aike volcanics provided sand for fluvial and marginal-marine deposits that formed the Springhill Formation, the oldest unit in the Magallanes Basin. The Springhill Formation is the primary oil and gas reservoir in the Magallanes Basin. This was followed by the deposition of the Río Mayer and Río Belgrano formations, a thick sequence of Early Cretaceous black marine shales. The shales, deposited across the basin in an anoxic environment, are the main potential hydrocarbon source rock. These were in turn overlain by the Late Cretaceous Punta Barrosa and Cerro Toro formations, deep-marine turbidite deposits separated by thick sections of dark shale. Turbidites are the result of turbidity currents, rapid downslope flows of sediment-laden water. After the flow stops in the deep basin, the sediments settle out in layers with coarse-grained sand at the base and fine mud on top. They often occur in sequences of multiple flow deposits. Turbidity currents are caused by the collapse of unstable slope deposits.

Uplift and deformation in the Cordillera is marked by a change to shallow marine coarse clastic sedimentation (sandstones and conglomerates) in the Magallanes Basin (e.g., the La Anita, Cerro Fortaleza, and Dorotea formations; Gorring, 2008).

Eocene and Miocene Andean east-directed shortening resulted in a north-south-oriented fold-thrust belt. Deformation is characterized by north-south to north-north-east-trending, east-vergent folds, reverse faults, and minor thrusts. Some areas contain west-vergent reverse faults and north-northeast-oriented open folds. Other areas have tight folds with box and kink geometries. Deformation becomes less intense as one proceeds eastward, where gentle folds deform the Upper Cretaceous sediments (Giacosa et al., 2012).

Uplift and deformation in the Patagonian Cordillera coincided with subsidence of the western Magallanes Basin and deposition of up to 5,000 m (16,400 ft) of sediment in the deepest part of the basin. This foreland basin sedimentation was dominated by continental and shallow-marine sediments eroded off the growing Andes. Angular unconformities and growth strata (units that thicken against faults) indicate active and ongoing deformation (Gorring, 2008). Sedimentary units during the foreland basin stage were transported south and east across the basin. (Giacosa et al., 2012).

Age data from folded gabbros suggest an Oligocene age for regional folding at 29.4 Ma. The east-northeast to east-west shortening direction was caused by convergence between the Nazca and the South American plates (Altenberger et al., 2003). Intense Neogene uplift and deformation occurred in the southern Cordillera starting around 25 Ma. The collision of the Chile Ridge spreading center with South America began around 15 Ma and may be responsible for another episode of intense deformation. This ridge collision caused arc magmatism to shut down and shift eastward. Neogene uplift may have kick-started glaciation in the southern Cordillera around 6 Ma (Gorring, 2008).

There were two major compressional deformation phases of the Andean orogeny, a deformational phase in the Eocene (Incaic, 56–34 Ma) and another in the late Miocene to early Pliocene (Quechua, 11.6–3.6 Ma). Rapid uplift and denudation of the Incaic phase started along the Pacific coast and migrated 200 km (120 mi) eastward until the Quechua phase. Radiometric ages constrain the age of the more pervasive deformation and uplift to before 8.6 Ma (Giacosa et al., 2012). Episodes of intense deformation have been linked to periods of rapid plate convergence.

Emplacement of granitic plutons occurred between 18 and 3 Ma. The Cerro Fitz Roy granitic pluton is 18 ± 3 Ma; the El Chaltén Adakite (silicic magma derived from partial melting of basalt in subduction zones) intrusion is dated at 14 Ma; the Torres del Paine granitic pluton is 13 to 12 Ma. The relatively young age and present elevation of these plutons indicate significant uplift and erosion since emplacement (Baumgartner, 2006; Gorring, 2008).

The spectacular Torres del Paine massif may be the best-exposed example of a laccolith in the world. A laccolith begins as a sill, a shallow magmatic injection parallel to bedding. As the injection pressure ramps up the sill begins to expand, lifting and arching the overlying sediments. The Torres del Paine laccolith eventually grew to 10 by 20 km (6 by 12 mi) wide and up to 2,000 m (6,560 ft) thick. Heat from the intrusion metamorphosed the surrounding sedimentary rocks, creating a contact metamorphic aureole. Sedimentary rock exposed at the Torres is mainly the dark, metamorphosed Cerro Toro Formation both above and below the intrusion. The Cretaceous Cerro Toro Formation was originally deposited as deep marine turbidite sandstones and conglomerates encased in dark shale in the Magallanes Basin. The intrusion consists of an early mafic (dark) unit and a later granitic (light) phase. Metasediments exposed above the granite intrusion are "roof pendants." Dark blocks of metasediment that broke off and began to sink can be seen embedded in the white granite (Baumgartner, 2006; Leuthold et al., 2014; Galland and Sassier, 2015; Abbott and Cook, 2016).

Basin Evolution and Stratigraphy

The Jurassic to Neogene basin evolution of southern Patagonia consists of two phases, an early back-arc rift phase (the predecessor Rocas Verdes Basin) and a later retro-arc/foreland-basin phase (Magallanes-Austral Basin). The southern Patagonian Andes region was characterized by back-arc extension during Jurassic to Early Cretaceous time. Extension resulted in bimodal volcanism including the basalt of the Sarmiento ophiolite complex and widespread silicic volcanism of the Jurassic Chon Aike volcanics (El Quemado, Ibañez, and Tobifera Formations). The resulting syn-rift Rocas Verdes Basin was filled by Chon Aike volcanics and by black shale of the Río Mayer and Zapata Formations.

The transition from back-arc extension to compression is recorded by the deepening of the basin axis from 100–500 m (330–1,640 ft) water depth during mid-Cretaceous time to up to 1,000–2,000 m (3,280–6,560 ft) by Late Cretaceous time. Then, as now, a subduction zone and trench lay west of the Chilean coast. Mountain building loaded the South American Plate on the east side of the Andes, creating a deep-marine foreland basin. Up to 4,000 m (13,000 ft) of sediments were deposited during roughly 30 million years of basin filling. The stratigraphic succession consists of three deep-water formations capped by a deltaic/ shallow-marine unit. The foreland-basin phase begins with deep-marine sandstone deposition in the 92-million-year-old Punta Barrosa Formation. Submarine landslides delivered sand and mud to the basin, gradually depositing a series of turbidite deposits. The 1,000 m- (3,300 ft) thick Punta Barrosa Formation consists of interbedded sandstone and mudstone. Paleocurrent direction was mainly toward the south. Because it is adjacent to the Andes, the Punta Barrosa Formation is pervasively deformed (see, for example, the Lago Grey stop).

Deep-marine sedimentation continued during the Late Cretaceous. Deepening of the foreland basin along with thrusting and uplift in the Andes resulted in the 80-million-year-old Cerro Toro Formation deep-marine channel system that delivered turbidites and filled channels with coarse sand and conglomerate from north to south and west to east. Thinly bedded shale and sandstone comprise most of the western Cerro Toro Formation.

Eastern exposures have thicker conglomerates (Abbott and Cook, 2016). Turbidites may have been mobilized by Upper Cretaceous earthquakes and ran out along the ocean floor to this deep marine basin (Ranney, 2016). The Cerro Toro Formation accumulated within a 4–5 km (2.5–3.1 mi) wide channel belt that occupied the axis of the Magallanes Basin during Late Cretaceous. The north-south elongated basin was the foredeep to the fold-thrust belt. The Andes were actively uplifting and were the source of the coarse-grained channel fill in the Cerro Toro Formation (Hubbard et al., 2007).

This, in turn, is overlain by the Tres Pasos Formation, which records the southward deposition of deep-marine slope deposits. Channel fill was deposited on a high-relief slope, and the fill is mostly sandy (McCauley and Hubbard, 2013). Tres Pasos slope channel deposits are 6–15 m (20–50 ft) thick and composed of stacked turbidites. Channels are roughly 300 m (1,000 ft) wide and typically 25–70 m (82–230 ft) thick. The oldest channels are widely spaced, reflecting a broad basin center. Later channels began to stack and amalgamate (merge together) as the basin became more confined. Out-of-channel deposits are mainly fine-grained shales (McCauley and Hubbard, 2013).

Ultimately, deep-marine sediments merge upward into shallow, marginal-marine deposits of the Late Cretaceous-Paleogene Dorotea Formation (Fildani et al., 2007; Romans et al., 2011; Malkowski et al., 2015; Abbott and Cook, 2016).

Glaciation

The southern Andes have been carved by glaciers into some of the most impressive and stunning landscapes on Earth. Deep U-shaped valleys, breathtaking spires, and glacial lakes surround you. The Andean Ice Cap, the largest outside of Antarctica, Greenland, and Iceland, spawned 47 glaciers, 13 of which flow towards the Atlantic. Perito Moreno Glacier is one of the few glaciers in the world that is advancing: It is in a state of dynamic equilibrium, with the rate of calving generally matching the rate of advance, about 2 m (6.5 ft) per day.

Brilliant blue-green icebergs float in glacial Lago Grey and Lago Argentino. The glacier-polished and striated Punta Barrosa Formation can be seen on the banks of Lago Grey. In 2014, the retreating Tyndall Glacier in Torres del Paine exposed a world-class dinosaur graveyard containing 46 nearly complete ichthyosaur skeletons.

Human History

Archaeology indicates that this region was inhabited by indigenous groups as early as 12,000 to 14,000 years ago. Several caves appear to have been occupied continuously since 10,000 BC. These were nomadic hunters of guanacos and the rhea. Several groups lived scattered across Patagonia and the southern Andes. Towards the end of the 1500s, Mapuche-speaking agriculturalists entered the area and spread across into the eastern plains and to the far south. Through confrontation and technological ability, they quickly dominated the earlier Tehuelche peoples of the region. The Mapuche are the main indigenous community today.

When Ferdinand Magellan first landed on the Patagonian coast in 1520, he encountered large native footprints. This led him to name the area Patagones, or "big feet." He encountered Tehuelche natives, taller than average Europeans of that time, and reported that the land was inhabited by giants. Later explorers repeated these stories, and for the next 200 years, it was commonly believed to be a land of giants.

In the mid-1800s Argentina and Chile began aggressively settling Patagonia. In 1860, the French adventurer Orelie-Antoine de Tounens proclaimed himself king of the Kingdom of Araucanía and Patagonia of the Mapuche. Tounens was arrested by the Chilean authorities on January 5, 1862, imprisoned, declared insane, and later expelled to France.

Argentina became concerned that the strong connections between indigenous tribes and Chile would give Chile influence over the pampas. They feared an eventual war in which the natives would side with the Chileans. The Conquest of the Desert program was probably triggered by the 1872 attack of 6,000 natives on the towns of General Alvear, Veinticinco de Mayo, and Nueve de Julio, where 300 inhabitants were killed, and 200,000 cattle taken. The Conquest of the Desert was a campaign by the Argentine government, led by General Julio Argentino Roca, to subdue or exterminate the native peoples of the south. It was largely successful.

Europeans only explored the southern Cordillera for the first time in the 1870s. Then German and British (mainly Welch) colonists settled the area, founding large sheep and cattle farms. The main industries today are energy, livestock, and tourism.

Stop 1 Torres del Paine National Park

Paine means "blue" in the native Tehuelche language and is pronounced "PIE-nay." The park, established in 1959 as Parque Nacional de Turismo Lago Grey (Grey Lake National Tourism Park), became Torres del Paine National Park in 1970. It was designated a UNESCO World Reserve of Biosphere in 1978 (Galland and Sassier, 2015). The park averages around 252,000 visitors a year.

The park centers on the peaks known as the Torres, or Towers. The highest summit of the range is Cerro Paine Grande. In 2011 it was ascended, measured using GPS, and found to be 2,884 m (9,462 ft).

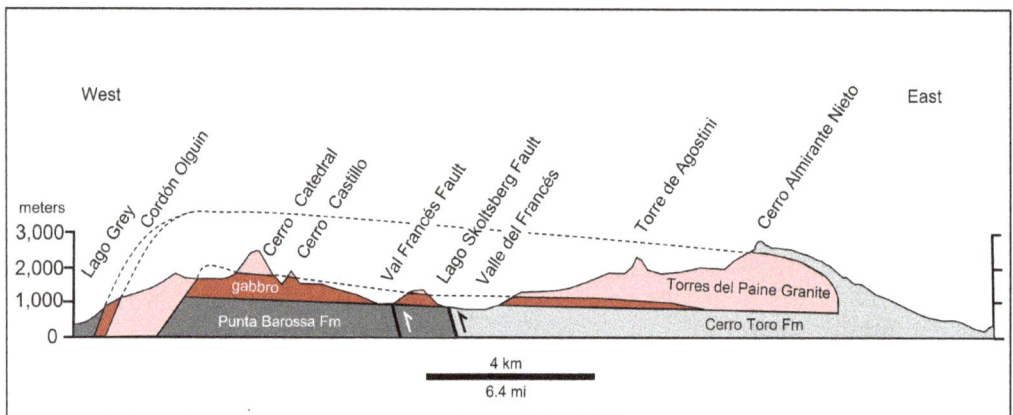

West-east cross-section through the Torres del Paine laccolith. Modified after Altenberger et al., 2003.

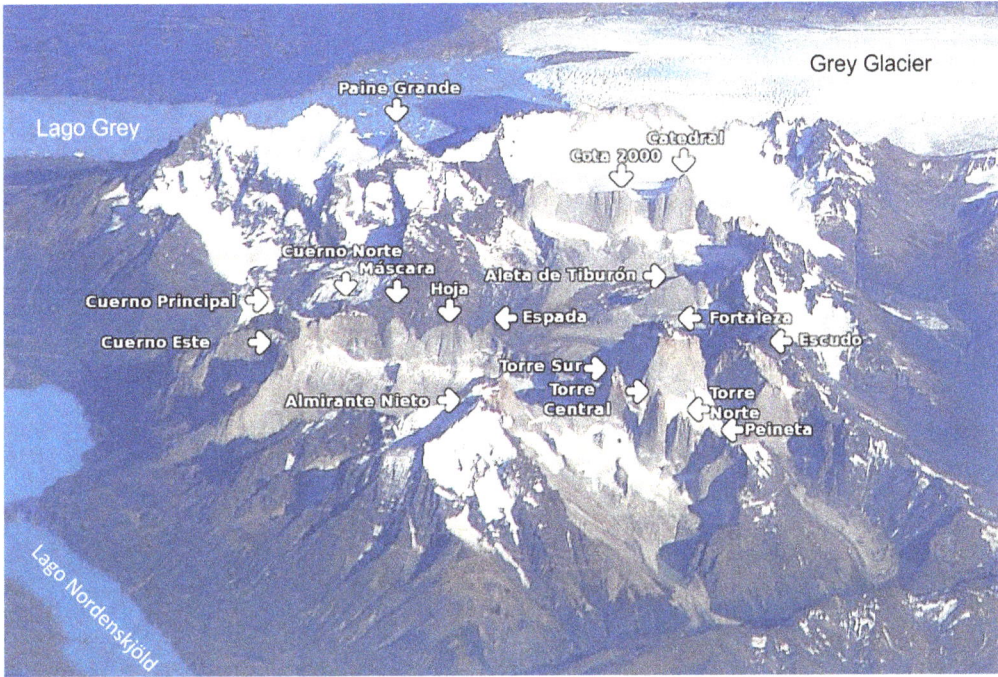

NASA satellite image looking southwest over Torres del Paine. https://commons.wikimedia.org/wiki/File
:Cordillera_del_Paine_annotated.jpg.

In 1976, British mountaineer John Gardner and two Torres del Paine rangers, Pepe
Alarcon, and Oscar Guineo pioneered the trail which circles the Paine massif. A certified
guide is required to access some parts of the park. Hiking and camping arrangements
should be made prior to entering the park. Hotel and hostel space is also limited and
should be arranged in advance.

Guido Monzino was an Italian mountaineer who climbed many of the world's most
challenging peaks. In 1957–58 he, along with Pierino Pession, Camilo Pelissier, Leonard
Carrel, and Jan Bich completed the first ascent of the North Tower (about 2,260 m, or 7,415
ft), the first of the three towers to be climbed successfully. In 1977, he donated 12,000 ha to
the Chilean government to help create the national park. The tower is now named after
him.

Chris Bonnington, a British mountaineer, and fellow Briton Don Whillans were the first
to ascend the Central Tower (about 2,460 m or 8,100 ft) in 1963. In 2017, three Belgian climb-
ers, Nico Favresse, Siebe Vanhee, and Sean Villanueva O'Driscoll, made the first free ascent
up the rock face (about 1,200 m or 4000 ft). The South Tower of Paine, about 2,500 m (8,200
ft), was first climbed by the Italian alpinist Armando Aste in the same year.

Visit

Visiting the parks is recommended between September and April, during the southern
spring, summer, and early autumn. During summer, daylight hours are long, given the

southern latitude. But the wind is always blowing, making it comfortable to cool. During southern winter the weather becomes extreme.

The park can be reached by driving on Chile Route 9, which is paved and connects Punta Arenas and Puerto Natales and continues as an asphalt road for 100 km (62 mi) before becoming a gravel road. Tire chains are recommended during winter. The park can also be reached by bus from Puerto Natales (Wikipedia). Alternatively, the park can be reached by car from El Calafate in Argentina by driving south on (mostly) paved RN-40 to the Chilean border, then Y-200 and Y-290 to the park.

Our tour begins in Torres del Paine National Park at Lago Nordenskjöld. This stop is 100 km (62 mi; 1 hr 50 min) north of Puerto Natales airport and 347 km (215 mi; 4 hr 30 min) south of the El Calafate airport by way of La Esperanza. We are in the eastern foothills of the Andean fold-thrust belt and on the margin of the Torres del Paine laccolith.

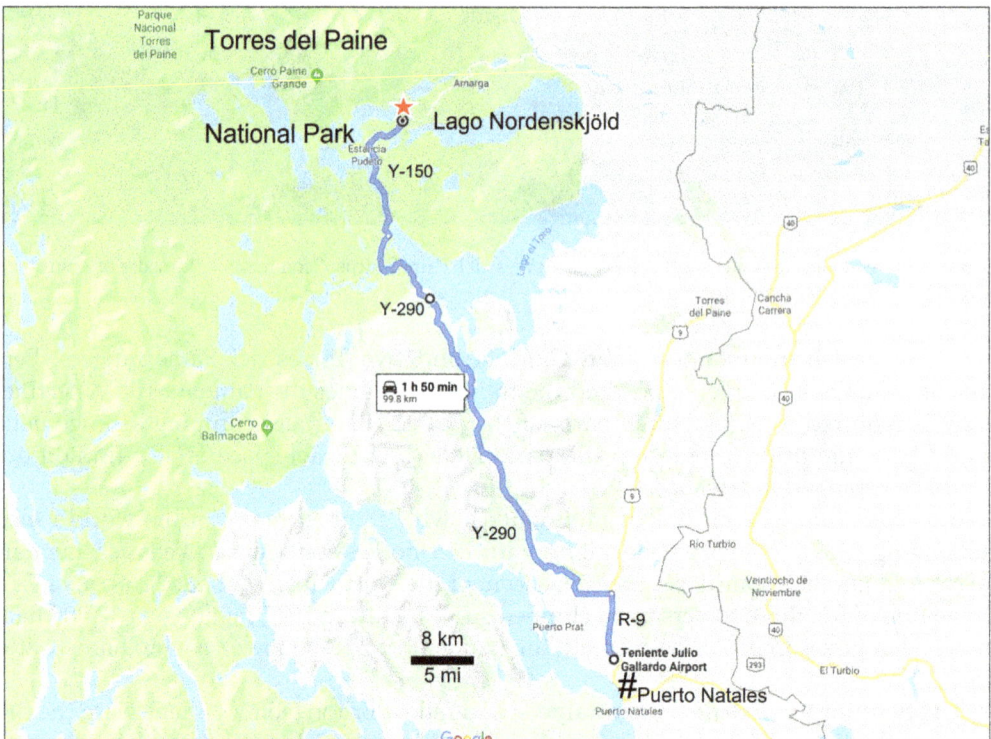

Map to Torres del Paine from Puerto Natales airport.

Begin Puerto Natales to Lago Nordenskjöld: *Take Ruta-9 north from Puerto Natales; after 17 km (10.5 mi) turn left (west) onto Y-290, a well-graded gravel road; at 87.5 km (54.4 mi) Y-290 turns right and becomes Y-150; continue north and east on Y-150 to km 108 (67.1 mi; 1 hr 50 min) from the start. This is* **Stop 1.1, Lago Nordenskjöld Overlook** *(–51.04385, –72.93213). The roads here are partly paved and partly graded gravel. The paved portions contain large potholes, so drive carefully.*

Map to Torres del Paine from El Calafate airport.

Begin El Calafate to Lago Nordenskjöld: Drive east out of town on RP-11, a nicely paved road; continue straight onto RN-40; continue straight onto RP-5 toward La Esperanza (this road is paved; the shortcut on R-40 is not well-maintained gravel); in La Esperanza circle around and turn left (southwest) onto RP-7; at Estancia Tapi Aike turn left (southwest) onto R-40; turn right (west) at sign to "Cancha Carrera 6" and drive on a graded gravel road to the Argentine Customs station at the border; pass through Argentine Customs, cross the border into Chile, and continue straight on Y-205 toward Cerro Castillo; at the roundabout take the second exit onto R-9 West; continue straight onto Y-150; stay on Y-150 north and west to **Stop 1.1, Lago Nordenskjöld Overlook** *(−51.04385, −72.93213). The roads here are partly paved and partly graded gravel. The paved portions contain large potholes, so drive carefully.*

Stop 1.1 Lago Nordenskjöld Overlook

This stop is in the turbidites of the Upper Cretaceous Cerro Toro Formation on the east flank of the north-south-trending Silla Syncline. The parking lot has great exposures of west-dipping graded beds typical of turbidite deposits, that is, repeating layers of thin-bedded sandstone fining upward to shale.

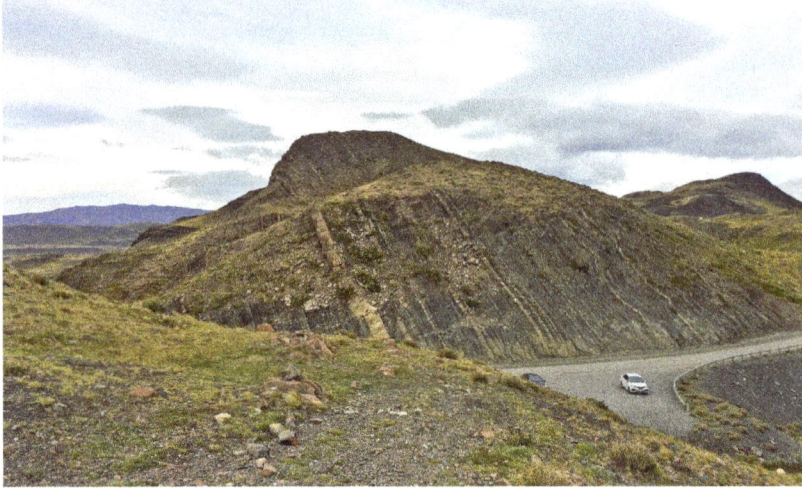

View south across a parking lot at west-dipping thin-bedded turbidites of the Cerro Toro Formation.

Looking west across the lake in the foreground you will get fine views of northwest-dipping beds near the center of the Silla Syncline. In the background are the towering spires of the Cuernos del Paine (Horns of Paine).

The turquoise color of the lake is a result of glacial flour, a rock powder formed by glaciers grinding against the bedrock. The suspended rock flour scatters blue-green light, creating the stunning color of the water.

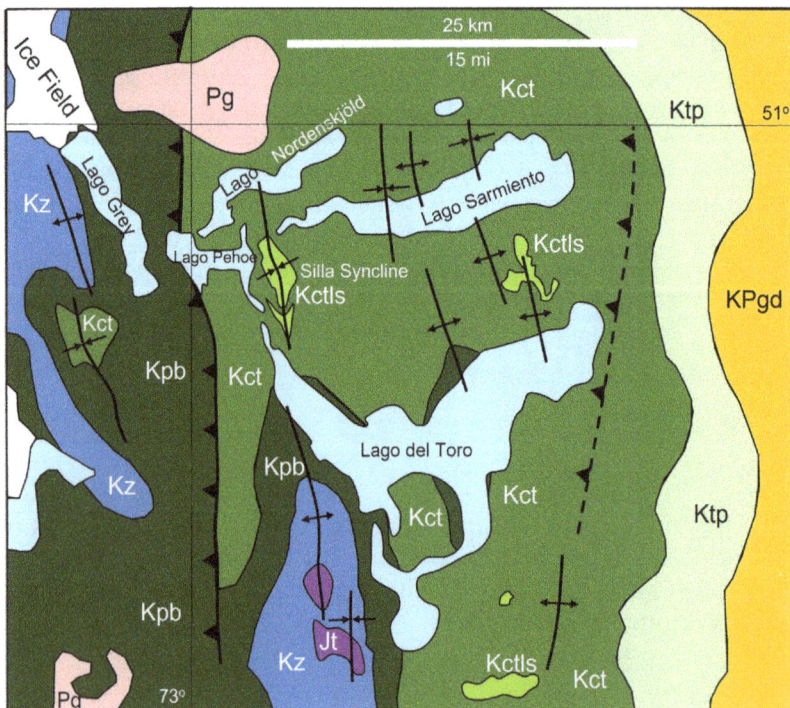

Geologic map of the Torres del Paine area. KPgd = Dorotea Formation; Ktp = Tres Pasos Formation; Kct = Cerro Toro Formation; Kctls = Lago Sofia Conglomerate, Cerro Toro Formation; Kpb = Punta Barrosa Formation; Kz = Zapata Formation; Jt = Tobifera Formation. Modified after Romans et al. (2011).

Age		Stratigraphic Unit	Tectonics
Pg		**Dorotea Fm** Shallow marine, delta, & shoreline sandstone	
Cretaceous	**Upper**	**Tres Pasos Fm** Deep marine slope mud & channel deposits	**Compression** Retro-arc & Foreland Basin
		Cerro Toro Fm Deep marine black shale & turbidites	
		Lago Sofia Conglomerate	
		Cerro Toro Fm	
		Punta Barrosa Fm Deep marine turbidites	
	Lower	**Zapata Fm** Deep marine black shale	**Extension** Rifting & Backarc Basin
Jurassic	**Upper**	**Sarmiento Ophiolite** **Tobifera Fm** Silicic Volcanics	

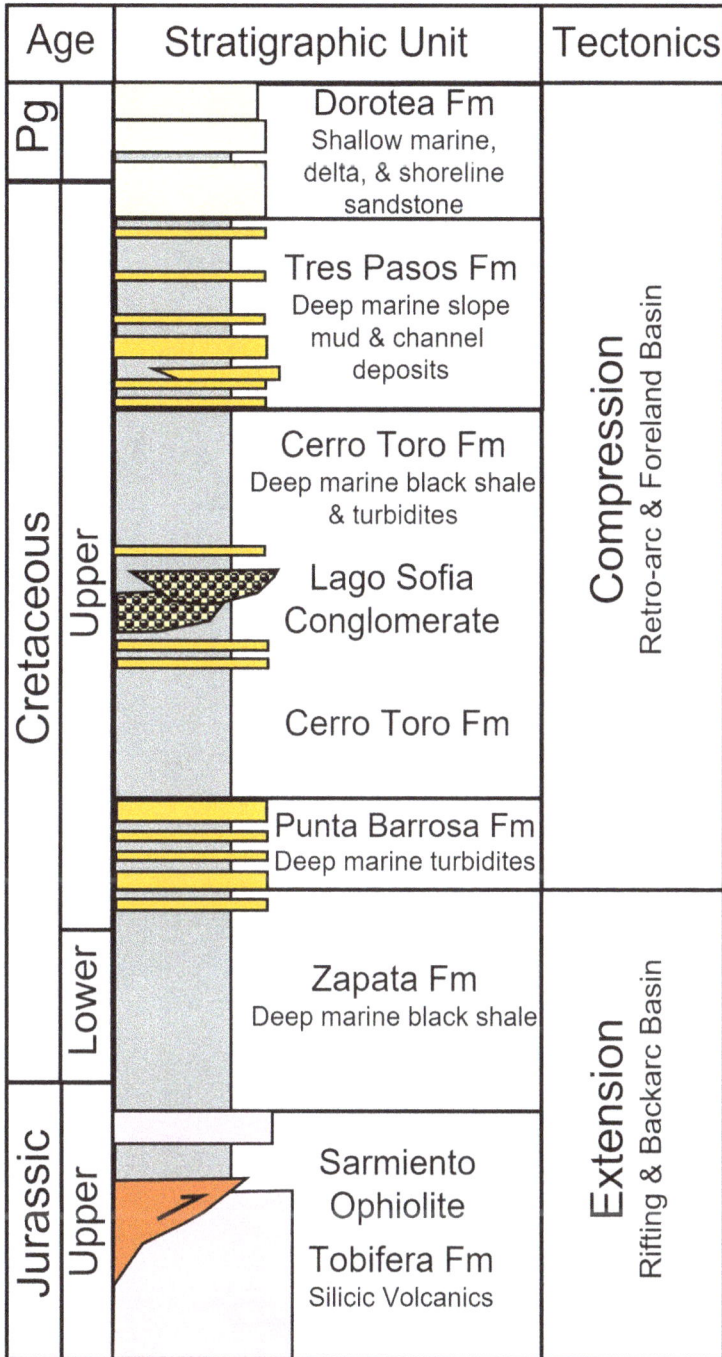

Stratigraphy of the Torres del Paine area, southern Andes and Magallanes-Austral Basin. Stratigraphic information drawn from Romans et al., 2011.

View west across Lago Nordenskjöld at northwest-dipping beds of the Cerro Toro Formation.

> *Lago Nordenskjöld Overlook to Turbidite: Continue west on Y-150 for 3.9 km (2.4 mi; 5 min) and pull over at the abandoned gravel quarry on the north side of the road. This is* **Stop 1.2, Cerro Toro Turbidite Channel** *(−51.06004, −72.97079).*

As you leave Lago Nordenskjöld Overlook and drive west, check out the excellent exposures of the Cerro Toro Formation graded-turbidite beds on the south side of the road 2.3 km (1.4 mi; 3 min) away at (−51.05821, −72.95069). There are pullouts on both sides of the road.

Stop 1.2 Cerro Toro Turbidite Sand Channel

This stop provides an example of a coarse clastic-dominated (mainly sand) channel deposit in the Cerro Toro Formation.

View northeast to channel in the Cerro Toro Formation.

> ***Cerro Toro Turbidite to Mirador Salto Grande****: Continue west on Y-150 to Y-158; turn left (north) on Y-158 and drive to **Stop 1.3, Mirador Salto Grande** parking lot (–51.067323, –73.005787) for a total of 2.9 km (1.8 mi; 4 min). Walk 335 m (1,100 ft) to the overlook.*

Stop 1.3 Mirador Salto Grande

The Salto Grande (Big Waterfall) is a 10 m (33 ft) drop in the Paine River between Lago Nordenskjöld and Lago Pehoe. Looking south toward Lago Pehoe, you will notice a small syncline developed in thin-bedded shales and sandstones of the Cerro Toro Formation. These scenic falls are a result of more resistant layers in the Cerro Toro.

Salto Grande on the Paine River. View to the west.

View south to small syncline developed in the Cerro Toro Formation just below the falls.

Mirador Salto Grande to Mirador Cuernos: *A 2.5 km (1.5 mi each way) trail leads to* ***Stop 1.4, Mirador Cuernos*** *(–51.048278, –73.012293). This is an easy walk and should take about 2 hours round trip.*

Stop 1.4 Mirador Cuernos

This overlook provides a spectacular view of the Cuernos del Paine (Horns of Paine). Notice there are glaciers perched on the peaks and coming down the valleys. Closer examination of the valley walls reveals east-directed thrust faults and east-verging folds.

A pervasive cleavage can be seen in the Cerro Toro shales outcropping at this stop. Cleavage is a type of deformation resulting mostly from high pressure metamorphism.

View north from Mirador Cuernos to the Horns of Paine.

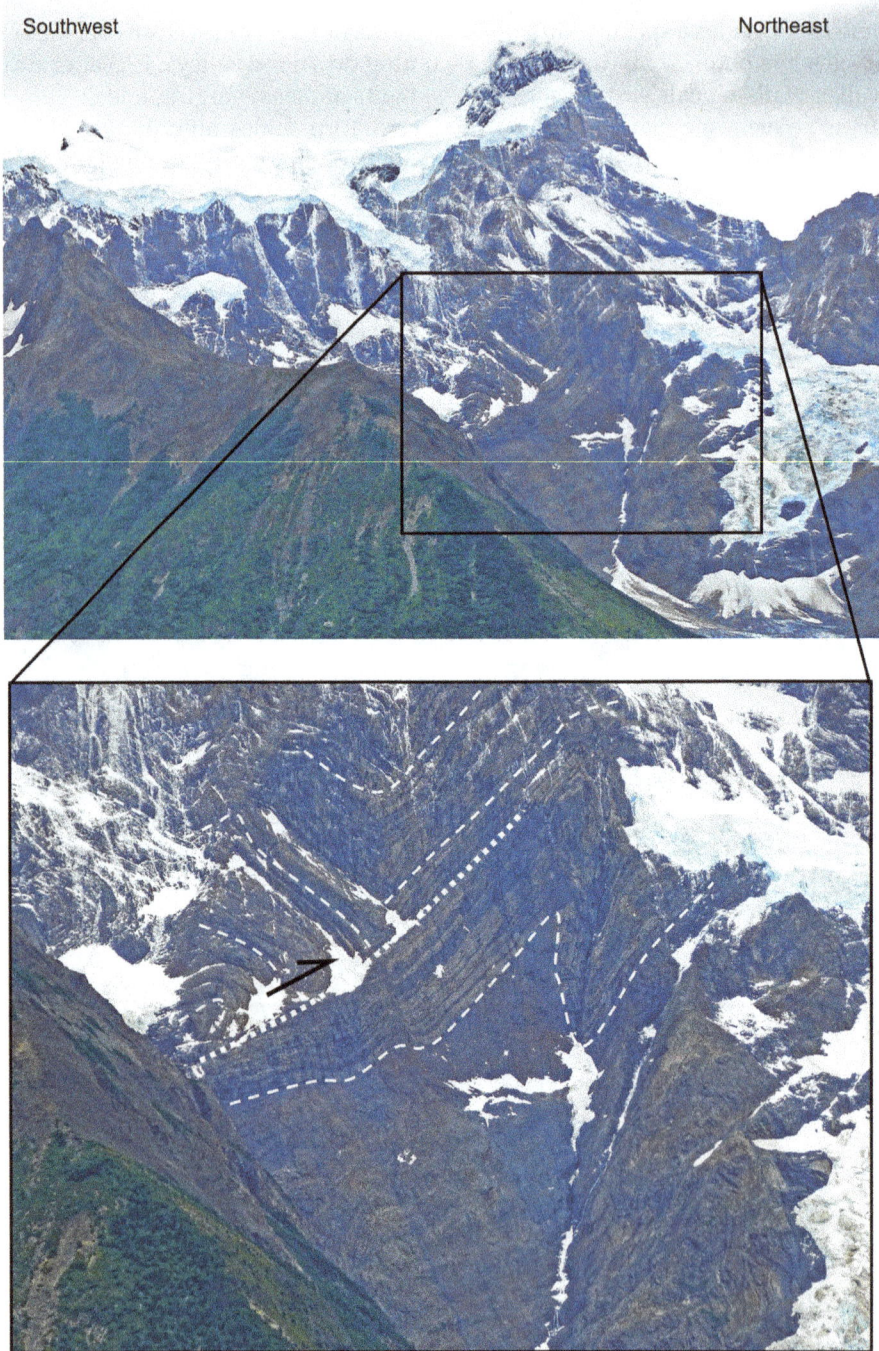

Thrusting and folding exposed on the northeast wall of Cuernos del Paine.

Pencil cleavage (platy aligned fabric) in the Cerro Toro shale. Chilean 100 p coin for scale.

Mirador Cuernos to Mirador Condor: Return east on Y-158 to Y-150; turn right (south) on Y-150 and drive a total of 7.6 km (4.8 mi; 11 min) to Stop 1.5, Mirador Condor (−51.106119, −72.985257), pullout on the left.

Stop 1.5 Mirador Condor

This short but steep hike, about an hour round trip, takes you from the lakeshore to the top of a ridge east of Lago Pehoe. The ridge is capped by the massive Lago Sofia Conglomerate member of the Cerro Toro Formation. Small caves in the ledge are used as nesting areas by condors. Even if you do not happen to see a condor, the views of Lago Pehoe and Cuernos del Paine are magnificent.

The massive Lago Sofia conglomerate channel in the Cerro Toro Formation forms a ledge along the ridge top. Condor caves have white guano streaks below.

> **Mirador Condor to Lago Grey:** *Continue south and then west on Y-150 for 26.7 km (16.6 mi; 36 min) to* **Stop 1.6, Lago Grey** *parking lot (−51.124649, −73.127980).*

Stop 1.6 Lago Grey

Take a moment to look at the outcrop south of the parking lot. A small east-directed thrust can be seen in the folded Upper Cretaceous Punta Barrosa Formation, a thin-bedded turbidite sequence.

Deformed Punta Barrosa Formation on the south side of the Lago Grey parking lot. Dashed lines are the trace of bedding; dotted lines are faults.

A 5 km (3 mi; about a 2-hour walk) round trip from the parking lot takes you over a bridge and across a beach to the Mirador Lago Grey lookout. Looking northwest from here across the lake you can see Grey Glacier, the source of the remarkable turquoise icebergs floating in the lake. The color of the icebergs is a result of air bubbles trapped within the ice. Longer wavelengths of light are absorbed in the ice; the blue and green wavelengths are reflected and refracted, creating the dazzling colored ice.

Looking northeast you get breathtaking views of the Torres del Paine and can clearly see the dark metamorphic roof pendant and bits of the roof rock that foundered and sank into the underlying granite.

View northwest toward Grey Glacier.

Bergy bits bejewel Lago Grey while the Torres clearly show the roof pendant arching over the granite laccolith. Foundered blocks of the dark roof rock can be seen suspended within the granite.

*Lago Grey to Mirador Río Serrano: Return east on Y-150 to Y-290; turn right (south) on Y-290 and drive past the park entrance; cross the Río Serrano and take the next road on the right (south) to **Stop 1.7, Mirador Río Serrano** (−51.232515, −72.968153) pullout on the right, for a total of 23.4 km (14.5 mi; 30 min).*

Stop 1.7 Mirador Río Serrano

This stop provides an excellent vista of the Torres del Paine in the background, Cerro Toro Formation hills in the middle distance, and the glacial till-filled valley in the foreground.

Torres del Paine and Río Serrano. The park entrance is just beyond the bridge.

*Mirador Río Serrano to Mirador Lago del Toro: Return north to Y-290 and turn right (east); drive east and south on Y-290 to **Stop 1.7, Mirador Lago del Toro** (−51.264079, −72.866370), pullout on the left for a total of 13.6 km (8.5 mi; 17 min).*

Stop 1.8 Mirador Lago del Toro

Many of the roadcuts along this road show excellent exposures of the Cerro Toro Formation turbidite deposits. This particular stop provides a safe pullout from which to examine the rocks. There is also an expansive view of Lago del Toro to the east.

Upper Cretaceous Cerro Toro Formation outcrop, Mirador Lago del Toro.

Mirador Lago del Toro to Frontal Folds: *Continue south on Y-290 for 16.9 km (10.5 mi; 19 min) to the intersection with Y-200 and pull over on the right. This is **Stop 2, Frontal Folds of Main Ranges** (–51.383729, –72.790355).*

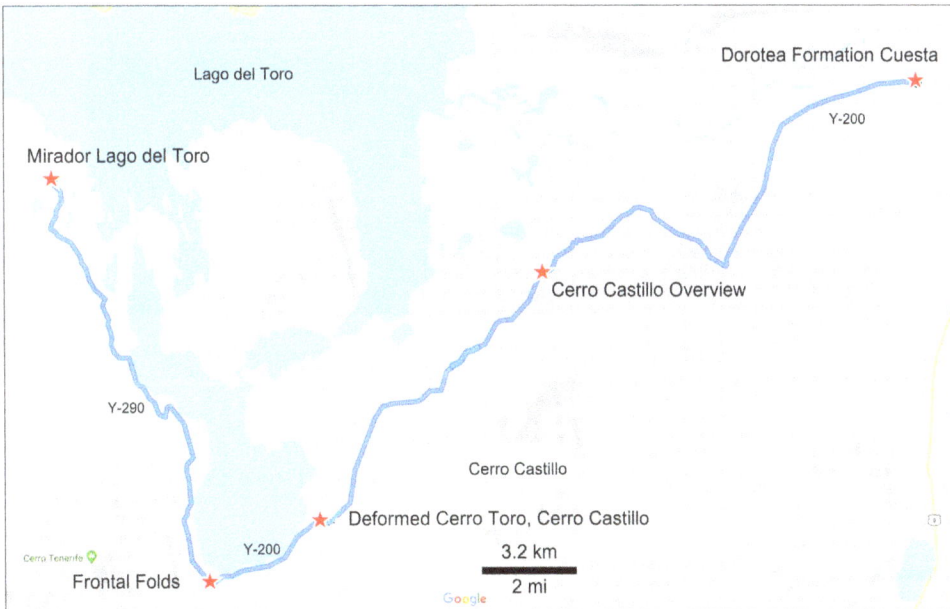

Stop 2 Frontal Folds of Main Ranges

Existing geologic maps are not definitive in this area, but it appears that the Jurassic Tobifera Formation is thrust over the Lower Cretaceous Zapata Formation here. The deformed layers exposed on the mountainside are all Tobifera Formation carried above the thrust.

View south to frontal folds in the Andean fold-thrust belt south and west of the Y-290/Y-200 intersection. Dashed lines are the trace of bedding; dotted line is a thrust fault.

> ***Frontal Folds to Cerro Castillo****: Turn left (east) onto Y-200 and drive 4.6 km (2.8 mi; 6 min) to* **Stop 3.1, Deformed Cerro Toro Formation, Cerro Castillo** *(−51.36581, −72.73606).*

Stop 3 Cerro Castillo

Cerro Castillo is an impressive erosional remnant, a mountain of shale held up by a cap of resistant conglomerate. It is developed entirely within the Cerro Toro Formation.

Stop 3.1 Deformed Cerro Toro Formation, Cerro Castillo

Cerro Castillo (Castle Hill), the large mountain south of the highway, consists entirely of Upper Cretaceous Cerro Toro Formation turbidite deposits. In addition to some folding, there appears to be normal faulting that may indicate the reactivation of Jurassic to Lower Cretaceous extensional structures.

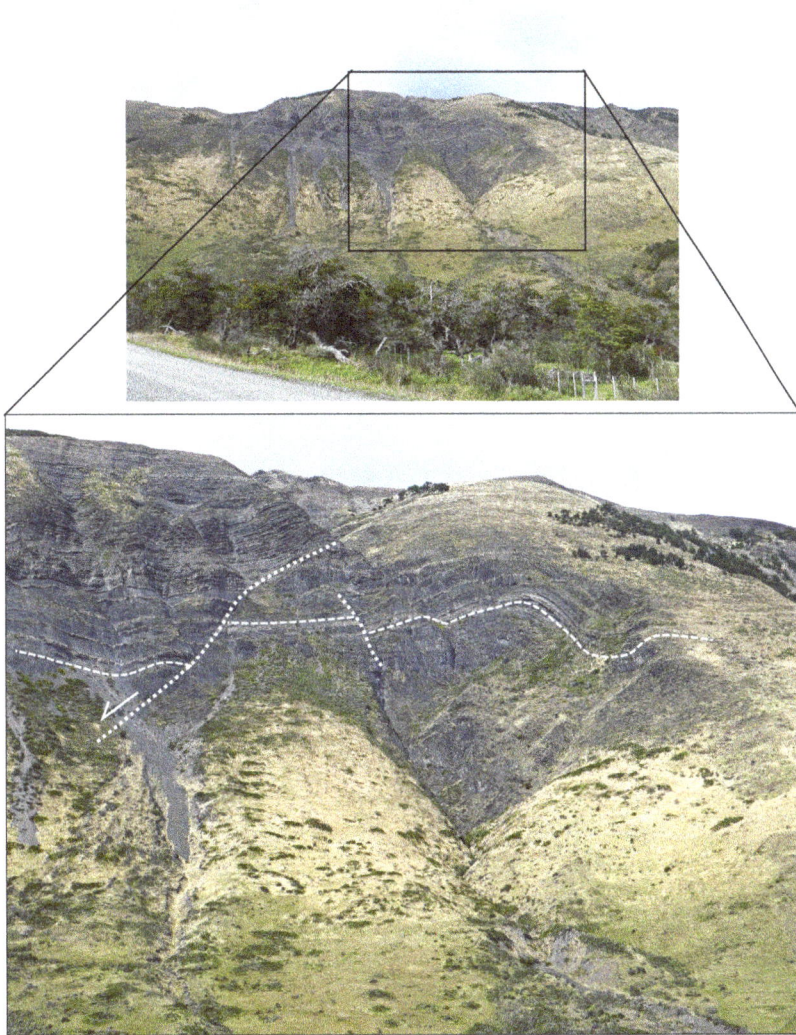

Detail of folded and normal-faulted Cerro Toro Formation on Cerro Castillo. Dashed lines are the trace of bedding; dotted lines are faults.

*Deformed Cerro Toro Formation, Cerro Castillo to Cerro Castillo Overview: Continue driving east on Y-200 for 12.4 km (7.7 mi; 14 min) to **Stop 3.2, Cerro Castillo Overview** (−51.29221, −72.62634) and pull over on the right.*

Stop 3.2 Cerro Castillo Overview

From this vantage, it is possible to view the entire Cerro Castillo. The mountain appears to be located along the axis of a broad, gentle, roughly north-south syncline developed in Upper Cretaceous Cerro Toro Formation turbidites.

Cerro Castillo looking southwest.

> ***Cerro Castillo Overview to Dorotea Formation***: *Continue driving east on Y-200 for 18.1 km (11.3 mi; 18 min) to* **Stop 4, Dorotea Formation Cuesta** *(–51.235173, –72.443008).*

Stop 4 Dorotea Formation Cuesta

At this stop, you can see a cuesta, or ridge of the Dorotea Formation above the mudstone-dominated Tres Pasos Formation. Both are dipping east, away from the Torres del Paine uplift. The Tres Pasos Formation, along with the Cerro Toro Formation, underlies the lowlands in this area.

The Magallanes-Austral retro-arc/foreland Basin of Patagonian Chile and Argentina is linked to the Late Cretaceous–Neogene uplift of the southern Andes. The Dorotea Formation, consisting of shallow-water basin shelf, delta (prodelta to delta-plain), and river deposits, records the final filling of the Late Cretaceous-Paleogene foredeep that developed in front of the Andean fold-thrust belt. Sediment was delivered to the shelf edge by a braided river system flowing south down the basin axis to the shelf edge. The shelf edge was the initiation point for turbidity flows into the deep basin. Abundant tidal features suggest that the Magallanes-Austral Basin was the result of an impinging fold-thrust belt to the west and a broad foreland bulge to the east (Schwartz and Graham, 2015). The Dorotea Formation contains marine invertebrates, plants, and marine reptiles including several partially articulated hadrosaurs found at the El Puesto site, Río de las Chinas Valley (Vogt et al., 2014). Detrital zircon dating and vitrinite reflectance data yield ages ranging from Late Cretaceous to as young as 65–63 Ma (George et al., 2019).

View east at a ridge of the Upper Cretaceous-Paleogene Dorotea Formation sandstone dipping away from Torres del Paine uplift.

Dorotea Formation Cuesta to Cerro Calafate, Lago Argentino, and the Glaciarium:
Drive east on Y-200 to the junction with Ruta 9; turn right (southeast) on R-9 and drive to
the village of Torres del Paine; from here continue driving straight (east) on Y-205 to the Chile-
Argentina border. Expect to spend at least an hour going through paperwork, having your lug-
gage x-rayed, and your vehicle inspected at the border. At the time of writing, fresh food was not
allowed across the border. Have vehicle import/export papers available. Expect to see guanacos
along and possibly on the highway. Once in Argentina, drive east on Ruta Nacional RN-40 to
Tapi Aike. It is recommended that from here you take the longer route and better highway to La
Esperanza rather than taking the shorter but more difficult RN-40 north. La Esperanza also
has a gas station, rare in these parts. Continue east on Ruta Provincial RP-7 to La Esperanza;
turn left (north) and take RP-5 to the intersection with RN-40; continue straight (north) on
RN-40 to the intersection with RP-11; continue straight (west) on RP-11 through the town of
El Calafate and turn left (south) on the gravel road marked "Glaciarium;" drive a few hundred
meters to **Stop 5, Cerro Calafate, Lago Argentino, and the Glaciarium** *(−50.336145,*
−72.338784), for a total of 311 km (193 mi; 3 hr 51 min).

Stop 5 Cerro Calafate, Lago Argentino, and the Glaciarium

This rather long drive takes you across the Patagonian pampas. These vast, nearly flat grasslands are the equivalent of the Great Plains in North America, the central Asian steppe, and the African savanna. There are not many outcrops, but there is a lot of wildlife. Watch for herds of guanacos and flocks of Darwin's Rhea.

The camel-like guanaco. They have no problem jumping livestock fences and seem to know to avoid vehicles.

Darwin's Rhea is a smaller cousin of the ostrich. This flightless, grassland bird can reach 170 cm (67 in) tall and can weigh up to 40 kg (88 lb).

We are now in the gently deformed foreland of the Andes. Lago Argentino is a glacially scoured "finger lake" formed during the retreat of the ice-age glaciers in the past 10,000 years. With an area of 1,415 km² (546 mi²), it is the largest freshwater lake in Argentina. The maximum depth is 500 m (1,640 ft). The lake lies partially in Los Glaciares National Park and is fed by numerous glaciers as well as several rivers. Lago Viedma, another glacial lake to the north, empties into Lago Argentino via the La Leona River. Lago Argentino drains into the Atlantic by way of the Santa Cruz River. The stunning turquoise color of the lake, once again, is due to suspended glacial flour.

The Andes mountain front can be seen west of the lake. South of us is Cerro Calafate, a ridge consisting of gently east-inclined layers of continental sediments shed off the Andes uplift. The bluffs show Eocene over Cretaceous rocks separated by an unconformity. The oldest rocks are Upper Cretaceous braided-river-channel sandstone and conglomerate of the La Irene Formation. Above this are river sandstones of the Chorrillo Formation. Overlying the Chorrillo Formation is a cliff-forming greenish sandstone, the Calafate Formation. The Calafate Formation contains glauconite, a green mineral indicative of a marine-shelf depositional environment. Above the Calafate Formation, there was a period of non-deposition or erosion, then the Man Aike Formation was deposited. The Man Aike consists of middle Eocene shallow-marine, shoreline, and fluvial sandstone and conglomerate.

View northwest toward Lago Argentina and the Andes from the Glaciarium.

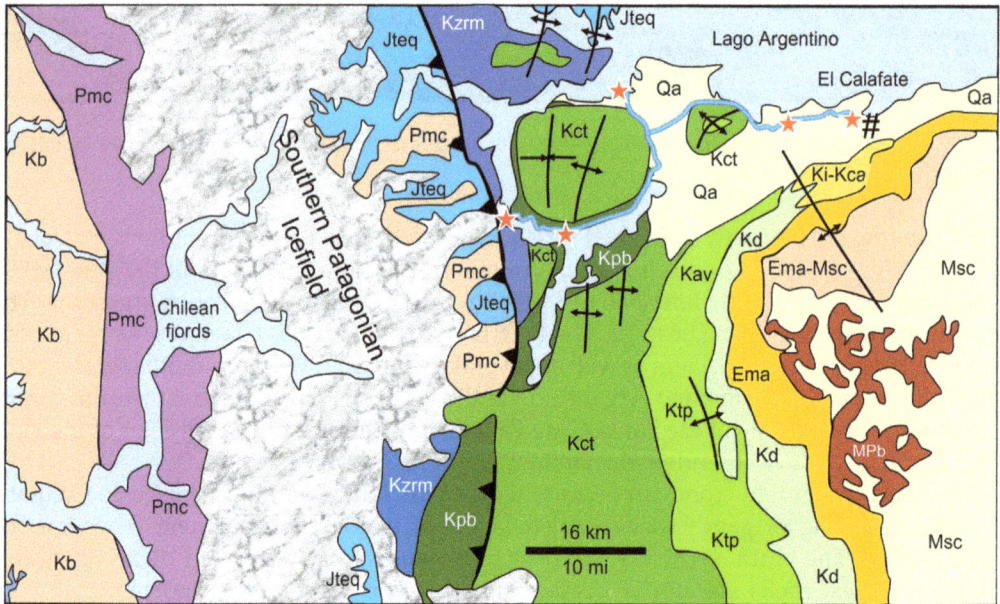

Geologic map of the El Calafate-Perito Moreno area. Pmc = Paleozoic metasediments; Jteq = Tobifera/El Quemado Formation; Kb = Southern Patagonian batholith; Kzrm = Zapata/Río Mayer Formation; Kpb = Punta Barrosa Formation; Kct = Cerro Toro Formation; Ktp = Tres Pasos Formation; Kd = Dorotea Formation; Ki-Kca = La Irene-Calafate Formation; Ema = Man Aike Formation; Msc = Santa Cruz Formation; MPb = basalt; Qa = alluvium. Modified after Malkowski et al., 2015.

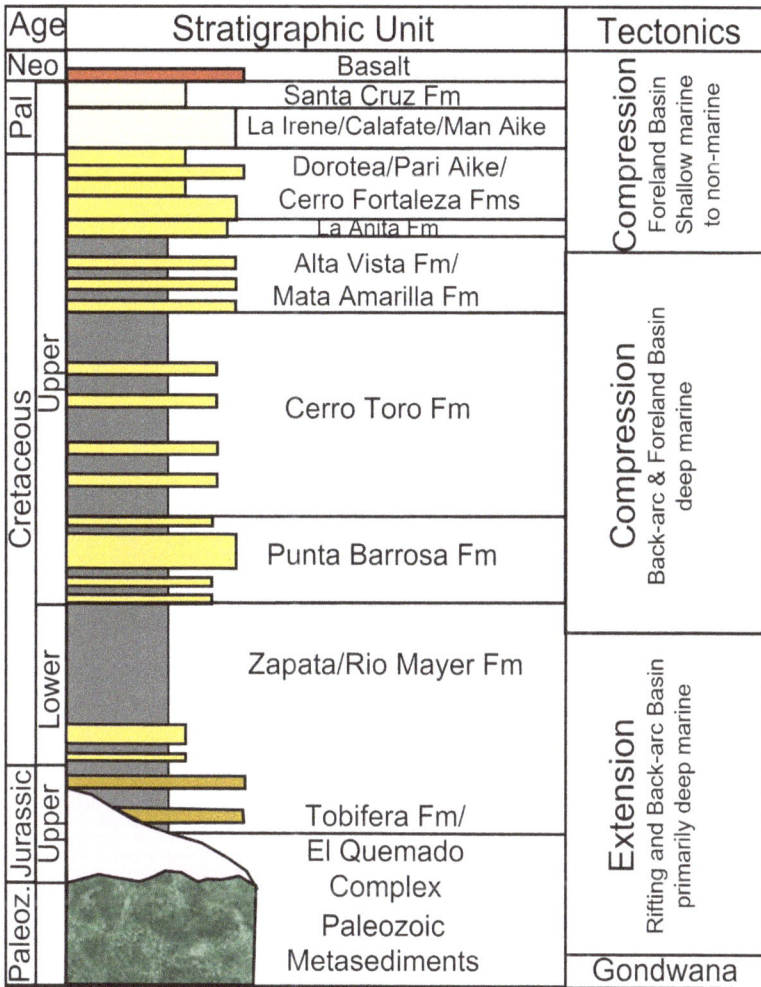

Age	Stratigraphic Unit	Tectonics
Neo	Basalt	Compression — Foreland Basin Shallow marine to non-marine
Pal	Santa Cruz Fm	
	La Irene/Calafate/Man Aike	
Cretaceous (Upper)	Dorotea/Pari Aike/ Cerro Fortaleza Fms	
	La Anita Fm	
	Alta Vista Fm/ Mata Amarilla Fm	Compression — Back-arc & Foreland Basin deep marine
	Cerro Toro Fm	
	Punta Barrosa Fm	
Cretaceous (Lower)	Zapata/Rio Mayer Fm	Extension — Rifting and Back-arc Basin primarily deep marine
Jurassic (Upper)	Tobifera Fm/ El Quemado Complex	
Paleoz.	Paleozoic Metasediments	Gondwana

Stratigraphy and tectonic regimes for the El Calafate-Perito Moreno region.

View south to Cerro Calafate. Bluffs consist, from bottom to top, of southeast-dipping La Irene, Chorrillo, Calafate, and Man Aike formations.

The Glaciarium, a museum that explains the glacial history of the region.

The Glaciarium is one of the few museums in the world dedicated to glaciology. Facilities include displays, a theater, a café, an ice bar, and an Ecoshop. The Glaciobar Branca is a bar that uses only glacier ice. In it, you will experience freezing temperatures, so the maximum stay is 20 minutes.

Visit

Hours: 9:00 am to 8:00 pm September to April, and 11:00 am to 8:00 pm May to August.

Address: Take Ruta 11 for 6 km west from El Calafate city center and turn left (south) on the gravel road by the sign for the Glaciarium.

Entrance fee: See the website for current entrance fees.

Website: https://www.patagonia-argentina.com/en/glaciarium-ice-museum-calafate/.

*Cerro Calafate to La Anita Sandstone: Return to RP-11 and turn left (west); drive 9.6 km (6.0 mi; 8 min) to **Stop 6, La Anita Sandstone** (–50.339542, –72.463153), and pull over on the right.*

Stop 6 La Anita Sandstone

This is a west-dipping ridge of the Upper Cretaceous La Anita Formation. The La Anita Formation consists of coarse-grained sandstones and conglomerates with occasional inter-bedded mudstones. South of Lago Argentino these marine deposits show clinoform development, that is, sloping depositional surfaces that indicate a shelf margin leading to a deep-marine setting. The La Anita Formation is stratigraphically below (older than) the units seen at Cerro Calafate.

View west to a ridge formed by the La Anita Formation.

The units at Cerro Calafate were dipping east-southeast, indicating that we just crossed the axis of a broad, roughly north-south-trending anticlinal foreland fold.

> ***La Anita Sandstone to Puerto Bandera****: Continue west on RP-11 to the intersection with RP-8; turn right (north) on RP-8 and drive to the end of the road. This is **Stop 7, Puerto Bandera** (−50.300439, −72.798196), for a total of 32.6 km (20.3 mi; 25 min).*

Stop 7 Puerto Bandera

As you drive to Puerto Bandera (Flag Harbor), the large mountain to the south is Cerro Frias. Cerro Frias is a large structural dome developed in the Cerro Toro Formation. The Cerro Toro Formation lies below the La Anita Formation.

Google Maps satellite image showing the large dome at Cerro Frias. Imagery ©2023 TerraMetrics.

At Puerto Bandera, in addition to being able to hire a boat to take you on a lake tour, you can look north across the lake to an anticline developed in the Jurassic El Quemado Formation. The El Quemado Formation consists of massive rhyolites, part of the Chon Aike volcanic event, deposited in northwest-oriented rift basins. Above the El Quemado Formation are the Jurassic to Early Cretaceous black shales of the Río Mayer Formation.

View north across Lago Argentino from Puerto Bandera. A large, broad anticline is developed in the Jurassic to Lower-Cretaceous units.

> **Puerto Bandera to Bajo Las Sombras**: *Return south on RP-8 to the intersection with RP-11; continue straight (south) on RP-11; about 3.7 km (2.3 mi) after the intersection with RP-11 is the entrance to Los Glaciares National Park; continue driving west on RP-11 for a total of 32.7 km (20.3 mi; 36 min) to **Stop 8.1, Bajo Las Sombras** (–50.4878, –72.9166); pull over on the right. Cross the road and climb down to the lake shore.*

Stop 8 Los Glaciares National Park, Perito Moreno

Los Glaciares National Park (Parque Nacional Los Glaciares), at 726,927 ha (2,807 sq mi), is the largest national park in Argentina. Established on May 11, 1937, it was declared a UNESCO World Heritage Site in 1981. The park's name refers to the Southern Patagonian Ice Field that feeds multiple glaciers. The north half of the park contains Lago Viedma, the Viedma Glacier, and a number of mountains popular for climbing and trekking, including Mount Fitz Roy and Cerro Torre. We will visit this area when we drive north to El Chaltén. The southern part has the major glaciers, including Perito Moreno Glacier, Upsala Glacier, and Spegazzini Glacier. Whereas the Perito Moreno glacier is accessible by road, the other two are reached only by tour boats.

As with Torres del Paine National Park, visiting the parks is recommended between September and April, during the southern spring, summer, and early autumn. During summer, daylight hours are long given the southern latitude. Summer temperatures are mild, but the wind blows hard most of the time. During southern winter the weather becomes extreme.

Stop 8.1 Bajo Las Sombras (Below the Shadows)

Once again, we are in the Upper Cretaceous turbidites of the Cerro Toro Formation. We will be in this unit all the way to Perito Moreno Glacier. The thin-bedded sandstone and shale were deposited in a deep-marine basin as a result of turbidity flows. Much later they were uplifted and folded as a result of Andean compression and shortening. We have now left the foreland and entered the zone of Andean thrusting, the foothills belt. Some maps show this unit as the Upper Cretaceous Punta Barrosa Formation, also a turbidite-dominated section.

Folded Cerro Toro or Punta Barrosa Formation thin-bedded turbidite sandstone at Bajo Las Sombras. Dashed lines are the trace of bedding.

Bajo Las Sombras to Bridge Roadcut: *Continue west on RP-11 for 1.4 km (0.9 mi; 2 min) to **Stop 8.2, Bridge Roadcut** (–50.48770, –72.9342), and pull over on the right either before or after the bridge.*

Stop 8.2 Bridge Roadcut

Here you see steeply-east-dipping alternating bands of thin sandstone and shale deposited from turbidity currents during Cerro Toro sedimentation.

Alternating beds of sandstone and shale in the Cerro Toro Formation at the bridge roadcut.

> **Bridge Roadcut to Mirador Los Suspiros**: *Continue driving west on RP-11 for 2.8 km (1.8 mi; 4 min) and pull into the parking area on the right. This is **Stop 8.3, Mirador Los Suspiros** (−50.486611, −72.964966).*

Stop 8.3 Mirador Los Suspiros (The Sighs Lookout)

From this vantage point, you have your first great view of the Perito Moreno Glacier to the west.

Perito Moreno Glacier and Miter Peak as seen from Mirador Los Suspiros.

You also have great views of the thrusted and folded Cerro Toro Formation to the south and north. Thrusts and fold vergence are all east-directed.

East-directed thrusting and east-verging folds in the Cerro Toro Formation looking south from Mirador Los Suspiros. Dashed lines are the trace of bedding; solid or dotted lines are faults.

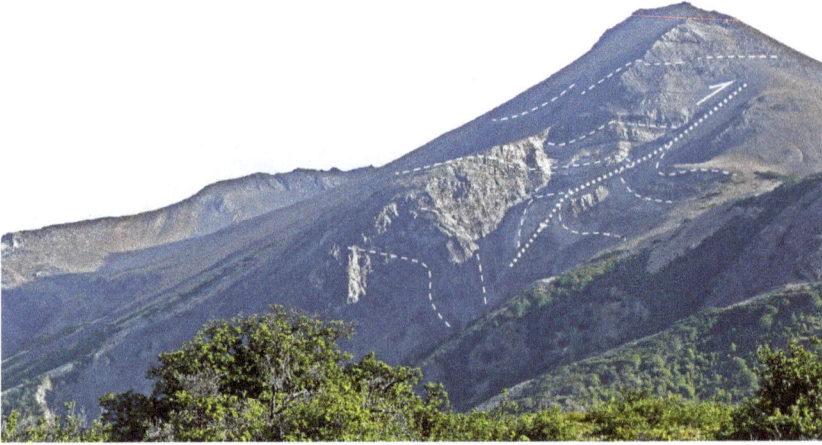

East-directed thrusting in the Cerro Toro Formation looking north from Mirador Los Suspiros.

> *Mirador Los Suspiros to Mirador Velo de Novia: Continue driving west on RP-11 for 2.2 km (1.4 mi; 3 min) to the parking area on the left. This is **Stop 8.4, Mirador Velo de Novia** (–50.475365, –72.980505).*

Stop 8.4 Mirador Velo de Novia (Lover's Veil Lookout)

The views of the glacier just keep getting better. Though it looks modest from here, the top of the glacier looms 238 m (780 ft) above the lake. That is just the thickness of ice above the surface.

Front of Perito Moreno glacier. Note the tour boats (circles) for scale.

Velo de Novia to Perito Moreno Glacier Overlook*: Continue driving west on RP-11 for 6.8 km (4.2 mi; 11 min) to* **Stop 8.5, Perito Moreno Glacier Overlook** *(–50.469132, –73.029761) and park in the parking area. Walk to the multi-level overlook.*

Stop 8.5 Perito Moreno Glacier Overlook

This is the pride of the park, a typical temperate-zone glacier with high accumulation and ablation rates. The ablation zone is near the toe of the glacier where there is a net loss due to melting, evaporation, and calving. The accumulation zone is where snow falls on the glacier and adds to the glacial mass.

In 1899 the glacier was 750 m away from the east shore. Then it advanced until 1917 when it reached the shore of the Magallanes Peninsula. Since then, it has been in a near-steady state due to local topography and climate. In other words, enough snow falls in the accumulation zone to keep up with melting and calving.

From this point, you will see mostly the ablation area of the glacier. On clear days you can see some of the peaks in the accumulation area. The glacier covers an area of 254 km² (98 mi²) and sits 185 to 3,000 m (600 to 9,850 ft) above sea level. It has a maximum length of 31 km (18.6 mi) and a thickness of 700 m (2,300 ft) at the centerline 8 km (5 mi) up from here. At that point, the ice velocity is more than 2 m/day (6.6 ft/day). At the terminus at Lago Argentino, it is 5 km (3 mi) wide.

There are restrooms and a small gift shop and café here. The viewing platforms extend over several levels, and there are excellent interpretive panels. You *will* see the ice calving. Listen for the sound of thunder.

Perito Moreno Glacier, a wall of ice at least 170 m (560 ft) thick at its termination.

Looking north at the valley walls above the glacier, you can see east-directed thrusting in the Cerro Toro Formation.

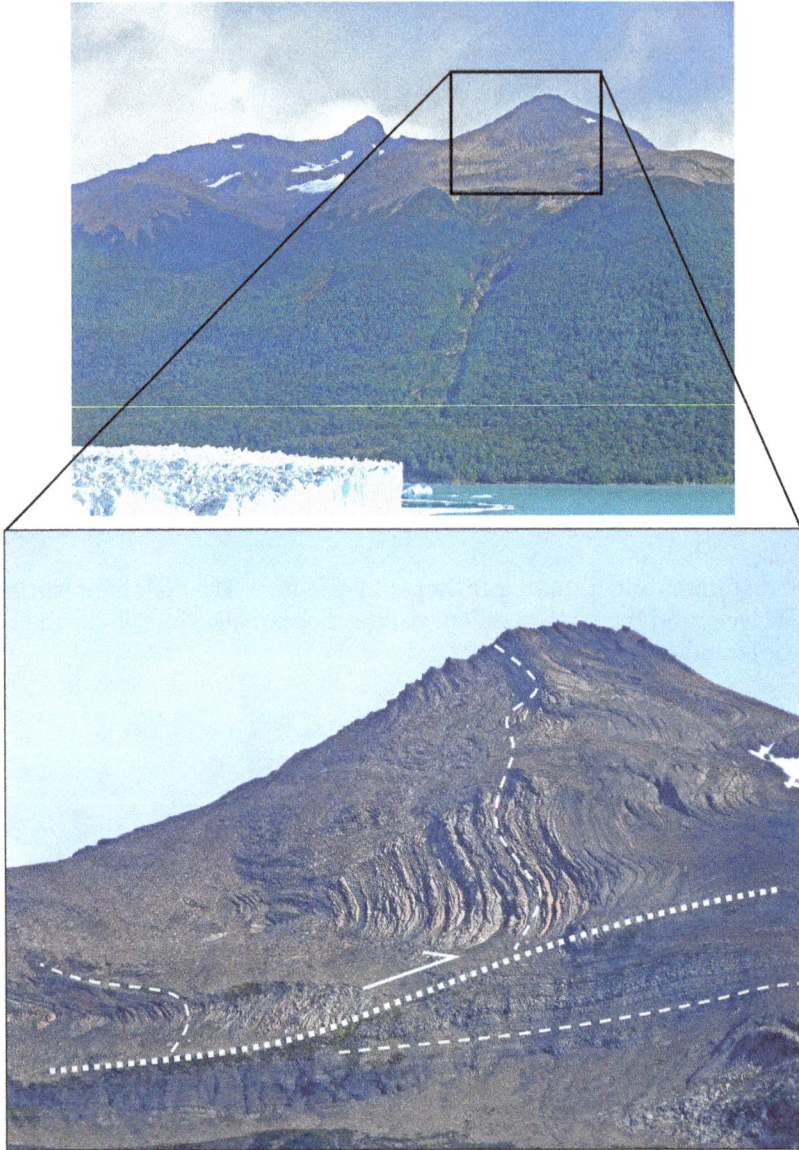

View north from Perito Moreno Glacier overlook. Thrusting puts the Jurassic El Quemado volcanics over Cretaceous Cerro Toro Formation. Dotted line is a thrust fault.

Perito Moreno Glacier Overlook to Cerro Fortaleza: *Return east on RP-11 through El Calafate to the intersection with RN-40; turn left (north) on RN-40 and drive to **Stop 9, Cerro Fortaleza and Mata Amarilla Badlands** (−49.9888, −72.09748) and pull over on the right for a total of 159 km (98.7 mi; 2hr 16 min).*

Stop 9 Cerro Fortaleza and Mata Amarilla Badlands

The Upper Cretaceous Mata Amarilla Formation consists of alternating white sand-stone and gray-to-black shale. The sandstone occurs as river channels within the shale. The formation was deposited in a coastal environment and includes shoreline sandstone, estuarine and lagoonal shale, and delta-floodplain and river-channel deposits (Poiré and Franzese, 2010; Malkowski et al., 2015). It typically weathers to a badlands topography. The formation is known for its abundant petrified conifer trunks (Martínez et al., 2017).

Cerro Fortaleza (Fortress Hill), a small mesa east of this location, is capped by resistant shallow-marine sandstone of the Upper Cretaceous La Anita Formation (which we saw previously at Stop 6). We are back in the undeformed Andean foreland.

View east to Cerro Fortaleza and the Mata Amarilla badlands. Río La Leóna is in the foreground.

*Cerro Fortaleza to La Anita and Mata Amarilla Formations: Continue driving north on RN-40 for 10.3 km (6.4 mi; 7 min) and pull over on the right. This is **Stop 10, La Anita and Mata Amarilla Formations** (−49.902538, −72.060058).*

Stop 10 La Anita and Mata Amarilla Formations

This excellent roadcut exposes both the Mata Amarilla and the La Anita formations. As we saw on the ridge west of El Calafate, the La Anita Formation consists of coarse-grained sandstone and conglomerate with some interbedded shale. This unit is indicative of a shal-low-marine environment during Late Cretaceous times in this area (Poiré and Franzese, 2010; Malkowski et al., 2015).

The La Anita Formation sandstone lies unconformably over the Mata Amarilla shales and sandstones.

> ***La Anita and Mata Amarilla Formations to Mirador Lago Viedma***: *Continue driv-*
> *ing north on RN-40 for 25.8 km (16.0 mi; 17 min) and pull over on the left. This is **Stop***
> ***11.1, Mirador Lago Viedma** (−49.717357, −71.966097).*

Stop 11 Mount Fitz Roy and Los Glaciares National Park

The Cerro Fitz Roy pluton has been dated at 18 ± 3 Ma (Gorring, 2008) and between 19 and 14 Ma (Giacosa et al., 2012). The plutonic complex consists of ultramafic, mafic, tonalitic, and granitic rocks. In other words, it runs the gamut from very dark to intermediate to light-colored igneous rocks. The pluton was cut by andesitic dikes ranging in age from 19 to 16 Ma. Those dikes were in turn displaced by Andean thrusting (the east-vergent Torre and Fitz Roy thrusts), indicating that pluton emplacement was coincident with compression and shortening. During this time, foreland basin sedimentation was ongoing (Giacosa et al., 2012).

The present elevation of the Cerro Fitz Roy and other nearby plutons requires over 4,000 m (13,000 ft) of post-mid Miocene uplift and erosion of cover rocks. Andean deformation extending into the Pliocene (5.3 to 2.6 Ma) is indicated by gentle tilting of Pliocene plateau basalts north of Lago Viedma (Gorring, 2008).

Monte Fitz Roy (also known as Cerro Chaltén, Cerro Fitz Roy, or Mount Fitz Roy) lies on the Argentina- Chile border at an elevation of 3,405 m (11,171 ft). It served as the profile for the Patagonia Clothing logo. Argentine explorer Francisco Moreno saw the mountain in 1877 and named it Fitz Roy in honor of Robert Fitz Roy, captain of HMS Beagle. Moreno, Fitz Roy, and Charles Darwin had traveled up the Santa Cruz River in 1834 while mapping the Patagonian coast. The mountain was first climbed in 1952 by French alpinists Lionel Terray and Guido Magnone.

There is no entrance station and no fee (yet) upon entering Los Glaciares National Park just east of the village of El Chaltén.

Stop 11.1 Mirador Lago Viedma

Lago Viedma, the second largest lake in Argentina, is 80 km (50 mi) long, between 12 and 20 km (7 to 12 mi) wide, and has an average depth of 100 m (330 ft). Named after the Spanish explorer Antonio de Viedma, the first European to see it in 1783, it is fed mainly by the Viedma Glacier. This lake, like Lago Argentino and others extending east of the Andes, was gouged out by glaciers and filled by melting ice. The lake empties into the La Leona River, which flows to Lago Argentino and from there to the Atlantic.

View west from Mirador Lago Viedma toward the Fitz Roy massif.

Keep an eye out for the Patagonian fox, or South American gray fox (*Lycalopex griseus*). They seem to be common around this stop. While not a true fox, this fox-like canid is limited to Argentina and Chile.

Patagonian fox at Mirador Lago Viedma.

Mirador Lago Viedma to Outer Fold-Belt Syncline: Continue driving north on RN-40 to the intersection with RP-23; turn left (west) on RP-23 and drive to Stop 11.2, Outer Fold-Belt Syncline (−49.47258, −72.62716) for a total of 69.4 km (43.1 mi; 1 hr 27 min). Pull over on the right.

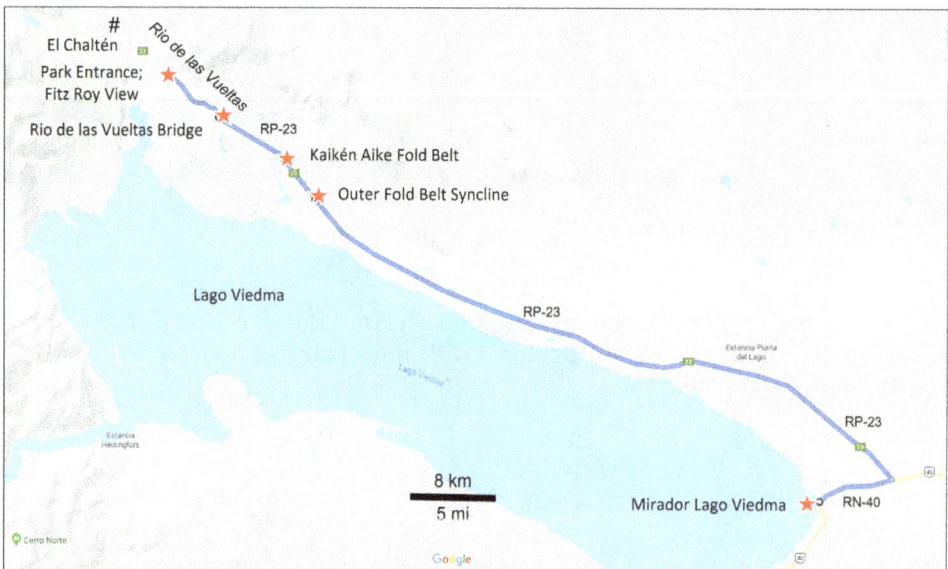

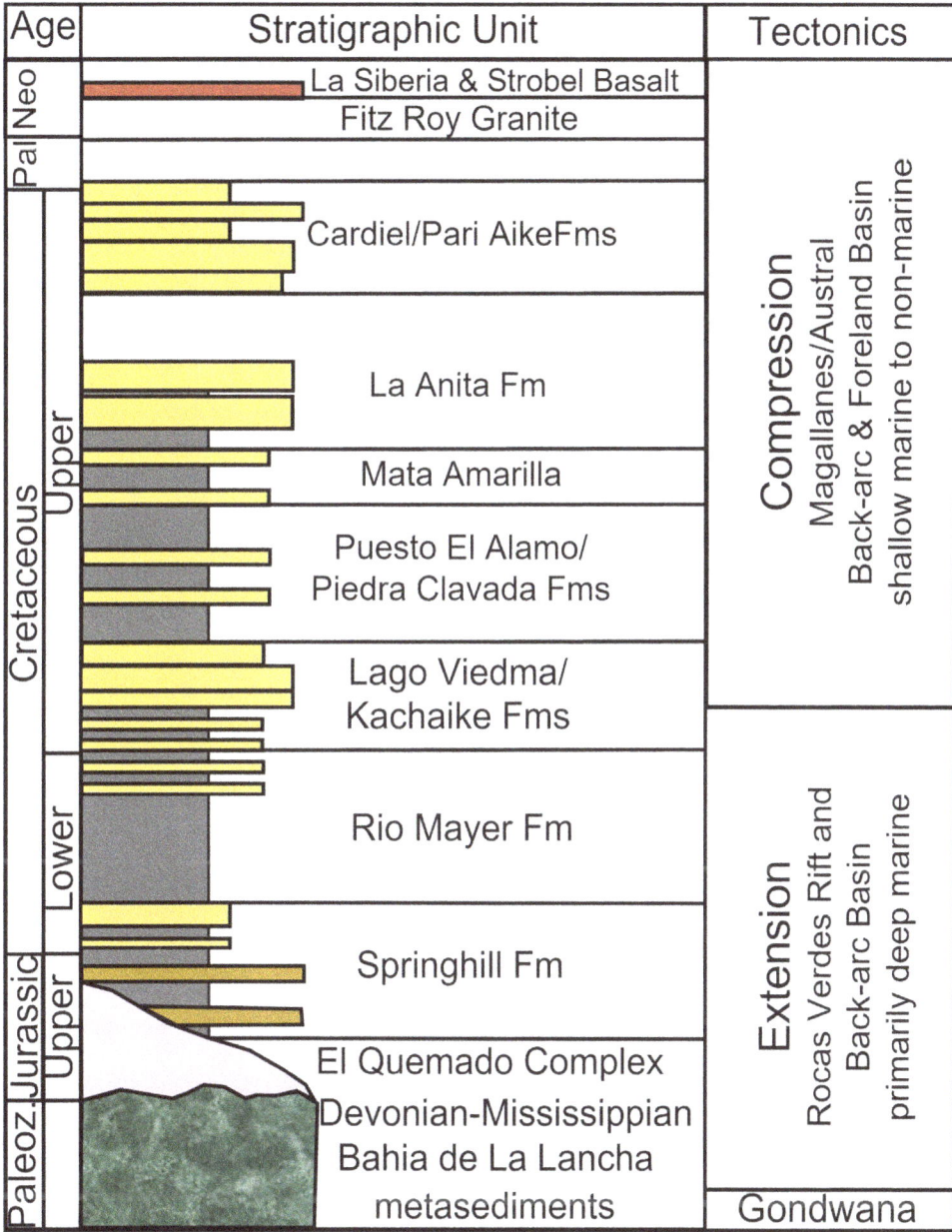

Age	Stratigraphic Unit	Tectonics
Pal Neo	La Siberia & Strobel Basalt	
	Fitz Roy Granite	
Cretaceous — Upper	Cardiel/Pari AikeFms	Compression — Magallanes/Austral Back-arc & Foreland Basin shallow marine to non-marine
	La Anita Fm	
	Mata Amarilla	
	Puesto El Alamo/ Piedra Clavada Fms	
	Lago Viedma/ Kachaike Fms	
Cretaceous — Lower	Rio Mayer Fm	Extension — Rocas Verdes Rift and Back-arc Basin primarily deep marine
Jurassic — Upper	Springhill Fm	
Paleoz.	El Quemado Complex Devonian-Mississippian Bahia de La Lancha metasediments	Gondwana

Stratigraphy and tectonic regimes of the Fitz Roy-El Chaltén area.

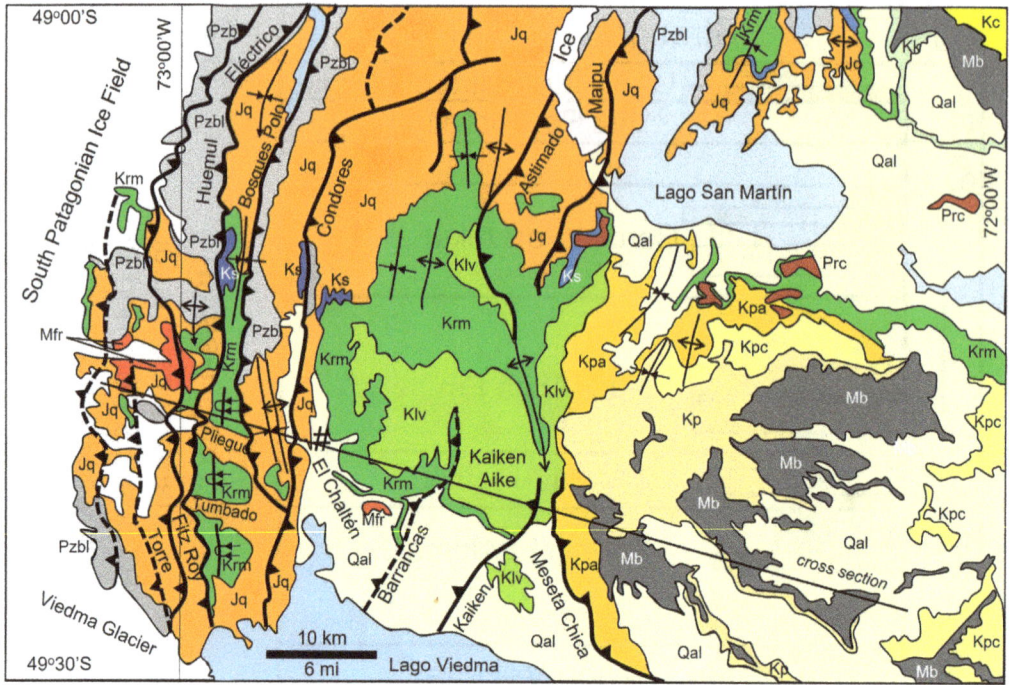

Geologic map of the El Chaltén and Fitz Roy area. Qal = Alluvium; Mb = Miocene La Siberia and Strobel basalt; Mfr = Miocene Fitz Roy Complex, including the Chaltén Adakite; Prc = Paleogene Río Carbón Essexite intrusive; Kc = Upper Cretaceous Cardiel Formation; Kp = Upper Cretaceous Pariaike Formation; Kpc = Upper Cretaceous Piedra Clavada Formation; Kpa = Upper Cretaceous Puesto Alamo Formation; Klv = Lower Cretaceous Lago Viedma Formation; Kk = Lower Cretaceous Kachaike Formation; Krm = Lower Cretaceous Río Mayer Formation; Ks = Lower Cretaceous Springhill Formation; Jq = Upper Jurassic Quemado Complex; Pzbl = Devonian-Mississippian Bahía Lancha Formation. Modified after Giacosa et al., 2012.

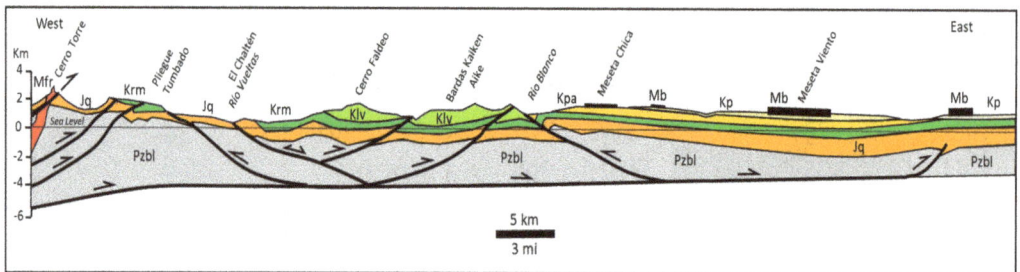

Cross-section from geologic map of the El Chaltén and Fitz Roy area. Unit color scheme is the same as on the map. Modified after Giacosa et al., 2012.

Stop 11.2 Outer Fold-Belt Syncline

Quick note: The best light for seeing detail on these south-facing slopes is in the evening during the summer.

The flat-topped or gently-inclined mesas north of RP-23 on the way to this stop are capped by Pliocene (3.5 Ma) basalt flows (Goring, 2008; Malkowski et al., 2015).

From this stop, looking at the slopes to the north we see a syncline developed in the Lower Cretaceous Lago Viedma Formation overlying the Lower Cretaceous Río Mayer Formation (Giacosa et al., 2012). It should be noted that other workers have mapped other units here: Gorring (2008) believes these cliffs are made of the Upper Cretaceous Cerro Toro Formation over the Río Mayer Formation, whereas Malkowski et al. (2015) has mapped this as the Upper Cretaceous Puesto El Alamo Formation over the Lower Cretaceous Lago Viedma Formation. Perhaps more important is the gentle deformation of these Cretaceous units into broad folds. We are going through a transition from gently-deformed foreland foothills to intensely-deformed units as we approach the thrust belt proper.

Deltaic deposits of the Lago Viedma Formation interfinger with shallow-marine platform sediments of the Río Mayer Formation. The Lago Viedma is time-equivalent to the Cerro Toro Formation south of here. These formations represent the final infill of the Jurassic Rocas Verdes rift basin.

We are now in the western part of the Magallanes-Austral foreland basin, and near the eastern edge of Andean deformation. In this Kaikén Aike fold belt, also called the "outer fold and thrust belt," there are few, relatively small thrusts and broad, low amplitude folds. Driving west, the basin fill becomes progressively older and more deformed (Giacosa et al., 2012; Malkowski et al., 2015).

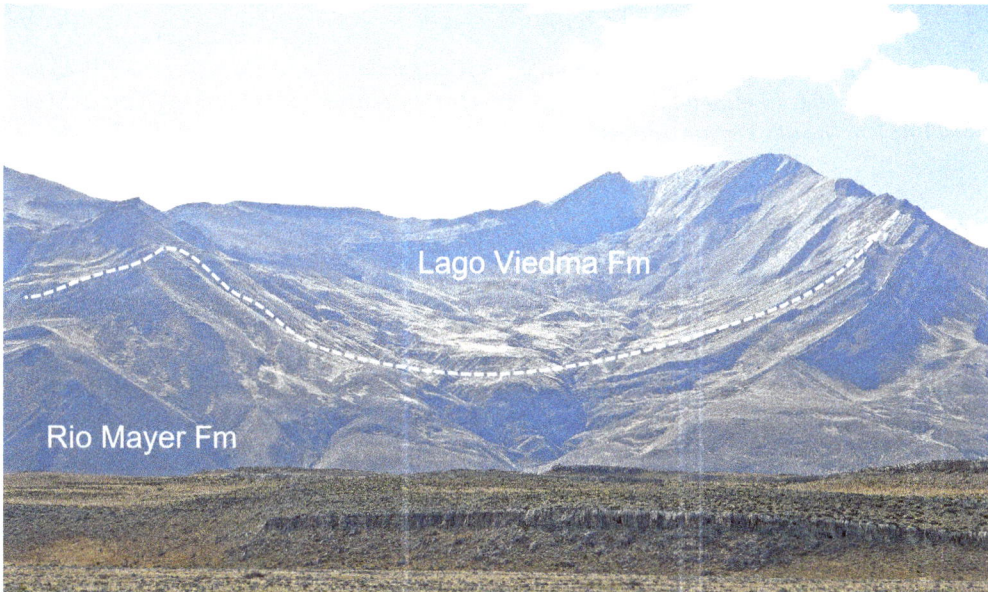

Syncline developed in the Lago Viedma Formation (upper unit) and the Río Mayer Formation (base of cliffs). Units are based on Giacosa et al., 2012. Dashed lines are the trace of bedding.

Outer Fold-Belt Syncline to Kaikén Aike Fold Belt: *Continue driving west on RP-23 for 4.5 km (2.8 mi; 6 min) and pull over on the right. This is* **Stop 11.3, Kaikén Aike Fold Belt** *(−49.4411, −72.66677).*

Stop 11.3 Kaikén Aike Fold Belt

Notice how the intensity of deformation has increased. These structures are related to the easternmost thrusts (both east and west-verging) of the Andean fold-thrust belt. Fault propagation folds in the Lago Viedma Formation marine sandstones along the Bardas de Kaiken Aike have chevron/kink geometries typical of flexural slip in thin layered turbidite sequences.

View north toward Bardas de Kaiken Aike showing one possible structural interpretation that includes a west-directed backthrust. The syncline seen at the previous stop is on the right (east). Units are based on Giacosa et al., 2012. Dashed lines are the trace of bedding; dotted lines are faults.

> ***Kaikén Aike Fold Belt to Río de las Vueltas Bridge***: *Continue driving west on RP-23 for 7.6 km (4.7 mi; 10 min) and pull over on the right (southeast side of bridge). This is **Stop 11.4, Río Mayer Formation, Río de las Vueltas Bridge** (−49.404977, −72.754586).*

Stop 11.4 Río Mayer Formation, Río de las Vueltas Bridge

The Río de las Vueltas, east of the bridge, has eroded a nice exposure of the Early Cretaceous black shales of the Río Mayer Formation. The Río Mayer Formation here is cut by several small basaltic dikes. Río Mayer shales form the décollement (detachment surface) for many of the thrusts in the Patagonian fold-thrust belt. Looking north, the slopes of Cerro Faldeo contain folds and thrusts within the Río Mayer Formation.

Glacial till covers most of the surface here.

Black shale of the Río Mayer Formation is cut by basaltic dikes at the Río de las Vueltas bridge. Google Street View looking east.

Río Mayer Formation, Río de las Vueltas Bridge to Park Entrance, Fitz Roy Massif Viewpoint: *Continue driving west on RP-23 for 6.6 km (4.1 mi; 7 min) and pull over on the right. This is* **Stop 11.5, Park Entrance, Fitz Roy Massif View** *(–49.36949, –72.81930).*

Stop 11.5 Park Entrance, Fitz Roy Massif Viewpoint

This stop affords an impressive view of the Fitz Roy massif, glacier-filled valleys, and the dark Jurassic Quemado Complex silicic volcanics in the foreground.

Fitz Roy massif as seen looking west from near the park entrance.

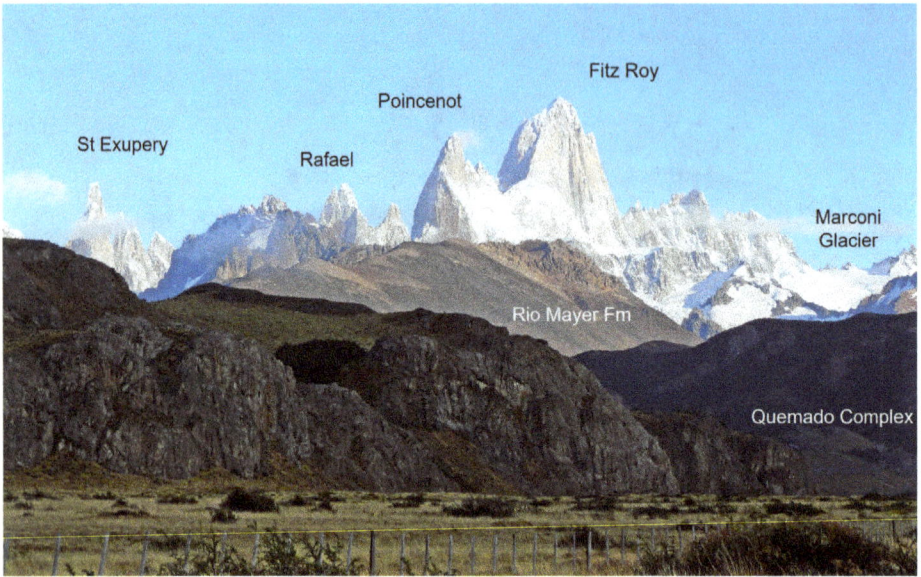

Iconic view west from near the park entrance.

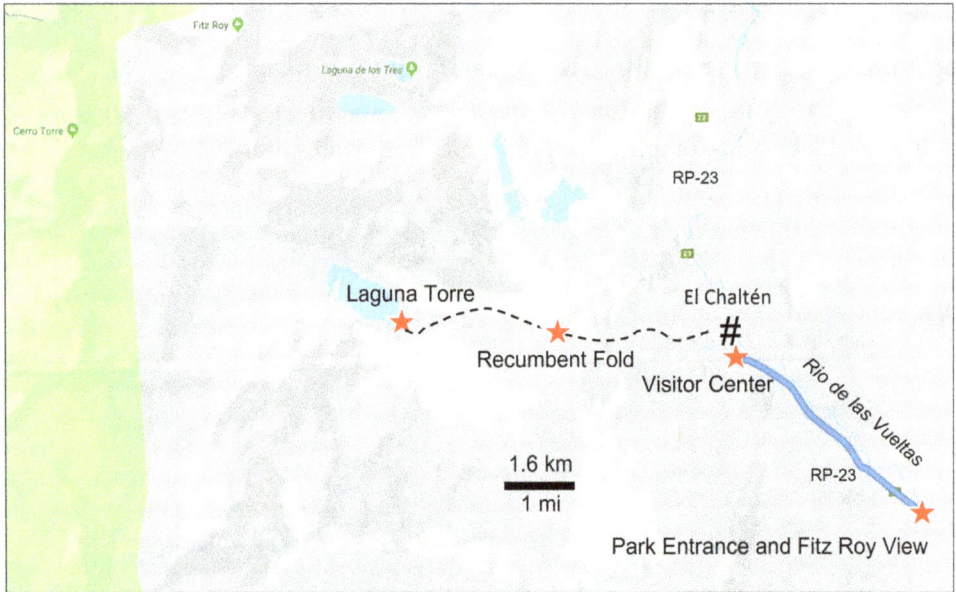

Park Entrance, Fitz Roy Massif Viewpoint to Visitor Center: *Continue driving west on RP-23 for 5.8 km (3.6 mi; 6 min) to the National Park Information sign and pull into the parking area on the left. This is* **Stop 11.6, Visitor Center** *(−49.337265, −72.880769).*

Stop 11.6 Visitor Center (Centro de Visitantes)

The park visitor center (and the adjacent town of El Chaltén) sits in a valley carved by the Río de las Vueltas into the Middle-Upper Jurassic Quemado Complex volcanics. These are the oldest rocks in the area we will be visiting.

This small and simple visitor center has displays that are mostly in Spanish. It does an adequate job explaining the geology, plants, animals, and climbing history of the region. Camping permits must be obtained here for remote areas; they are not required for areas close to town, although the park wardens request that you inform them of your plans.

Sunrise on the granite spires of Cerro Fitz Roy as seen from El Chalten.

> *Visitor Center to Recumbent Fold, Valle del Torre*: *Continue driving west on RP-23 into El Chaltén; continue straight on Miguel Martín de Güemes; turn right (northeast) on Lago del Desierto and drive to San Martín; turn left (northwest) on San Martín and drive to Las Loicas; turn left (west) on Las Loicas and drive to the third street on the left (unnamed); turn left (southwest) and drive to the end of the street where you will find a small unmarked parking area (−49.33113, −72.89542) for a total of 1.8 km (1.1 mi; 6 min). This is the trailhead for the Laguna Torre Trail. The trail sign is high on the hillside in a saddle; all you see from the parking area are footpaths heading up the hillside. Hike this moderate-difficulty trail about 3.8 km (2.4 mi; 1 hour 15 minutes) to a view south across the valley. This is* **Stop 11.7, Recumbent Fold in Río Mayer Formation, Valle del Torre** *(−49.332085, −72.938028).*

Stop 11.7 Recumbent Fold (Pliege Tumbado) in Río Mayer Formation, Valle del Torre

We are now fully back into the Andean Fold-Thrust belt. Looking south across the valley you can see an east-verging recumbent syncline developed in mudstones of the Río Mayer Formation. The fold lies below the Huemul-Eléctrico Thrust. It may have been part of an anticline-syncline fold pair cut by the thrust before erosion removed the upper part of the system.

Recumbent fold developed in the Río Mayer Formation. This syncline is overturned toward the east. Dashed line follows bedding. View south to Loma Pliege Tumbado from the Laguna Torre trail.

Same folds in the Río Mayer Formation, different angle. View south from the Laguna Torre trail.

Recumbent Fold, Valle del Torre to Torre Lagoon: *Continue hiking west on Laguna Torre trail for approximately 4.3 km (2.7 mi) to **Stop 11.8, Torre Lagoon, Moraine, and Glacier Torre** (Laguna Torre) and its moraine (−49.330167, −72.988986).*

Stop 11.8 Torre Lagoon, Moraine, and Glacier Torre

This, the last stop on our Patagonian geo-tour, provides a view of Torre Glacier, Torre Lagoon, and the moraine that dams the lagoon. Deformation related to the Huemul-Eléctrico Thrust can be seen in the canyon walls along the north side of the valley. You can see the Jurassic Quemado Complex volcanics thrust over Lower Cretaceous Río Mayer shale. The Río Mayer Formation in the footwall to the thrust is highly deformed into tight east-verging folds.

View west across Torre Lagoon. Meltwater from Torre Glacier fills the lagoon.

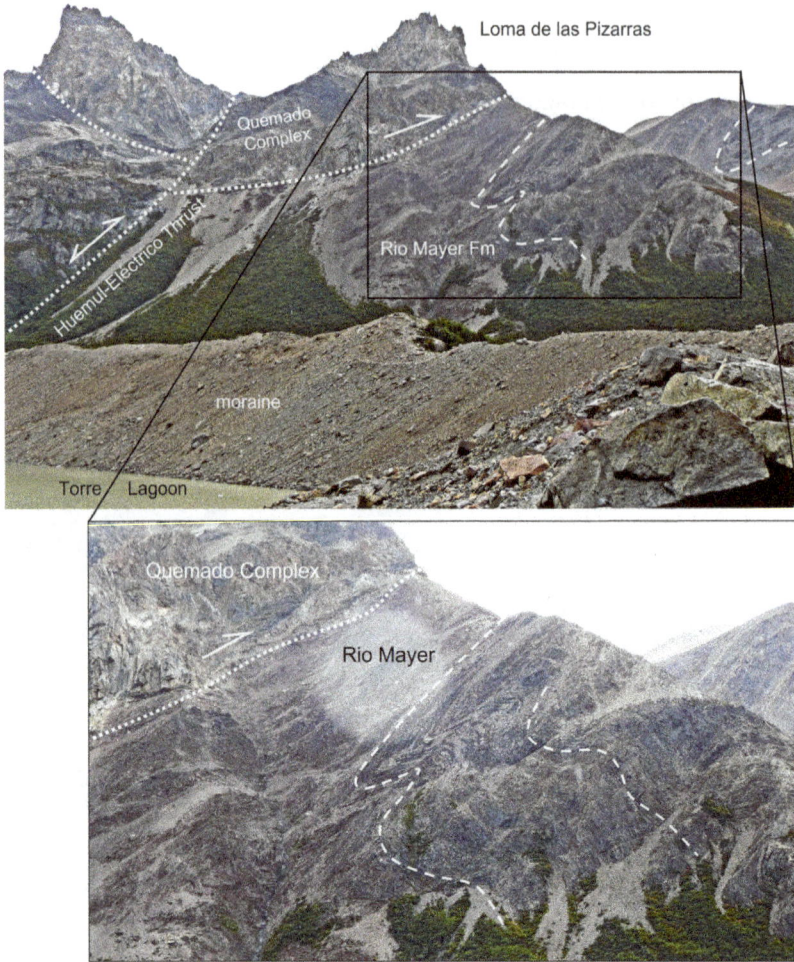

View north from Torre Lagoon to Loma de las Pizarras showing the Quemado Complex thrust over Río Mayer Formation. The moraine that dams Torre Lagoon is in the foreground. Dashed lines are the trace of bedding; dotted lines are faults. The subscene shows the details of subthrust deformation in the Río Mayer Formation.

Los Glaciares is the last stop on our tour of Patagonian National Parks and the southern Andes. It would be difficult to find another place on earth that has such a spectacular collection of landscapes in such a relatively small area. This region truly is a geologic wonder of the world.

References

Abbott, L., and T. Cook. 2016. Travels in geology: Exploring an icon of Patagonia: Chile's Torres del Paine National Park. *Earth Magazine*. Accessed 29 September 2019, https://www.earthmagazine .org/article/travels-geology-exploring-icon-patagonia-chiles-torres-del-paine-national-park.

Altenberger, U., R. Oberhänsli, B. Putlitz, and K. Wemmer. 2003. Tectonic controls and Cenozoic magmatism at the Torres del Paine, southern Andes (Chile, 51°10′S). *Revista Geológica de Chile* v. 30 no. 1, p. 65–81.

Baumgartner, L.P. 2006. The geology of the Torres del Paine Laccolith, S-Chile (abs.). Geological Society of America Paper 4-9, Abstracts with Programs, Speciality Meeting No. 2, p. 50.

Fildani, A., M.R. Shultz, S.A. Graham, and A.L. Leier. 2007. A deep-water amalgamated sheet system, Punta Barrosa Formation, Marina's Cliff, Chile. American Association of Petroleum Geologists Studies in Geology #56, p. 125–127, DOI: 10.1306/1240920St563291.

Fosdick, J.C., B.W. Romans, A. Fildani, A. Bernhardt, M. Calderón, and S.A. Graham. 2011. Kinematic evolution of the Patagonian retroarc fold-and-thrust belt and Magallanes foreland basin, Chile and Argentina, 51°30′S. *Geological Society of America Bulletin* v. 123 no. 9/10, p. 1679–1698.

Galland, O., and C. Sassier. 2015. Torres del Paine – The Patagonian diamond. *GEOExPro* v. 12 no. 1, 16 p.

George, S.W.M., S.N. Davis, R.A. Fernández, L.M.E. Manríquez, M.A. Leppe, B.K. Horton, and J.A. Clarke. 2019. Chronology of deposition and unconformity development across the Cretaceous-Paleogene boundary, Magallanes-Austral Basin, Patagonian Andes. *Journal of South American Earth Sciences* v. 95, 18 p.

Giacosa, R., D. Fracchia, and N. Heridia. 2012. Structure of the southern Patagonian Andes at 49°S, Argentina. *Geologica Acta* v. 10 no. 3, p. 265–282, DOI: 10.1344/105.000001749.

Gorring, M.L. 2008. Field trip guide: Ridge-trench collision – The southern Patagonian Cordillera east of the Chile Triple Junction. *Geological Society of America Field Guide* 13, 22 p.

Hubbard, S.M., B.W. Romans, and S.A. Graham. 2007. Deep-water channel margin architecture, Cerro Toro formation, Cerro Mocho, Chile. *American Association of Petroleum Geologists Studies in Geology*, #56, p. 128–131, DOI: 10.1306/1240921St563265.

Leuthold, J., O. Müntener, L.P. Barumgartner, and B. Putlitz. 2014. Petrological constraints on the recycling of mafic crystal mushes and intrusion of braided sills in the Torres del Paine mafic complex (Patagonia). *Journal of Petrology* v. 55 no. 5, p. 917–949.

Macauley, R.V., and S.M. Hubbard. 2013. Slope channel sedimentary processes and stratigraphic stacking, Cretaceous Tres Pasos Formation slope system, Chilean Patagonia. *Marine and Petroleum Geology* v. 41, p. 146–162, DOI: 10.1016/j.marpetgeo.2012.02.004.

Malkowski, M., T. Schwartz, and Z. Sickmann. 2015. Jurassic-cretaceous stratigraphic evolution of the Magallanes-Austral deep-water foreland basin, Argentine Patagonia. Field Trip Guidebook, Stanford Project on Deep-water Depositional Systems (SPODDS), Department of Geological and Environmental Sciences, Stanford University, 63 p.

Martínez, L.C.A., A. Iglesias, A.E. Artabe, A.N. Varela, and S. Apesteguía. 2017. A new Encephalarteae trunk (Cycadales) from the Cretaceous of Patagonia (Mata Amarilla Formation, Austral Basin), Argentina. *Cretaceous Research* v. 72, p. 81–94.

Poiré, D.G., and J. Franzese. 2010. Mesozoic clastic sequences from a Jurassic rift to a Cretaceous foreland basin, Austral Basin, Patagonia, Argentina. *In* del Papa, C. and R. Astini (Eds.), Field Excursion Guidebook, 18th International Sedimentological Congress, Mendoza, Argentina, FE-C13, p. 1–53.

Ranney, W. 2016. Torres del Paine National Park, Chile. Earthly Musings – Wayne Ranney's Geology Blog. Accessed 29 September 2019, https://earthly-musings.blogspot.com/2017/03/the-ever-incredible-torres-del-paine.html.

Romans, B., A. Fildani, S. Hubbard, J. Covault, S. Graham, and J. Fosdick. 2011. The influence of tectonic evolution on deep-water stratigraphic architecture: Upper Cretaceous, Magallanes Basin, Chile. American Association of Petroleum Geologists Search and Discovery Article #50323, 25 p.

Schwartz, T.M., and S.A. Graham. 2015. Stratigraphic architecture of a tide-influenced shelf-edge delta, Upper Cretaceous Dorotea Formation, Magallanes-Austral Basin, Patagonia. *Sedimentology* v. 62, p. 1039–1077, DOI: 10.1111/sed.12176.

Vogt, M., M. Leppe, W. Stinnesbeck, T. Jujihara, H. Mansilla, H. Ortiz, L. Manríquez, and E. González. 2014. Depositional environment of Maastrichtian (Late Cretaceous) Dinosaur-bearing Deltaic Deposits of the Dorotea Formation, Magallanes Basin, Southern Chile (abs.). *In* Ifrim, C., F.J.C. Berciano, and W. Stinnesbeck (Eds.), The 23rd Latin American Colloquium on Earth Sciences, p. 156.

2

Neuquén Basin and Andean Foothills: Neuquén to Mendoza, Argentina

Tupungato volcano as seen looking west from the vineyards of Mendoza.

Overview

This is a land of oil and dinosaurs, majestic volcanos and dry salt lakes, magnificent foothill folds, and abundant ammonites. The intersection of the Andean foothills fold-thrust belts with the Neuquén Basin provides breathtaking vistas of rolling hills and broad valleys with a backdrop of snow-capped volcanic peaks. There is no shortage of interesting geology. Andean compressional thrusts die eastward into large anticlinal folds that contain much of Argentina's oil and gas. Erosion of volcanic, granitic, and sedimentary materials provides the raw ingredients for rich, well-drained soils that nurture the vineyards of Argentina's key wine-growing region. We start with the dinosaur museum in Plaza Huincul, also a thriving oil town. Driving north from Zapala we pass through the Agrio Fold-Thrust Belt of the Andean foothills, passing spectacular outcrops of the Vaca Muerta Formation, a key source rock and at the heart of the latest "resource play." A resource play is a hydrocarbon exploration concept where the oil is contained in the same rock it is generated in. At Chos Malal we enter a vast volcanic province and cross the Tromen and Payunia volcanic plateaus. North of Malargüe we traverse the Malargüe Fold-Thrust Belt. We pass through the colorful Cañon del Atuel before entering the world-famous wine-growing region (think Malbec) of Valle de Uco near Mendoza. During

DOI: 10.1201/9781351168281-2

our 1,300 km (780 mi) tour, we will encounter fossil beds, discuss dinosaur discoveries, review the development of the local energy industry, and appraise the Mendoza wine region.

Main stops on the Neuquén Basin geo-tour.

Itinerary

Start: Neuquén Airport

Stop 1 Carmen Funes Museum, Plaza Huincul

Stop 2 Los Catutos member, Vaca Muerta Formation

Stop 3 Arroyo Covunco

Stop 4 Lotena Formation

A Brief Review of the Neuquén Basin

Setting

At 160,000 km² (62,000 mi²), the Neuquén Basin covers most of Neuquén Province and part of Mendoza Province. The basin is bounded on the west by the Andes volcanic arc, on the southeast by the North Patagonian Massif, and on the northeast by the San Rafael Block and the Sierra Pintada Massif (Howell et al., 2005).

The basin contains between 7 and 8 km (23,000 and 26,000 ft) of strata (Kurcinka et al., 2018) that chronicle a continuous geological record from Triassic to Quaternary times. The basin began filling with Jurassic synrift deposits and continued with a series of Jurassic-Cretaceous marine transgressions and regressions that deposited sediments in a subduction-related retro-arc basin which evolved into an orogenic-foreland basin setting (Howell et al., 2005; Rojas Vera et al., 2015).

The basin is usually divided into a western sector containing roughly north-south-trending fold-thrust belts, and an eastern Neuquén Embayment, essentially flat and undeformed. Late Cretaceous-Cenozoic Andean deformation resulted in the development of a series of north-south-oriented fold-thrust belts (Marlargüe, Chos Malal, Agrio). Hydrocarbons are found throughout the basin, but most of the production is from the eastern sector.

Basin Development

The stratigraphic basement consists of Permo-Triassic igneous units of the Choiyoi Group. The Late Triassic to Early Jurassic synrift phase involved back-arc extension that led to the formation of rift basins that were filled with the pre-Cuyo Group volcanics and terrestrial deposits (Kurcinka et al., 2018). A main source of sediment was igneous and metamorphic rocks from the area of the present Frontal Cordillera and San Rafael-Las Matras blocks (Borghi et al., 2019).

The Early Jurassic to Early Cretaceous phase involved high-angle subduction along the western margin of Gondwana, allowing a thick succession to accumulate in the back-arc basin created by thermal subsidence (Howell et al., 2005; Horton et al., 2016). The arc contained gaps through which the basin maintained a connection with the Pacific, although tectonic uplift of the arc or sea level changes may have cut the basin off from the ocean at times (Kurcinka et al., 2018

Three major phases of the Andean Orogeny have been defined relative to the evolution of the basin: the Peruvian (Late Cretaceous), Incaic (Paleogene), and Quechua (Neogene to Recent) events (Cobbold and Rossello, 2003). Late Cretaceous shortening across the region was due to plate reorganization produced by the opening of the Southern Ocean. Westward movement of the South American Plate exceeded the rate of slab rollback in the subduction zone, triggering compressional uplift of the Central Andes and eastward thrusting towards the foreland in the Neuquén Basin. The retro-arc basin evolved into a foreland basin at this time, and the arc became the main source of sediment. Sedimentation during this time was primarily continental and included evaporites (Kurcinka et al., 2018). The Late Cretaceous Neuquén Group is considered to be a nonmarine sedimentary unit 50–1,500 m (164–4,920 ft) thick consisting of fluvial, alluvial, aeolian, and lacustrine deposits eroded from actively growing structures. This section unconformably overlies the Bajada del Agrio Group and is unconformably overlain by marine and continental deposits of the Campanian to Oligocene Malargüe Group (Borghi et al., 2019). The younger Malargüe Group represents an Atlantic marine transgression presumably produced by tectonic loading by the Andes in conjunction with global sea level rise, or perhaps flat-slab-related subsidence.

Another extensional event affected both the arc and retro-arc areas during Oligocene to early Miocene. Finally, a major shortening event, characterized by magmatism and foreland migration of deformation, began in early Miocene and continues to the present. Magmatism produced during subduction is represented by the intrusion of arc granitic rock in the Main Cordillera, or Cordillera Principal, by the emplacement of stratovolcanoes and in the foreland, and by the development of basaltic plateaus such as the Payunia Volcanics (Borghi et al., 2019).

Structure

Thrusts and folds along the western edge of the basin trend north-south, indicating that the principal stress was directed toward the east. Jurassic rift-related normal faults were reactivated during the Andean Orogeny, creating basement-cored structures mainly in the western part of the basin. As the faults propagated up-section during later phases of compression, the detachment surfaces/thrust faults accommodated shortening along evaporite (salt, gypsum) and shale horizons. Thrust movement along detachment surfaces caused detachment folds, fault-bend folds, and fault-propagation folds. Thrusts are directed mainly to the east, and structures verge mainly to the east, but backthrusts do occur. Folds tend to be older and of higher amplitude in the west and become younger and more gentle to the east.

The Huincul basement high (Huincul Arch, or "dorsal de Huincul") divides the Neuquén Basin into northern and southern sub-basins. The basement high is about 250 km (160 mi) long and essentially east-west. Originally thought to be a transpressive fault zone, it is now thought to be an inverted half-graben with perhaps some strike-slip on reactivated deep normal faults (Wikipedia, Neuquén Basin).

Regionally, the Mesozoic sedimentary basin fill dips and thickens to the west until it encounters the deformation front, where it has been uplifted and eroded several kilometers (Borghi et al., 2019).

Stratigraphy

The Early Jurassic to Early Cretaceous paleocurrent data show that the sediments were derived mainly from igneous and metamorphic rocks of the Patagonian and Sierra Pintada massifs, with a minor component from the Andean volcanic arc. This sedimentary sequence begins with deep-marine to slope and shelf turbidite deposits and mudstones of the Los Molles Formation, followed by the deltaic, shallow-marine Lajas, and fluvial Challaco formations. Multiple transgressive-regressive cycles are represented by the Cuyo, Lotena, Mendoza, and Bajada del Agrio groups. The total thickness of these four groups exceeds 6,000 m (19,700 ft) in the center of the basin and includes both marine and nonmarine sedimentary environments (Borghi et al., 2019).

Jurassic and Early Cretaceous marine units are found in the foothills zone. Upper Cretaceous sedimentary rocks are up to 1,500 m (4,920 ft) thick and mainly of continental origin. Only in the latest Cretaceous and Paleocene did transgression of the Atlantic result in shallow-marine strata. Upper Cretaceous redbeds underpin the foreland plateaus. Tertiary sediment is no more than a thin veneer in the distal foreland (Cobbold and Rossello, 2003).

Volcanics

We will visit a number of volcanos, cinder cones, and lava flows, mainly between Chos Malal and Malargüe. Two of these are exceptional.

Tromen Volcano, at 4,114 m (13,498 ft), is a major landmark northeast of Chos Malal. The Tromen Volcanics include basalt flows as well as andesites and ignimbrites (ashfall tuffs and pyroclastic flows). These volcanics cover over 2,000 km^2 (800 mi^2) and all erupted in the past 2.3 million years (Galland et al., 2007).

The Payunia Volcanic Field covers over 5,200 km² (2,000 mi²) in the western and central part of the basin. The field is part of the back-arc volcanism of the Andes, and likely formed in the past 300,000 years. The main volcanic center, Payún Matrú, standing 3,715 m (12,188 ft) high, is a 15 km (9.3 mi) wide shield volcano. Northeast of Payún Matrú is a 9 by 7 km (5.4 by 4.2 mi) caldera formed after the magma chamber erupted and the summit collapsed. The entire volcanic field sits 2,000 m (6,600 ft) above sea level. The volcanic field contains over 800 cones and basalt flows. Payún Matrú produced the longest known individual Quaternary lava flow on earth – 181 km (112 mi). The Payunia Volcanic Field is a protected area and is a candidate to be a UNESCO World Heritage site (Galland and Sassier, 2016).

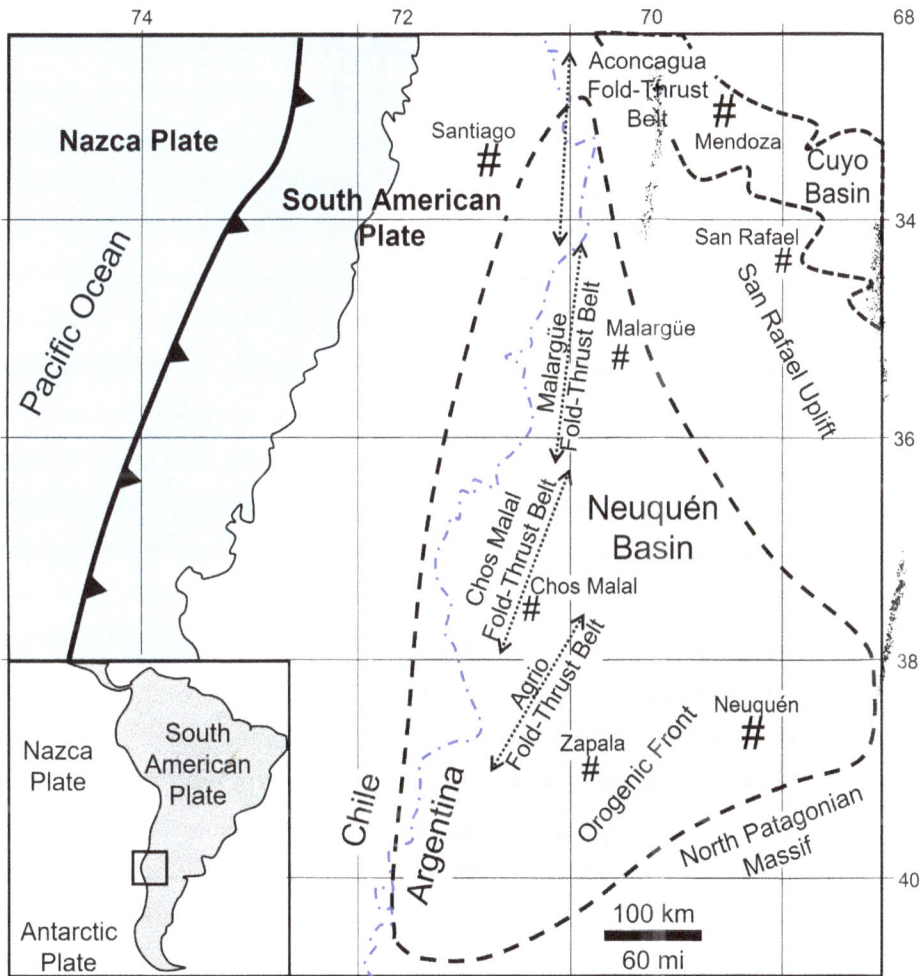

Present-day plate tectonic setting of the Neuquén Basin.

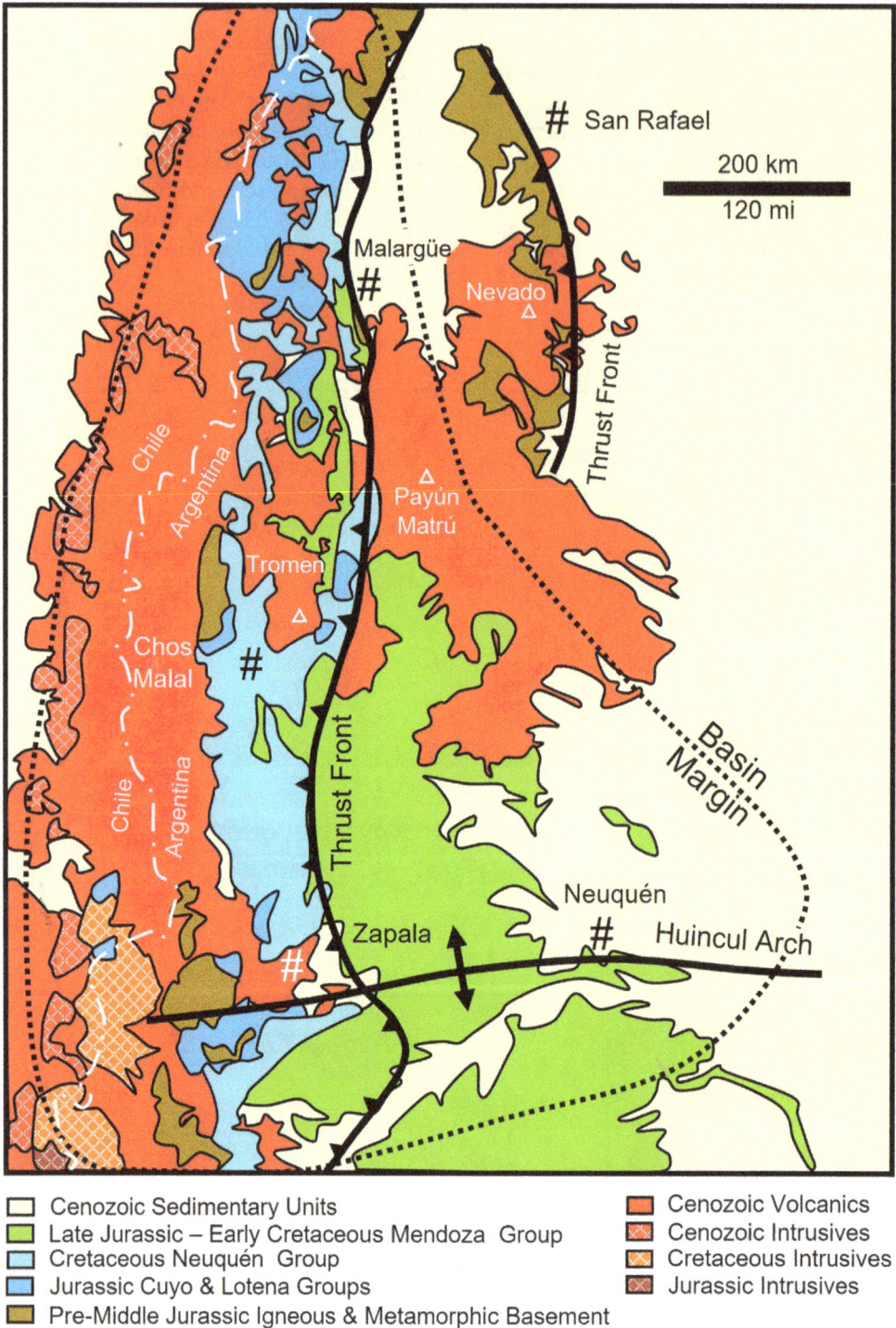

Simplified geologic map of the Neuquén Basin. Modified after Cobbold and Rossello, 2003; Naipauer and Ramos, 2016; Galland et al., 2019.

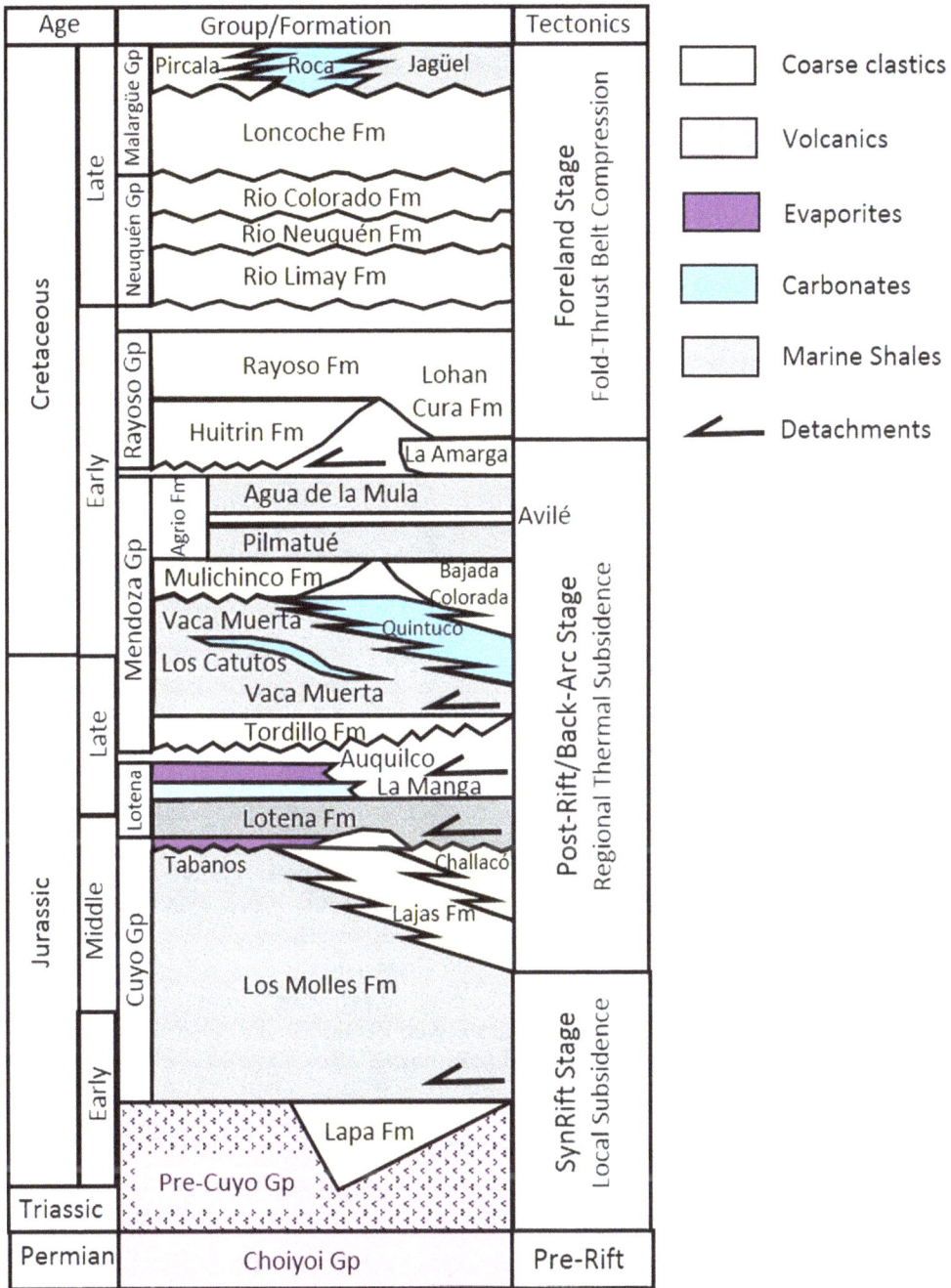

Stratigraphy of the Neuquén Basin. Stratigraphy derived from Cobbold and Rossello, 2003; Howell et al., 2005; Turienzo et al., 2012; Branellec et al., 2016; Kurcinka et al., 2018; Borghi et al., 2019.

Local History

The first inhabitants of the area were nomadic hunter-gatherers of the Puelches group. Hunting guanacos was the main means of subsistence, but they also hunted rheas and armadillos and fished near Llancanelo Lake. In the seventeenth, eighteenth, and nineteenth centuries, Mapuche groups migrated eastward into the Andes and pampas, establishing relationships and assimilating with the pre-existing groups. Largely driven out of the Neuquén area by settlers in the 1800s, the Mapuche today only occupy areas in south-central Chile and southwestern Argentina.

In 1604, Hernando Arias de Saavedra explored Patagonia. Starting in Buenos Aires, he reached what is now the city of Neuquén. In 1782, Basilio Villarino left Carmen de Patagones and traveled up the Río Negro. On January 23, 1783, he arrived at the confluence of the Limay and Neuquén Rivers (called Confluencia), camping on an island. He then followed the Limay to the confluence with the Collón Curá, and continued from there to the Chimehuin River. In 1885, the lands of Confluencia were auctioned to settlers. Shortly after, during the Conquest of the Desert campaign, the indigenous tribes that inhabited the province were either killed or driven out. The province became a center for cattle ranching, with some irrigated farming.

Founded in 1904, Neuquén is the capital of Neuquén province and the newest provincial capital in Argentina. It lies west of the confluence of the Limay and Neuquén rivers, which join here to form the Río Negro. The metropolitan area has a population of over 340,000, making it the largest city in Patagonia.

Neuquén today is both an agricultural center and a center for the petroleum industry, collecting oil and gas extracted from different points across the province.

As you drive from Neuquén to Stop 1, you are passing along the Huincul Arch, a roughly east-west-trending basement high that divides the Neuquén Basin into northern and southern sub-basins. The Huincul Arch is an inverted half-graben with perhaps some strike-slip on reactivated deep normal faults. The arch is bounded on the north by a northwest-inclined Triassic-Jurassic normal fault that had renewed movement in an opposite sense during the Peruvian phase (late Early Cretaceous through Late Cretaceous) of the Andean Orogeny. The existing normal fault was reactivated as a reverse fault that uplifted and inverted the Huincul Arch. In Late Cretaceous, the ridge became buried and less active. In 1897, soldiers discovered a large oil seep near Cerro Lotena on the arch. This was followed in 1918 by the first successful oil well in the basin drilled at Plaza Huincul (Cobbold and Rossello, 2003; Grimaldi and Dorobek, 2011).

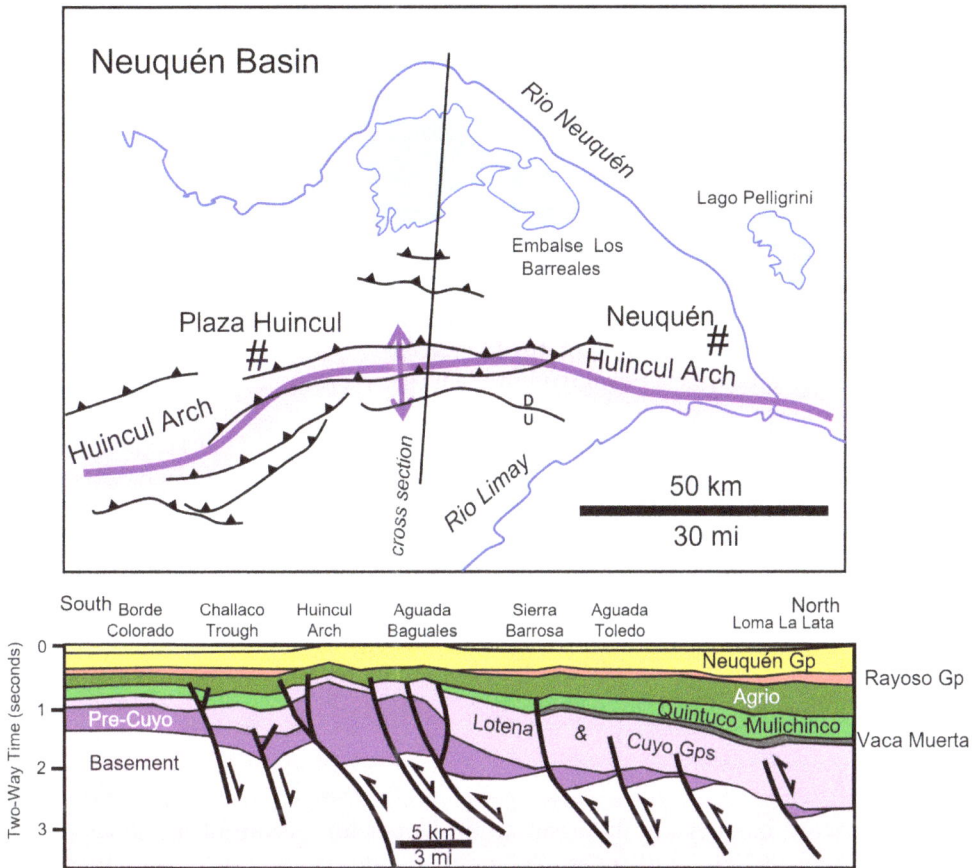

Map and cross-section of the Huincul Arch. The cross-section shows how the original normal faults were later inverted to form hydrocarbon traps. Field names are given. Modified after Grimaldi and Dorobek, 2011.

Neuquén Basin and the Energy Industry

The Neuquén Basin is the most important hydrocarbon-producing basin in Argentina and one of the most important in South America. In 2023, production was 51,000 m³ oil per day (320,000 barrels of oil per day) and 91 million m³ per day (3.2 billion ft³) of gas. Since the first discovery in 1918, close to 2.2 billion m³ (14 billion barrels) of oil have been found in the basin and 280 million m³ (1.8 billion barrels) produced (Howell et al., 2005). Proved and probable reserves in 2005 were estimated to be 300 million m³ (1.9 billion barrels) of oil and 495 billion m³ (17.5 trillion ft³) of gas (Legarreta et al., 2005; Smith Llinas, 2019; Saucier, 2019). The basin accounts for 45% of Argentina's oil production and 61% of its gas production (Wikipedia, Neuquén Basin). Hydrocarbons are found throughout the basin, although most are in the relatively undeformed eastern Neuquén Embayment (Howell et al., 2005; Galland and Sassier, 2016).

Source Rock

Three thick, high-quality marine organic-rich intervals cover most of the basin. Source rock formations include the Vaca Muerta, and to a lesser extent the Agrio and Los Molles formations (Fuentes et al., 2016; Ostera et al., 2016). As well, a Lower-Middle Jurassic non-marine source rock was deposited in anoxic lakes that developed within local half-grabens (Legarreta et al., 2005).

The Vaca Muerta Formation is the primary source rock in the basin. The Vaca Muerta Formation mainly consists of a succession of bituminous black shales deposited in an anoxic, restricted-circulation marine basin. The formation also contains occasional volcanics and limestones. The Vaca Muerta Formation ranges in age from latest Jurassic to Early Cretaceous. This formation was deposited in a northwest-southeast-elongated embayment bounded to the south by the Huincul Arch. The Vaca Muerta Formation represents the deepest and most distal part of the northwest-prograding Quintuco-Picún Leufú depositional system. The Vaca Muerta-Quintuco System ranges from up to 1,800 m (5,900 ft) thick at the basin depocenter to between 200 and 400 m (650 to 1,300 ft) at its margins (Cruset et al., 2021).

Total organic carbon and thermal maturity (ability to generate oil and gas) in the Vaca Muerta Formation appear best developed in the western basin; the unit becomes lean and immature in the eastern foreland. The Vaca Muerta-Quintuco System reached a maximum burial depth of around 6,000 m at the Andean front of the Agrio Fold-Thrust Belt during the Miocene, whereas the system only reached a maximum burial of around 3,700 m at the Chos Malal Fold-Thrust Belt during Late Cretaceous and reached a burial of up to 1,500 m during Early Cretaceous in Sierra de Vaca Muerta. Burial depth is directly related to thermal maturity as measured by vitrinite reflectance (Ro). Ro indicates that over much of the Neuquén Basin, this source rock is in the oil window (will generate oil), whereas in the western fold-thrust belt it is in the gas window (Cruset et al., 2021). In the Los Molles shale play, only the Lower Los Molles Formation seems to have appropriate conditions to generate (Fuentes et al., 2016; Ostera et al., 2016).

Reservoir Rock

The primary reservoir rocks in the basin are in Upper Jurassic and Cretaceous units. In the Jurassic, these include sandstones and conglomerates in the Lotena and Tordillo formations and limestones in the Barda Negra Formation. Cretaceous reservoirs are almost entirely sandstones and occur in the Troncoso member of the Huitrín Formation, the Avilé member of the Agrio Formation, and the Mulichinco Formation. Limestones of the Cretaceous Chachao also hold some oil.

Bitumen dikes are widespread in the Agrio Fold-Thrust Belt along the western side of the basin. The source of these hydrocarbons is thought to be organic-rich black shales of the Vaca Muerta and Agrio formations. (Cobbold and Rossello, 2003).

Seals

Regional seals occur in evaporites of the Auquilco and Huitrín Formations, with local seals in shales of the Vaca Muerta, Agrio, and Catriel formations (Wikipedia).

Traps

In the western part of the basin, structures are exposed at the surface in a series of essentially north-south fold-thrust belts (Galland and Sassier, 2016). The central and eastern basin has structural accumulations on broad upwarps over basement highs, including the Huincul and Chihuido highs. The central and eastern basin has accumulations that are stratigraphically trapped in updip reservoir pinchouts.

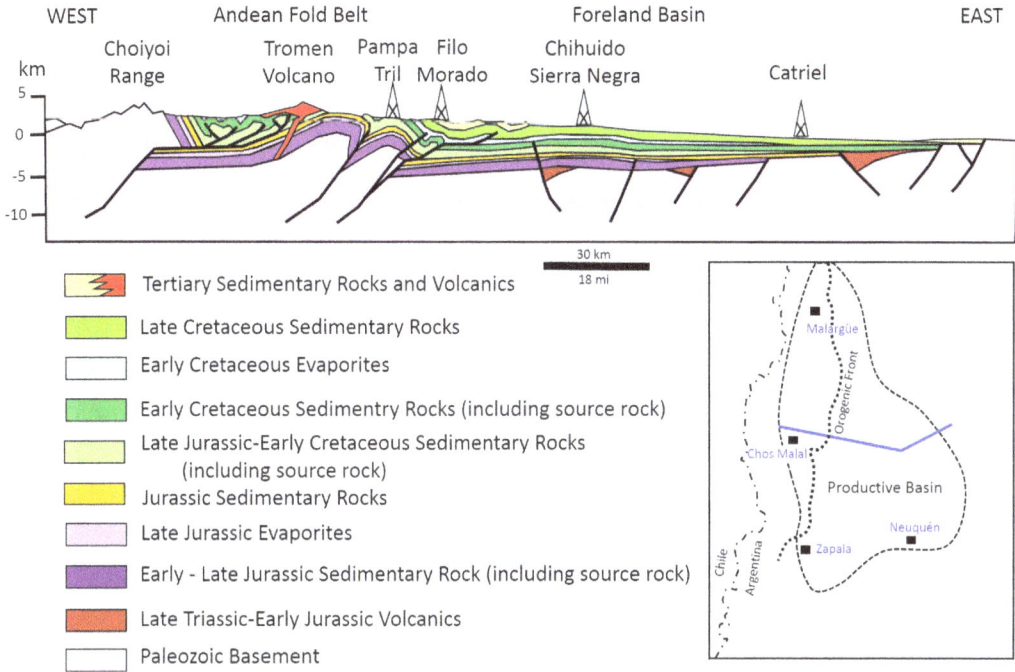

Basin cross-section showing structural domains. Modified from Legarreta et al., 2005.

Timing of Generation and Migration

The location of the source rock hydrocarbon kitchens is in the central and western deepest part of the basin. From there, hydrocarbons migrated updip to the basin margins, mostly in the east (Legarreta et al., 2005).

The timing of hydrocarbon generation affected the likelihood of accumulation and preservation. The Los Molles Formation's organic matter was almost entirely converted to hydrocarbons between Early Cretaceous and Early Tertiary time. The Vaca Muerta Formation generated hydrocarbons during Late Cretaceous to Miocene. The Agrio Formation generated hydrocarbons between Eocene and late Miocene (Zapata et al., 2003; Legarreta et al., 2005).

Early generated hydrocarbons, in the more mature western areas, had little chance of accumulating in later structures formed during Tertiary deformation. Thus, the fold belts have few fields sourced from the Los Molles and Vaca Muerta formations. Late generation and early (Late Cretaceous to Paleogene) trap development led to multiple accumulations along the eastern, updip margin of the Neuquén Embayment, including the Huincul High (Legarreta et al., 2005; Rojas Vera et al., 2015).

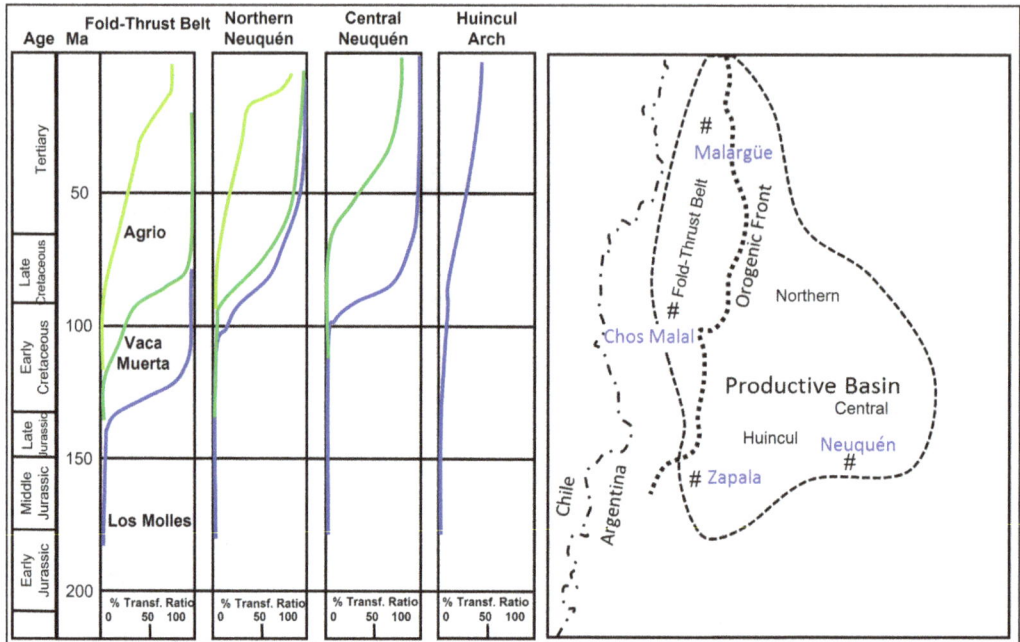

Transformation ratios show the timing of generation for the three primary source rocks in various parts of the basin. 50% transformation means half of the kerogen has produced hydrocarbons. Modified after Legarreta et al., 2005.

Unconventional Oil and Gas – the Vaca Muerta and Other Plays

Bituminous shale was discovered in the Río Salada Valley by Guillermo Bodenbender in 1892. The name Vaca Muerta was introduced in 1931 by the American geologist Charles Weaver. The Jurassic-Cretaceous unit in deeper parts of the basin consists of mature black shale, marl, and lime mudstone. This is the major source rock for all large oil accumulations in the basin. In 2010 Repsol-YPF opened the Vaca Muerta shale play with the first economic production at Loma La Lata Field. In addition to Repsol, the play is being actively explored/developed by Chevron, Total, Royal Dutch Shell, and ExxonMobil, among others (Fleming, 2018).

The Vaca Muerta Formation is the largest shale-oil producer in the Neuquén Basin. Its thickness ranges between 30 and 400 m (100 and 1,300 ft; Smith Llinas, 2019). The Vaca Muerta play extends over an area three times the size of the Permian Basin shale play in west Texas. The U.S. Energy Information Administration estimates total recoverable hydrocarbons from the Vaca Muerta at 2.58 billion m³ (16.2 billion barrels) of oil and 8.7 billion m³ (308 trillion cubic feet) of gas. Industry observers expect 140 to 150 frac'd wells to be drilled in the play in 2019, mostly high-density, long lateral horizontal wells, and up to 250 wells in 2021. Production in this play grew from 9,540 m³/day (60,000 barrels of oil per day) at the end of 2018 to 33,400 m³ per day (280,000 barrels per day) in 2022, mostly from the Loma Compana area operated by Repsol-YPF (Fleming, 2018). Other Vaca Muerta fields being developed are at Sierras Blancas/Cruz de la Lorena (Shell), El Orejano (YPF), and Fortin de Piedra (Smith Llinas, 2019). In August 2022, the Vaca Muerta reached an all-time production high of 91 million m³ (3.21 billion ft³) of gas per day.

Other shale oil and gas plays include the Agrio and Los Molles formations. Although lightly explored at the time of writing, they show promise.

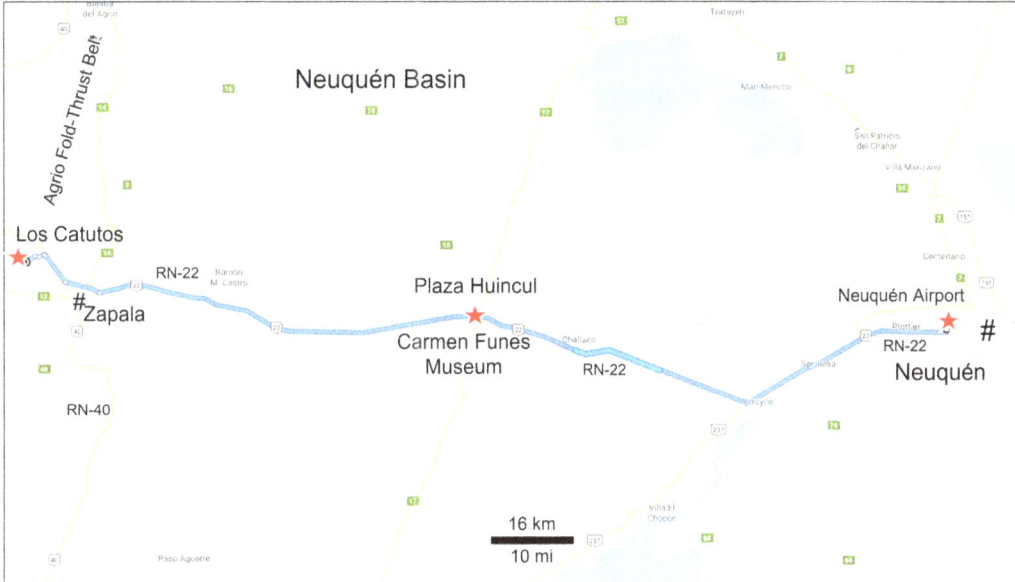

Start: Neuquén Airport
Most flights to Neuquén originate in Buenos Aires, Argentina.

Neuquén Airport to Carmen Funes Museum, Plaza Huincul: From Acceso al Aeropuerto drive south on Goya to Ruta Nacional-22; turn right (west) on RN-22 and drive to **Stop 1, Carmen Funes Museum, Plaza Huincul** *on the left (−38.931008, −69.193980), for a total of 98.4 km (61.1 mi; 1 hr 17 min).*

Stop 1 Carmen Funes Museum, Plaza Huincul

The Neuquén Basin is known locally as the "Valley of the Dinosaurs." The first dinosaur fossils were found here in 1883 and there has been a trove of specimens found since. Most dinosaur fossils come from Upper Cretaceous continental deposits of the Neuquén Embayment, the eastern, undeformed part of the basin. While excellent samples can be found at the Bernardino Rivadavia Museum of Natural Science in Buenos Aires, the best fossils are exhibited in the dusty oil town of Plaza Huincul in the middle of the flat Neuquén Basin.

The museum was established in 1984 and has become a significant tourist attraction. It is named after an early pioneering lady who offered lodging, food, and water to travelers at her post near here. At the museum, you will see *Argentinosaurus huinculensis*, an herbivore of the sauropod family, considered the largest dinosaur ever discovered. This giant is estimated, mainly based on 1.5 m (5 ft) high vertebrae, to have been 35 m (115 ft) long and weighed 100 tons (110 US tons). It was found at a site 8 km (5 mi) east of Plaza Huincul. As well, *Giganotosaurus carolinii*, one of the largest carnivores ever found, is displayed along with *Caypullisaurus bonapartei*, the most complete ichthyosaur (marine reptile) fossil found in South America. The museum features the only known sauropod embryos, found in a remarkable nesting site near Auca Mahuida, 150 km (90 mi) northwest of Neuquén. The embryos are so well preserved that you can see patches of delicate fossilized skin (Galland and Sassier, 2016).

Argentinosaurus huinculensis at the Museo Municipal Carmen Funes, Plaza Huincul, Neuquén, Argentina. Photo courtesy of W.I. Sellers, L. Margetts, R. Aníbal Coria, and P.L. Manning, https://commons.wikimedia.org/wiki/File:Argentinosaurus_skeleton,_PLoS_ONE.png.

The museum has 11 holotype specimens (specimens upon which the description and name of a new species is based). Field work, often in cooperation with other American museums, continues in this fossil-rich basin.

If you are interested in fossils, you should also visit the Ernesto Bachmann Paleontological Museum in Villa El Chocón, 75 km (47 mi) southwest of Neuquén, and fossil sites such as Lago Barreales, where excavations are currently ongoing.

Visit

The museum has an exhibition area with skeletons, dioramas, and videos. There are also exhibits on local history, archeology, and anthropology. The museum has workshops for preparing fossils and a gift shop.

Address: 55 Córdoba Avenue, Plaza Huincul, Neuquén Province (postal code 8318), east of the intersection of National Route 22 and Provincial Route 17.

Phone: +54 299 496-5486.

Hours: Mondays through Fridays from 9 am to 7:30 pm. Saturdays, Sundays, and holidays, from 9 am to 9 pm

Check the website or call for current entrance fees, and to be sure it is not closed for renovation.
    Website: https://www.interpatagonia.com/cutralco-huincul/carmen-funes-municipal-museum.html.

> **Carmen Funes Museum, Plaza Huincul to Los Catutos Member, Vaca Muerta Formation:** *Continue east on RN-22 to the town of Zapala; turn right (north) on RN-40; drive 11.1 km (6.9 mi) north on RN-40 to the sign for Los Catutos by an unpaved and unnamed road on the left; turn left (west) and drive 4.3 km (2.6 mi) to the village of Los Catutos. Park and walk about 700 m (2,400 ft) south up the hill to an abandoned quarry. This is* **Stop 2, Los Catutos Member, Vaca Muerta Formation** *(−38.837011, −70.197860), for a total of 92.4 km (57.4 mi; 1 hr 13 min).*

Stop 2 Los Catutos Member, Vaca Muerta Formation

As you drive west across the flat expanse of the basin, you will see multicolored shale and sandstone beds on the mesas north and south of the highway. This is the Lisandro Formation. The Lisandro Formation (Neuquén Group) varies between 35 and 75 meters (115 and 246 ft) thick. It is composed of red siltstones and claystones, which have been interpreted as deposited in a swampy to fluvial (river) environment. The red-striped Lisando Formation is generally easy to distinguish from the greenish or yellowish sandstones of the underlying Late Cretaceous Huincul Formation.

Redbeds in the Neuquén Group between Plaza Huincul and Zapala. View southwest from RN-22.

The Los Catutos stop is on a small rise just south of the village of the same name. A quarry has been cut into the rhythmic carbonate-marl sequence. The outcrop represents a Late Jurassic carbonate ramp with north-prograding clinoforms (sloping surfaces) developed in water depths estimated to be less than 50 m (160 ft). Abundant fossils indicate peak biological activity (Leanza, 2012).

The outcrops around the Sierra de la Vaca Muerta are good analogs for the distal portion of the depositional system and therefore are good analogs for exploration in the Vaca Muerta tight oil and tight gas plays. Brittle and permeable zones within the overall tight shale usually occur in the carbonate-rich sequence tops. These sweet spots for unconventional exploration are detectable on seismic based on rock properties (Zeller, 2013).

Google Maps satellite image showing the Los Catutos and Sierra Vaca Muerta stops. Imagery ©2023 TerraMetrics.

Quarry in the Los Catutos member of the Vaca Muerta. This represents a carbonate ramp deposit.

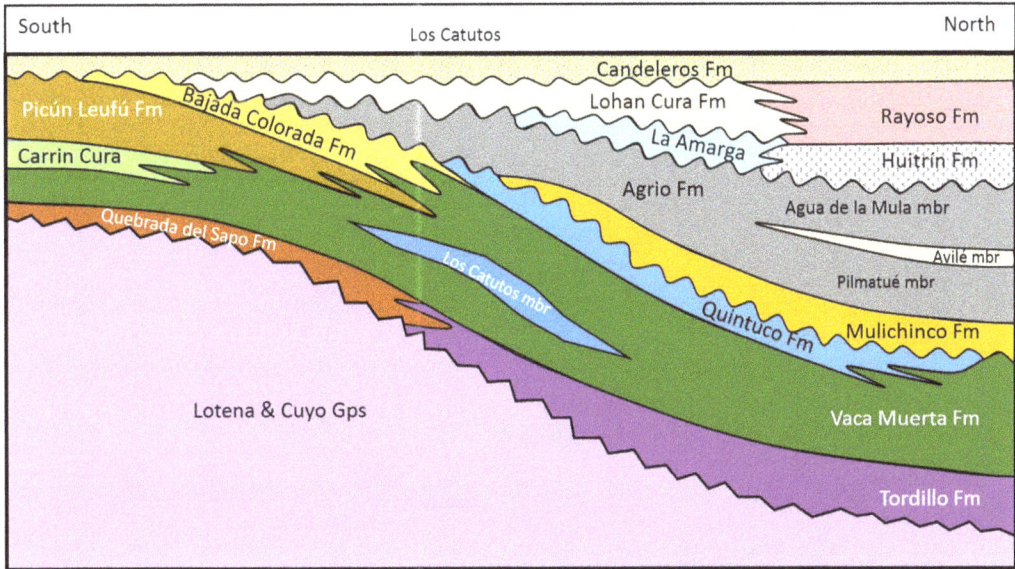

Schematic stratigraphic cross-section showing the Los Catutos member, Vaca Muerta Formation. Modified after Leanza, 2012.

Ammonite impression from the Los Catutos quarry.

Los Catutos Member, Vaca Muerta Formation to Arroyo Covunco: Return to RN-40 and turn left (north); drive to **Stop 3, Arroyo Covunco** (–38.79628, –70.20085), pullout on the left for a total of 10.6 km (6.6 mi; 12 min).

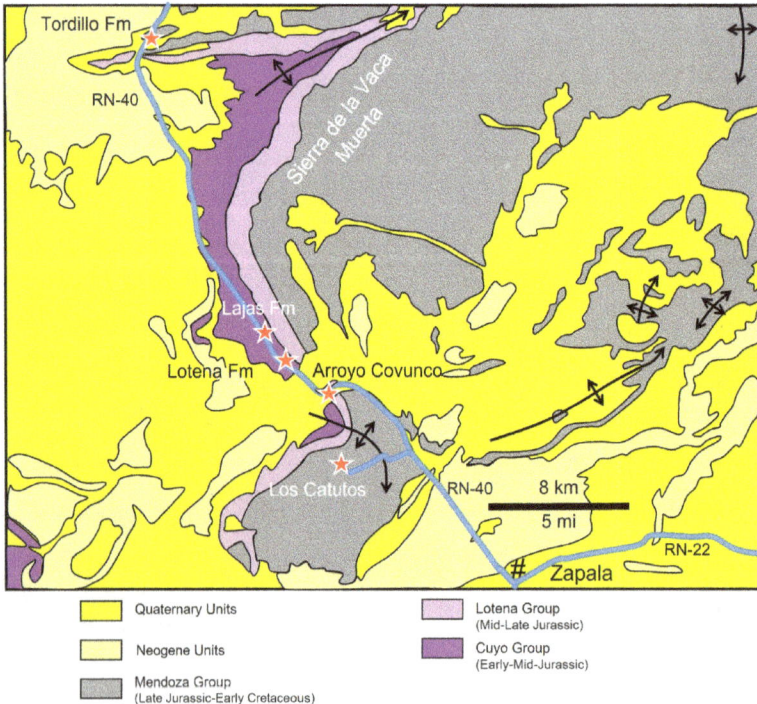

Geologic map of the Sierra Vaca de Muerta area north of Zapala. Modified after Leanza and Hugo, 2005.

Stop 3 Arroyo Covunco

This stop provides a roadcut through the east flank of a southeast-plunging anticline. Outcrops, from bottom (west) to top (east), include the Jurassic Lajas, Tabanos, Lotena, and La Manga formations (Canale et al., 2016). The mesa is capped by flat-lying Pleistocene gravel and conglomerate (Leanza and Hugo, 2001).

View northeast across RN-40 at east-dipping Jurassic units including the Lajas, Tabanos, Lotena, and La Manga formations from base to top, respectively. Dashed line indicates trace of bedding.

> ***Arroyo Covunco to Lotena Formation:*** *Continue north of RN-40 for 6.0 km (3.7 mi; 5 min) to **Stop 4, Lotena Formation** (−38.767661, −70.251222) and pull over on the right.*

Stop 4 Lotena Formation

Fair exposures of the Lotena Formation can be seen alongside RN-40 at this stop. The unit consists of Middle to Upper Jurassic fluvial conglomerates, calcareous sandstones, and marine limestones and shales. In places, the Lotena Formation can be a massive sandstone up to 45 m (150 ft) thick and has been mentioned as a potential hydrocarbon reservoir rock in the basin (Arcuri and Zavala, 2008).

View northeast toward the Sierra de la Vaca Muerta. Sandstone and shale in the foreground are of the Lotena Formation. The fence limits access to the outcrop.

Lotena Formation to Lajas Formation: Continue north on RN-40 for 3.0 km (1.8 mi; 2 min) to Stop 5, Lajas Formation (−38.747946, −70.274002) and pull over on the right.

Stop 5 Lajas Formation at km 2457

The 600 m (2,000 ft) thick Middle Jurassic sandstone and conglomerate of the Lajas Formation (Cuyo Group) has been considered primarily deltaic, perhaps tidally influenced (Mcilroy et al., 2005). The Jurassic basin was funnel-shaped, and narrow in the south, which accentuated any tidal influence. Recent work, however, suggests that the Lajas Formation is instead a river-dominated system, with only moderate amounts of tidal influence. Waves and locally strong tidal currents are present mainly during periods of sea advance (transgressions) when the shoreline changes from river-dominated deltas to wave-dominated barrier islands (Kurcinka et al., 2018).

Lajas Formation, west side of the Sierra Vaca Muerta, dipping southeast. On the side of RN-40 at km 2,457 .

> *Lajas Formation to Tordillo Hogback: Continue driving north on RN-40 for 17.3 km (10.7 mi; 11 min) to **Stop 6, Tordillo Hogback and Auquilco Formation** (−38.61060, −70.34212), and turn right (east) onto a dirt road; drive 300 m (0.2 mi) down the road to the outcrop.*

Stop 6 Tordillo Hogback and Auquilco Formation

This stop affords the opportunity to examine the Tordillo and, to a lesser extent, the Auquilco formations. Redbeds of the Late Jurassic Tordillo Formation, which lies above the Auquilco and just below the Vaca Muerta Formation, consist of coarse conglomerates, alluvial, fluvial, and aeolian sandstones, and volcaniclastic deposits that range from less than 40 m (130 ft) thick in the southern Neuquén Basin to as much as 300 m (980 ft) thick in the northern basin, although it is known to have abrupt lateral thickness changes. The Tordillo Formation forms one of the main commercial reservoirs in the Neuquén Basin.

The widespread Auquilco Formation (Lotena Group) represents a Late Jurassic marine regression (withdrawal) that deposited thick (up to 1,100 m thick) evaporites, mainly gypsum, gypsiferous sandstone, and gypsiferous limestone (Leanza and Hugo, 2001). It forms a major detachment surface in the Agrio and Malargüe fold-thrust belts (Lebinson et al., 2018). Here the Auquilco Formation lies in the depression below and south of the hogback and is indicated by white to light-colored soil deposits.

View northeast from RN-40 to the Tordillo Sandstone hogback, west side of the Sierra de la Vaca Muerta. The gently folded Tordillo Sandstone sits on Auquilco evaporites, mainly gypsum.

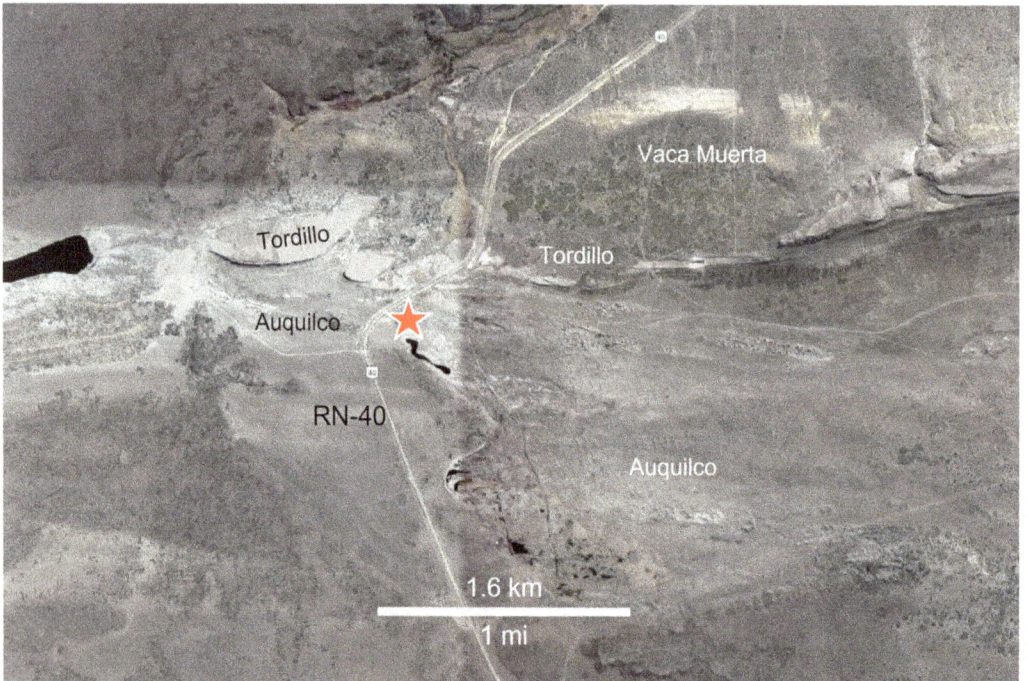

Google Maps image showing the Tordillo-Auquilco stop. Imagery ©2023 Airbus, CNES/Airbus, Landsat/ Copernicus, Maxar Technologies.

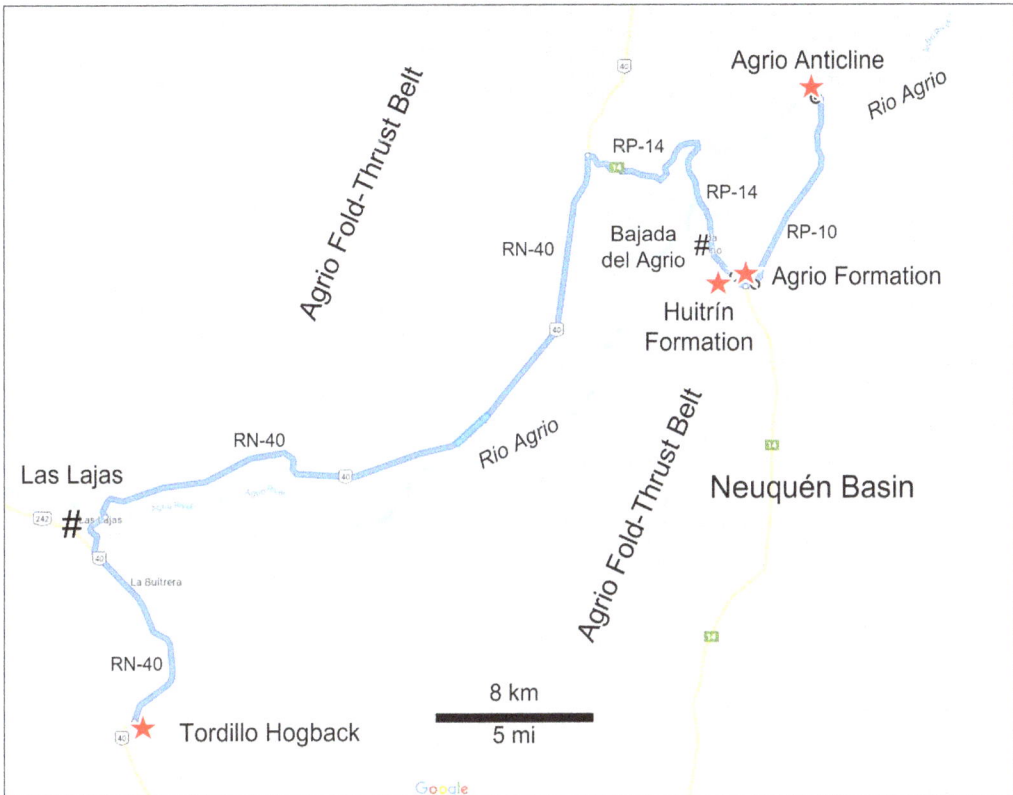

***Tordillo Hogback and Auquilco Formation to Huitrín and Agrio Formations,
Bajada del Agrio:*** *Continue north on RN-40 to Ruta Provincial-14 turnoff on the right
going to Bajada del Agrio; turn right (east) on RP-14 (nicely graded but unpaved) and
drive to* **Stop 7.1, Huitrín Formation over Agrio Formation, Bajada del Agrio**
(−38.421767, −70.012929), for a total of 61.4 km (38.1 mi; 51 min).

Stop 7 The Agrio Fold-Thrust Belt

Fold-thrust belts formed along the western margin of the Neuquén Basin as a result of
the Andean Orogeny. They include, from south to north, the Aluminé Fold-Thrust Belt
(FTB; from near Malleo to Zapala), the Agrio FTB (between Zapala and Chos Malal), the
Chos Malal FTB (around Chos Malal to Barrancas), the Malargüe FTB (from Barrancas
to around Sosneado; Rojas Vera et al., 2016), and the Aconcagua FTB around that peak.
Taken together, these fold-thrust belts all have a similar history and timing. They began to
develop in the eastern Andes-western Neuquén Basin in Late Cretaceous and progressed
eastward in at least two phases until Miocene. Much of what is said here about the Agrio
Fold-Thrust Belt also applies to the Chos Malal and Malargüe fold-thrust belts.

The Agrio Fold-Thrust Belt, which we entered near Zapala, is bounded to the east by the Los Chihuidos High and to the west by the Loncopué Trough. The fold-thrust belt is a roughly 50 km (36 mi)-wide zone consisting of Jurassic and Cretaceous units carried mainly on east-directed thrusts and folded into symmetric and east-verging, roughly north-northwest-trending anticlines and synclines. There are also west-directed back-thrusts and associated folds. The Agrio and Chos Malal Fold-Thrust Belts contain up to 8,000 m (26,000 ft) of Jurassic and Cretaceous sedimentary rocks that unconformably over-lie a Permo-Triassic volcanic basement, the Choiyoi Group (Rojas Vera et al., 2010; Rojas Vera et al., 2015; Borghi et al., 2019).

The fold-thrust belt is characterized by a combination of thin-skinned "sled-runner" thrusts similar to the Sevier deformation in North America, and thick-skinned basement-cored thrusts similar to Laramide deformation in North America. There is a difference in deformation style between the inner, western fold-belt and the outer, eastern part. Western structures are more affected by the reactivation of favorably oriented normal faults that originated during Middle to Late Jurassic rifting and breakup of Gondwana (140–170 Ma). East-directed compression and inversion of these normal faults resulted in uplifted and thrust basement blocks. Continuing displacement was transferred to detachments in Late Jurassic evaporites (Auquilco Formation) or Late Jurassic to Early Cretaceous shales (Vaca Muerta and Agrio formations). This continued shortening led to thin-skinned detachment folds, fault-bend folds, and fault-propagation folds in Upper Cretaceous-Tertiary units.

The eastern Agrio Fold-Thrust Belt consists of large, tight anticlines detached in Auquilco Formation evaporites that bound deep basement blocks. Detachments propa-gated up from the Auquilco Formation through the Mendoza Group until they reached Huitrín Formation evaporites of the Rayoso Group, which became a main detachment in the eastern sector. The anticlines are separated by broad synclines with shapes that appear to reflect basement blocks at depth (Zapata et al., 2008).

The Agrio Fold-Thrust Belt developed during two main compressive phases of the Andean Orogeny (Late Cretaceous-Paleocene and the middle-late Miocene) in response to convergence between the Nazca and South American plates. Restored structural cross-sections show tectonic shortening across the fold belt on the order of 11 km (18%; Lebinson et al., 2018). Progressive eastward deformation began in Late Cretaceous during a period of shallowing of the subduction zone between the Nazca and South American tectonic plates. Deformation began earlier and is more intense in the west; as it progressed eastward into the basin, the thrusts become shallower and involve younger strata; eventually, the eastern tip of thrusting stops in the subsurface and a reverse thrust from that tip point forms a triangle zone in an area characterized by gentle folding.

The earliest deformation in the thrust belt occurred at the Cordillera del Viento block north and west of Chos Malal. The first detachment developed in the Agrio Fold-Thrust Belt at the Naunauco-Loma Rayoso area. At the same time, tectonic inversion affected the inner (western) part of the Agrio Fold-Thrust Belt, as confirmed by thick synorogenic deposits of the Neuquén Group to the east. At that time, the back-arc region of the Neuquén Basin became a foreland province with sedimentation derived from the Andes to the west. The magmatic arc began migrating east toward the foreland at this time (Zapata et al., 2003; Rojas Vera et al., 2015). An early foreland basin formed in Late Cretaceous time and filled with syntectonic deposits of the Neuquén Group.

Late Miocene deformation, which dominated in the eastern part of the thrust belt, also partially reactivated the western structures (Rojas Vera et al., 2015). The main compres-sional deformation ended in late Miocene (Ramos and Folguera, 2005). Some structures, including the Cerro Negro, Cerro Mayal, and Cerro Caicayen folds were intruded by Eocene andesitic magmatism, thus nailing down their timing of development (Cobbold and Rossello, 2003; Pons et al., 2010). We will see the Collipilli intrusion into the La Mula Anticline. During the last compressional pulse, which began in middle Miocene, the entire fold-thrust belt moved toward the foreland. The thrusts probably used pre-existing Jurassic detachment surfaces (Zapata et al., 2008).

The last uplift and exhumation event is interpreted as late Miocene (10 Ma), and some workers feel it continues to the present. This last compressive event formed the El Cholar Anticline, the Loma Rayoso Anticline (we will examine an unconformity there), and the Chihuidos High. The triangle zone developed at this time. Miocene uplift of the Agrio Fold-Thrust Belt is indicated by the presence of the synorogenic Tralalhue conglomerates located in piggyback basins (Rojas Vera et al., 2015) and the Agua de la Piedra, Molle, Pincheira, Coyocho, and Huincán formations in the Malargüe Fold-Thrust Belt (Silvestro et al., 2005). We will see synorogenic deposits in the Pincheira Syncline west of Malargüe.

The magnitude of deformation can be evaluated based on the thickness of the syn-tectonic deposits. Late Cretaceous deformation is associated with more than 1,500 m (4,900 ft) of synorogenic deposits whereas Miocene deformation is associated with less than 150 m (490 ft) of synorogenic deposits (Rojas Vera et al., 2015).

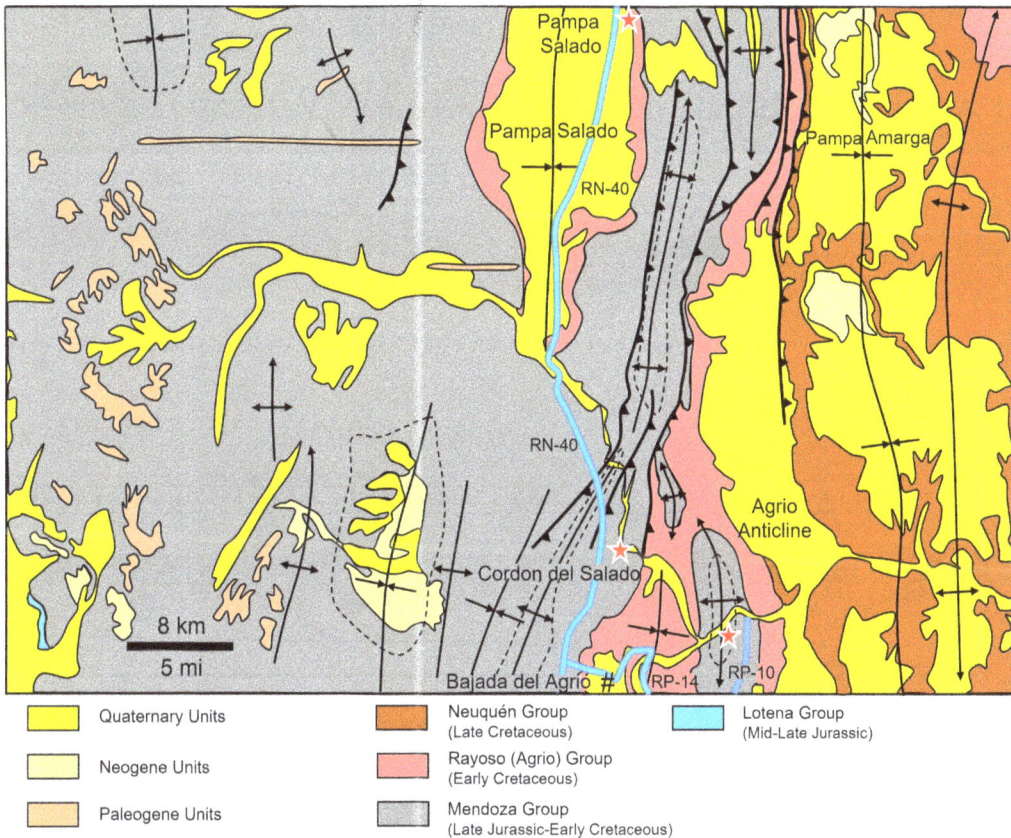

Geologic map of the Agrio FTB between Bajada del Agrio and Pampa del Salado. Modified after Leanza and Hugo, 2005.

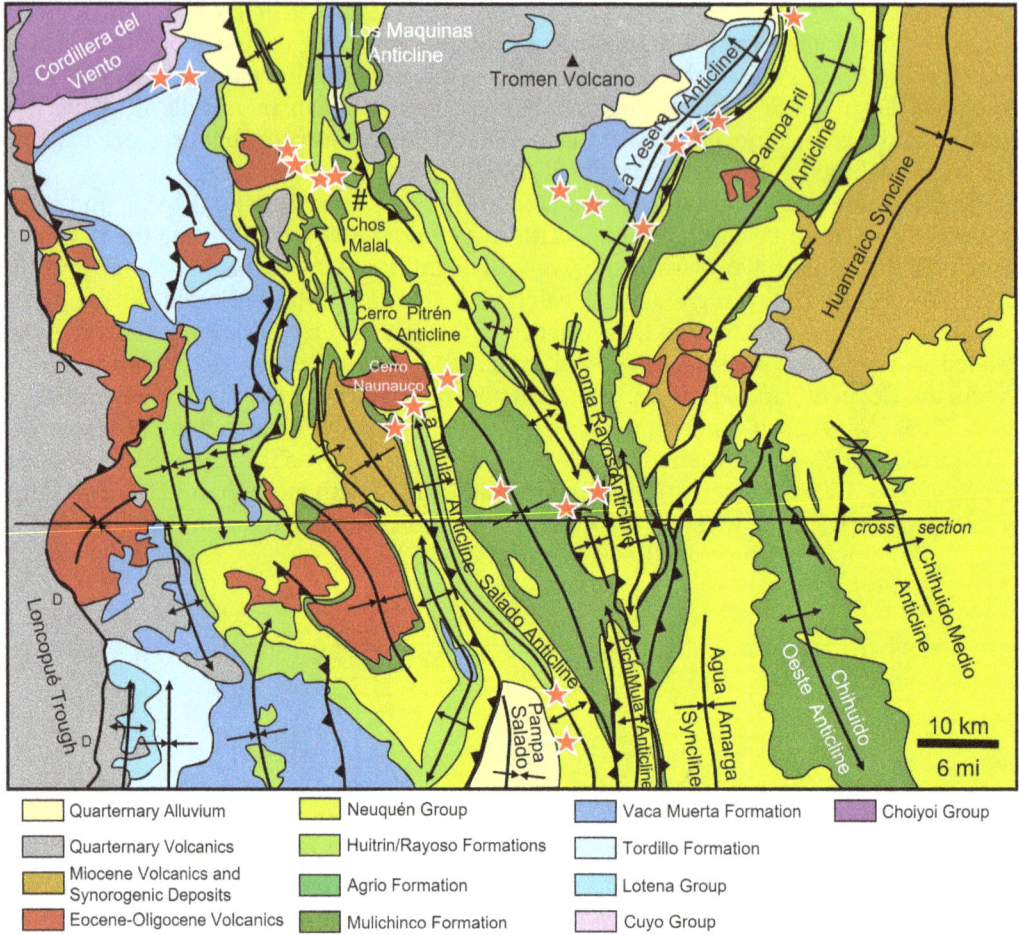

Geologic map of the Agrio FTB between Pampa del Salado and Chos Malal. Modified after Rojas Vera et al., 2015.

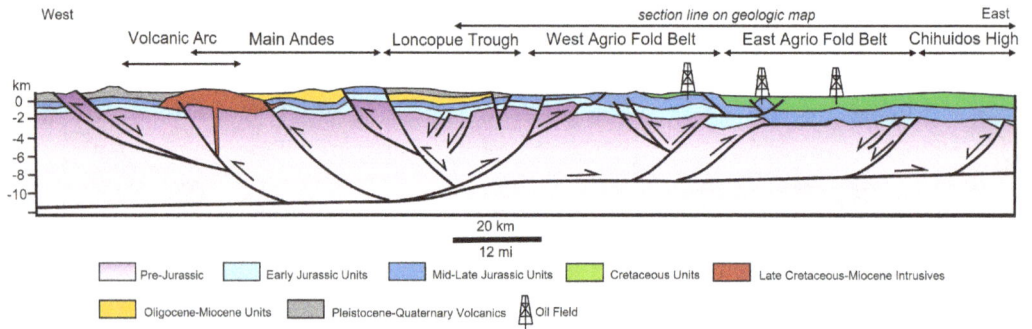

West-east cross-section through the Andes and Agrio Fold-Thrust Belt at about –37º45'. Modified after Rojas Vera et al., 2015.

Stop 7.1 Huitrín and Agrio Formations, Bajada del Agrio

The west flank of the Agrio Anticline exposes the Huitrín Formation over the Agrio Formation. The Early Cretaceous Huitrín Formation, Bajada del Agrio Group, can be up to 65 m (215 ft) thick and is divided into three members: (1) the Lower Troncoso member sandstone, composed of a lower fluvial and an upper eolian section; (2) the Upper Troncoso, containing evaporites deposited in a restricted marine environment; and (3) the La Tosca member consisting of interbedded marine limestones and shales (Zapata et al., 2008; Naipauer and Ramos, 2016). At this stop, we are looking at the Lower Troncoso Sandstone member. Sandstones of the Lower Troncoso member are some of the best hydrocarbon reservoir rocks in the basin (Lebinson et al., 2018). We discuss the Agrio Formation at the next stop.

The Upper Troncoso and La Tosca members indicate the final marine incursion from the Pacific Ocean into this area (Rojas Vera et al., 2015).

Huitrín Formation sandstone over Agrio Formation shale, Bajada del Agrio. Looking west along RP-14.

Huitrín and Agrio Formations, Bajada del Agrio to Agrio Formation, Bajada del Agrio: *Continue driving southeast on RP-14 to RP-10; turn left (northeast) on RP-10 (unpaved but well-graded) and drive to* **Stop 7.2, Agrio Formation, Bajada del Agrio** *(−38.425247, −70.001983) for a total of 1.2 km (0.8 mi; 2 min).*

Stop 7.2 Agrio Formation, Bajada del Agrio

We are driving across the west flank of the Agrio Anticline, moving to older rocks as we approach the core of the fold. At this stop, we see the Agrio Formation (Early Cretaceous Mendoza Group), which serves as a hydrocarbon source rock, reservoir rock, seal, and detachment surface in the basin. It consists of three members: (1) the lower Pilmatué member, cyclic deep-marine to near shore gray-green to dark gray shales up to 577 m (1890 ft) thick; (2) the Avilé member consisting of widespread anastomosing fluvial (river channel) medium-grained sandstone 10–90 m (33–295 ft) thick, and (3) the Agua de la Mula member, up to 489 m (1,600 ft) of black marine mudstone and dark gray-green shale rhythmically interbedded with limestone, muddy limestone, and limy sandstone (Leanza and Hugo, 2001; Borghi et al., 2019). A continental conglomeratic unit of the Agrio Formation has been recognized in the eastern part of the basin in both outcrops and wells. The conglomerates may indicate uplift of the eastern foreland (a foreland bulge) during Early Cretaceous (Naipauer and Ramos, 2016). Overall, the unit indicates a Pacific marine transgression during a period of tectonic quiescence (Zapata et al., 2008; Turienzo et al., 2012).

Agrio Formation, Bajada del Agrio. View southwest from RP-10.

The black shales of the lower Agrio Formation are good source rocks. The organic material is Type II kerogen (marine algal, oil-prone) deposited in an anoxic environment with minor terrestrial input. Total organic carbon (TOC) values vary between 1 and 3.5%. Thermal maturity varies from immature to mature (late oil stage) depending on the depth of burial. The upper Agrio Formation is also a potential source rock, as it contains bituminous black shales with TOC values that vary from 1 to 2%. The organic matter is kerogen type II and III (marine algal and gas-prone woody material). Thermal maturity varies from mature to immature (Ostera et al., 2016). Modeling suggests that the carbon-rich black shales of the Agrio Formation became mature and may have generated oil in Late Cretaceous or Paleogene (Cobbold and Rossello, 2003).

The Agrio Formation contains one of the main hydrocarbon reservoir units in the basin, the Avilé member. The Avilé consists of fluvial and eolian sandstones that reflect a major sea-level drop.

Shales within the Agrio Formation form widespread detachment horizons in the Agrio and Chos Malal fold-thrust belts in the western part of the Neuquén Basin (Lebinson et al., 2018).

Agrio Formation, Bajado del Agrio to Agrio Anticline: Continue driving north on *RP-10 for 9.8 km (6.1 mi; 17 min) to* **Stop 8, Agrio Anticline** *(−38.34639, −69.9675). Pull over on the right and walk up the hill to the left (west) for the best view.*

Stop 8 Agrio Anticline

This stop is on the east flank of the Agrio Anticline and in the outer, eastern sector of the Agrio Fold-Thrust Belt. The view north across the Agrio River looks up the axis of the anticline. The Agrio Anticline is at the thrust-front near the south end of the thrust belt. This north-south-oriented, relatively tight, doubly plunging and slightly west-verging anticline exposes the Pilmatué member of the Agrio Formation in the core, and Avilé/Agua de la Mula members dipping 40–60 degrees on the flanks. It is bounded on both sides by broad synclines: The Villa del Agrio Syncline on the west exposes the Rayoso Formation; the Agua Amarga Syncline on the east contains Huincul and Cerro Lisandro formations. The anticline is formed over an east-dipping backthrust (Leanza and Hugo, 2005).

The Agrio Anticline was the first fold in the fold belt to be drilled, by Standard Oil, in 1935. The RA x-1 well was located on the surface anticline near an oil seep. This and a following well, drilled by YPF in 1949, were both dry holes (Zapata et al., 2008). The fold is non-productive to this day.

Google Earth oblique view north over the Agrio Anticline and Bajada del Agrio. Stars indicate stops. Imagery ©2023 Maxar Technologies, Airbus, Landsat/Copernicus.

View north into core of the Agrio Anticline.

Bivalves weathering out of the Agrio Formation, Agrio Anticline.

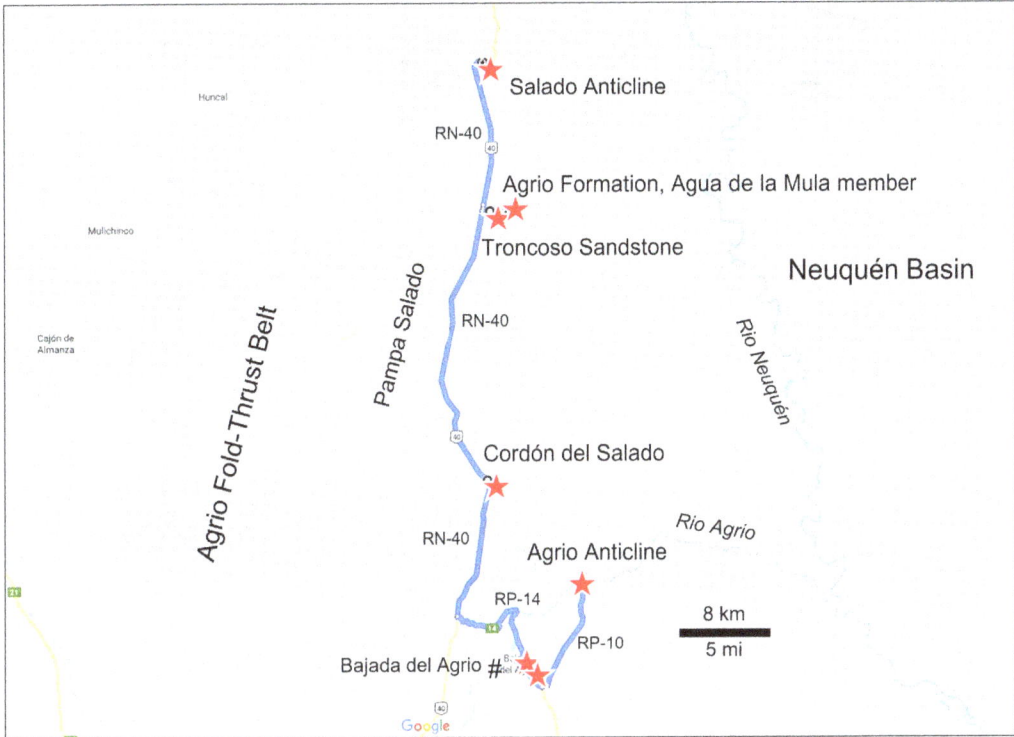

*Agrio Anticline to Cordón del Salado: Return south and west to RN-40; turn right (north) on RN-40 and drive to **Stop 9, Cordón del Salado** (−38.263946, −70.062655) and pull over on the right for a total of 37.9 km (23.5 mi; 46 min). Note that sections of RN-40 are graded gravel rather than paved.*

Stop 9 Cordón del Salado

This stop provides a view into the Cordillera (or Cordón) del Salado at the southern plunge of the Cerro La Mula-Naunauco Anticline (Zapata et al., 2008), also called the Salado Anticline (Rojas Vera et al., 2015). At 60 km (36 mi) long, this doubly plunging anticline is the longest continuous fold in the Agrio Fold-Thrust Belt. Complex internal deformation is marked by lateral ramps, minor thrusts, backthrusts, and layer-parallel shear (Zapata et al., 2008). The flanks dip steeply and are overturned in places. The Vaca Muerta Formation is exposed in the core of the anticline, and the flanks consist of resistant sandstone beds of the Mulichinco Formation.

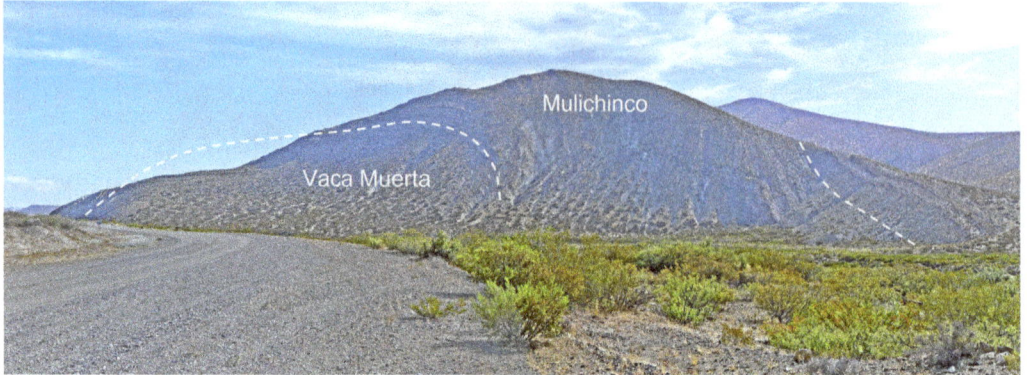

View north into the core of the Cerro La Mula-Naunauco Anticline from RN-40.

> *Cordón del Salado to Troncoso Sandstone, Pampa Salado: Continue driving north on RN-40 for 25.7 km (16.0 mi; 17 min) to an unnamed dirt road on the right; turn right (east) and drive to **Stop 10.1 Troncoso Sandstone, Pampa Salado** (−38.052432, −70.059242), for a total of 26.5 km (16.5 mi; 20 min) and pull over on the right.*

Stop 10 Pampa Salado

The purpose of this stop is to observe the stratigraphy of the upper Agrio and Huitrín formations. As we discussed at the Agrio Anticline, the Huitrín Formation has three members, the Lower Troncoso sandstone, Upper Troncoso evaporites; and La Tosca interbedded limestones and shales (Zapata et al., 2008). This stop provides an opportunity to walk up to and closely examine outcrops of this important reservoir sandstone.

Stop 10.1 Troncoso Sandstone, Pampa Salado

As we drive across the west flank of the Cerro La Mula-Naunauco/Salado Anticline we first approach the resistant, west-dipping Lower Troncoso Sandstone member of the Huitrín Formation.

Troncoso Sandstone member of the Huitrín Formation looking east from the access road.

Huitrín Formation Troncoso Sandstone to Agrio Formation, Agua de la Mula Member: *Continue driving east on the dirt road for 1.8 km (1.1 mi; 4 min) to* **Stop 10.2, Agrio Formation, Agua de la Mula Member** *(−38.05473, −70.03985) and pull over on the right.*

Stop 10.2 Agrio Formation, Agua de la Mula Member

After we cross the resistant Huitrín Formation Troncoso Sandstone, we enter the more easily eroded shales of the Agua de la Mula member of the Agrio Formation in the core of the structure. These black marine mudstones are a likely source rock for oil in the overlying Troncoso Sandstone and other reservoirs.

West-dipping shale of the Agua de la Mula member, Agrio Formation. View south, at Pampa Salado.

*Agrio Formation, Agua de la Mula Member, to Salado Anticline: Return to RN-40 and turn right (north); drive a total of 16.6 km (10.3 mi; 17 min) to **Stop 11, Salado Anticline** (−37.935494, −70.066578) and pull over on the right.*

Stop 11 Salado Anticline

You get a good view of the west flank of the Salado Anticline as you drive to the next stop.

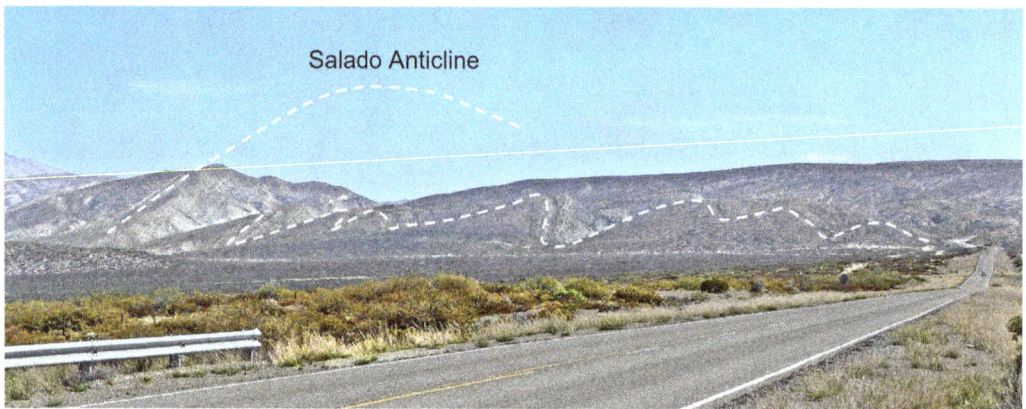

View north along RN-40 toward the Salado Anticline stop. Dashed lines indicate the trace of bedding.

The Salado Anticline roadcut exposes the steeply west-dipping Agua de la Mula member of the Agrio Formation on the west flank of the Salado Anticline. Here the member consists of thick black marine shale alternating with thin siltstone, sandstone, and limestone layers. Lower Cretaceous decapod crustaceans (lobster-like creatures) are commonly preserved in calcareous nodules in the shale here (Aguirre-Urreta et al., 2012).

The Salado Anticline and Pichi Mula Anticline immediately to the east are detached in Jurassic Auquilco Formation evaporites. These folds have wavelengths on the order of 2 to 4 km (1.2 to 2.4 mi). At the surface, the Pichi Mula Anticline is thrust west over the Salado Anticline. East-west shortening is estimated at about 9 km (5.5 mi) across the Agrio Fold-Thrust Belt here (Rojas Vera et al., 2015).

Agua de la Mula member of the Agrio Formation exposed on the west flank of the Salado Anticline, RN-40 roadcut.

The Salado Anticline is continuous with the La Mula-Nanauco Anticline to the north. The Salado Anticline and La Mula Anticline stops are indicated by the red stars. Google Earth oblique view north. Imagery ©2023 Maxar Technologies, Airbus, Landsat/Copernicus.

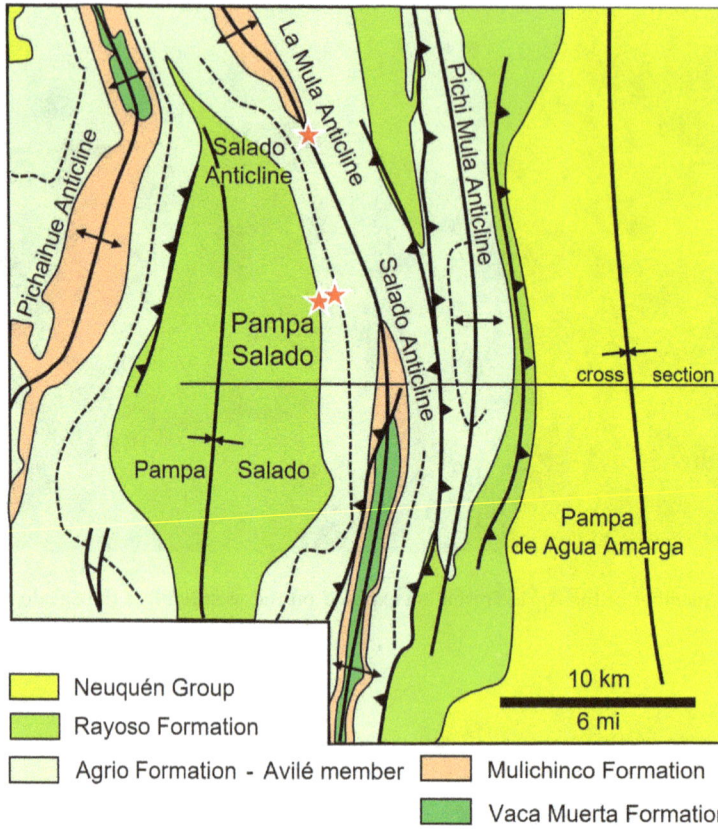

Geologic map of the Pampa Salado, Salado Anticline, and Pichi Mula Anticline triangle zone. Modified after Zamora Valcarce and Zapata, 2015.

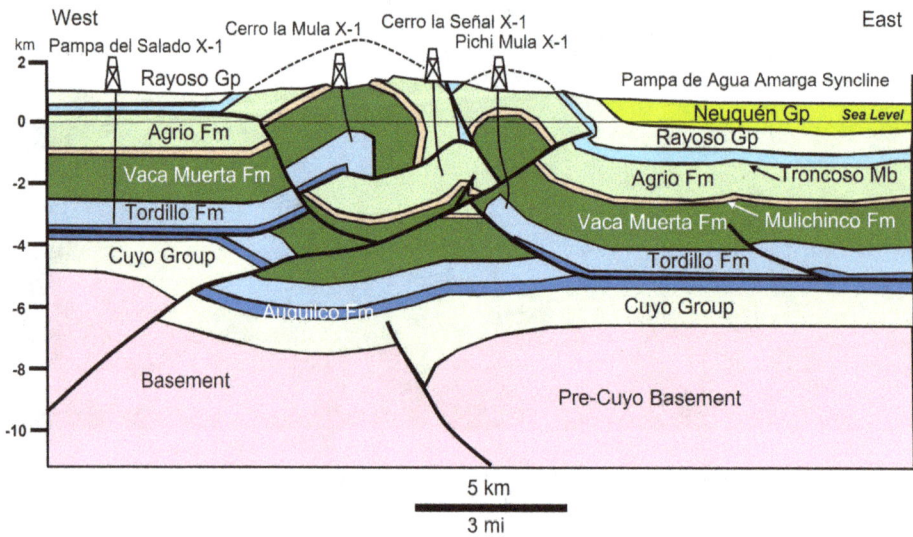

Cross-section through the Salado/Cerro La Mula and Pichi Mula anticlines showing the triangle zone. Modified after Zamora Valcarce and Zapata, 2015.

*Salado Anticline to Cerro Rayoso Anticline: Continue north on RN-40 to RP-9 on the right; turn right (east) onto RP-9 and drive on this graded gravel road to **Stop 12, Cerro Rayoso Anticline** (–37.717971, –70.007473) for a total of 32.8 km (20.4 mi; 30 min). Pull over on the right.*

Stop 12 Cerro Rayoso Anticline

This stop provides a nice overview of the Cerro Rayoso Anticline. This west-verging (tilted), doubly plunging breached fold exposes the Agrio Formation in the core and the Huitrín Formation on its flanks. A west-directed backthrust is mapped on the west flank of this structure (Rojas Vera et al., 2015).

West-verging Cerro Rayoso Anticline, view to the east from RP-9. Because the structure is west-verging, the core of the anticline is exposed on the west flank of this ridge.

Cerro Rayoso Anticline to Rayoso Formation, Pichi Neuquén Syncline: Turn around and drive back west on RP-9 for 1.1 km (0.7 mi; 2 min) to **Stop 13, Rayoso Formation, Pichi Neuquén Syncline** *(–37.72718, –70.01233).*

Stop 13 Rayoso Formation, Pichi Neuquén Syncline

Just west of Cerro Rayoso Anticline is the broad Pichi Neuquén Syncline that contains the Rayoso Formation at the surface. The syncline lies over the southern termination of the Cerro Rayoso Anticline. The contact between the two structures is an east-dipping backthrust. The detachment surface is in Rayoso Formation evaporites (Zapata et al., 2008).

The Early Cretaceous Rayoso Formation continental sediments in the Neuquén Basin were deposited after the basin was cut off from the Pacific. The distinct red-and-white banded unit contains over 1,200 m (3,900 ft) of alternating evaporites, carbonates, and sandstones. The sandstones are potential reservoirs, whereas the evaporites act as local detachment surfaces. Deposition is thought to have been in a fluvial-alluvial system in an arid climate with distinct seasonal precipitation (Barros et al., 2016).

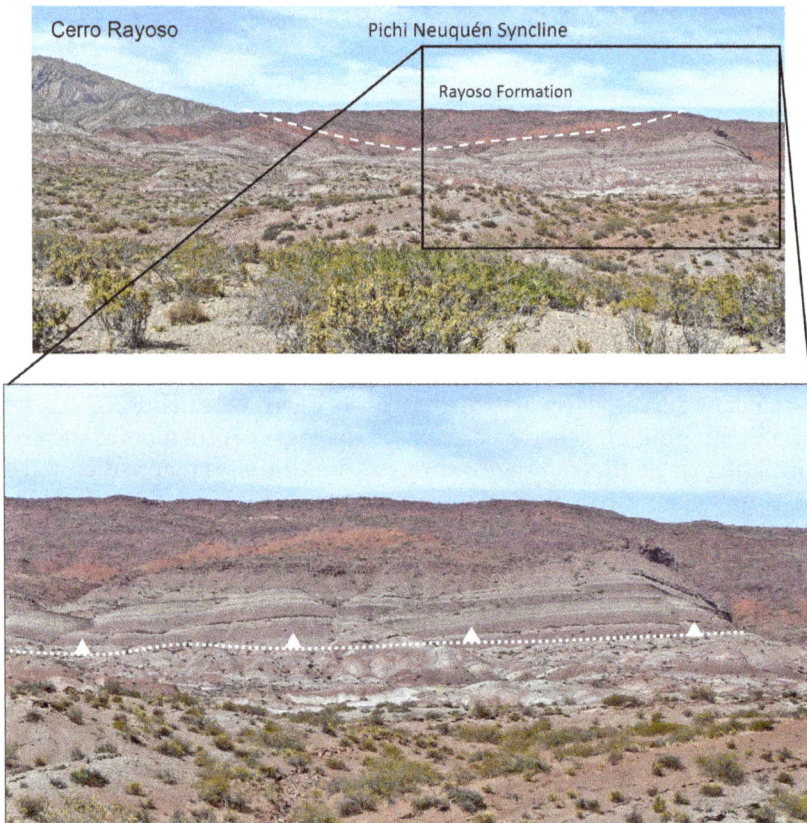

View south down the axis of the Pichi Neuquén Syncline and at the backthrust in the Rayoso Formation.

Rayoso Formation, Pichi Neuquén Syncline to La Mula Anticline: *Return to RN-40; turn right (north) on RN-40 and drive to* **Stop 14, La Mula Anticline Overview** *(−37.72701, −70.15267) for a total of 17.9 km (11.1 mi; 20 min) and pull over on the right.*

Stop 14 La Mula Anticline Overview

From this stop on RN-40, you can look southwest into the core of La Mula-Naunauco Anticline where it has been dissected by the Río Tralalhue. It appears to be a simple anticline capped by resistant Mulichinco sandstone and containing dark Vaca Muerta Formation shale in the core. As mentioned previously, this doubly plunging anticline is the longest fold in the Agrio Fold-Thrust Belt (the south end is sometimes called the Cerro La Mula Anticline, while the north end is called the Cerro Naunauco Anticline). This structure is continuous with the Salado Anticline that we saw earlier. The apparent simplicity belies the fact that the anticline's flanks are steep to overturned and are cut by thrusts and back thrusts (Zapata et al., 2008).

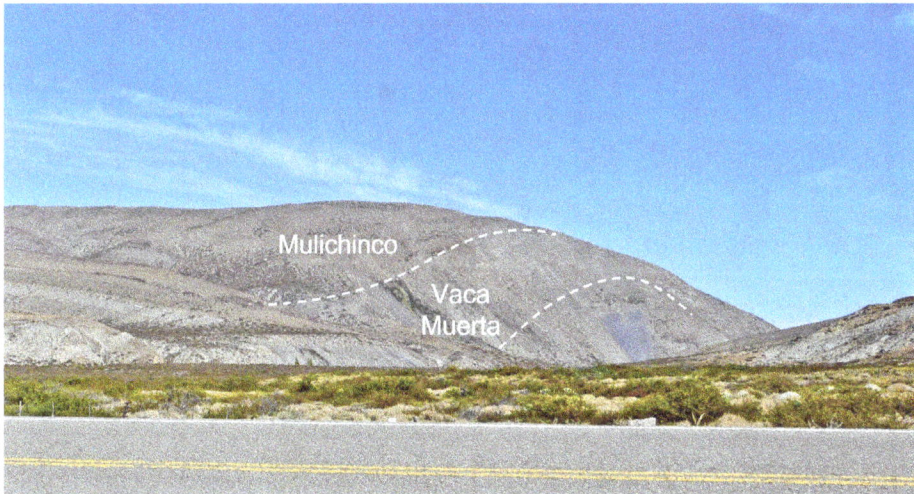

Looking west from RN-40 into the valley of the Río Tralalhue that cuts across the La Mula Anticline.

La Mula Anticline Overview to Agrio-Huitrín Contact, La Mula Anticline: *Continue north on RN-40 to RP-4 on the left near the village of Naunauco; turn left (west) on unpaved but graded RP-4 and drive through the anticline to* **Stop 15.1, Agrio-Huitrín Contact, La Mula Anticline** *(−37.65498, −70.23548) for a total of 18.7 km (11.6 mi; 18 min).*

Stop 15 La Mula-Cerro Naunauco Anticline

These stops cross the Cerro Naunauco Anticline at the north end of the regional Salado/ Cerro La Mula/Cerro Naunauco trend.

Stop 15.1 Agrio-Huitrín Contact, La Mula Anticline

Cross the La Mula-Naunauco Anticline to arrive at this stop. Looking south, you see the contact between the upper Agrio Formation (La Tosca member interbedded limestones and shales) and overlying Troncoso sandstone member of the Huitrín Formation along the west flank of the structure.

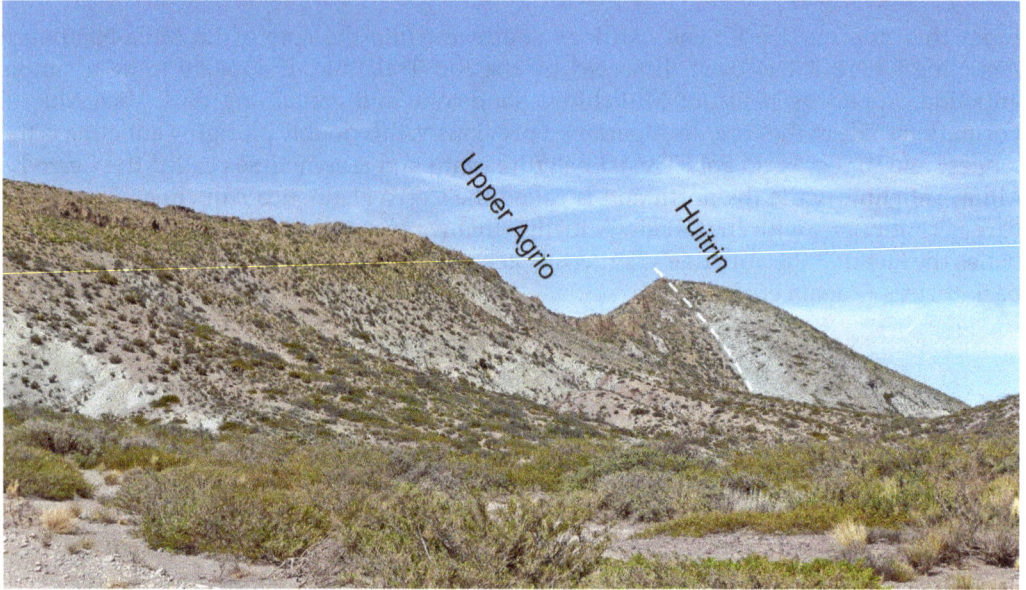

View south along the west flank of the La Mula-Cerro Naunauco Anticline.

Agrio-Huitrin Contact to Collipilli Laccolith and Anticlinal Axis: *Return east on RP-4 for 3.2 km (2.0 mi; 4 min) to **Stop 15.2, Collipilli Laccolith and Anticlinal Axis** (−37.63577, −70.21201) and pull over on the right.*

Stop 15.2 Collipilli Laccolith and Anticlinal Axis

Intrusives of the Collipilli Formation were emplaced between bedding as laccoliths and sills. The Collipilli Laccolith is intruded near the axis of the La Mula-Cerro Naunauco Anticline in the interval between the Agrio and the Rayoso formations. From here you can look north along the axis of the anticline to see the laccolith.

Whereas a sill is a relatively thin intrusion parallel to bedding, a laccolith is a sill that has thickened and domed-up the overlying layers.

The area had been uplifted and partially eroded prior to intrusion. Age dating correlates the Collipilli region igneous rocks with Paleogene units mapped in the Andean Cordillera. An early subvolcanic facies called the Collipilli Formation has ages ranging from 50 to 45

Ma, whereas a later, Eocene volcanic unit called the Cayanta Formation is dated around 39 Ma. Recent age dating indicates a much larger age range, from 65 to 12 Ma. The Cerro Naunauco laccolith is dated at 65.5 Ma.

The Collipilli and Naunauco intrusives are subalkaline low-potassium andesites indicative of a volcanic arc, which is typical for the Andes. Trace element analyses suggest they are arc to back-arc units (Zapata et al., 2008).

View north from RP-4 along the axis of the La Mula-Naunauco Anticline. This is Cerro Naunauco and the Collipilli Laccolith.

Collipilli Laccolith and Anticlinal Axis to Collipilli Laccolith, Cerro Naunauco: *Continue driving east on RP-4 to RN-40; turn left (north) on RN-40 and drive to **Stop** **15.3**, **Collipilli Laccolith, Cerro Naunauco** (–37.62664, –70.16010) for a total of 6.4 km (4.0 mi; 9 min).*

Stop 15.3 Collipilli Laccolith, Cerro Naunauco

This stop provides a view west from RN-40 toward Cerro Naunauco and the Collipilli Laccolith at the north end of the La Mula-Naunauco Anticline. The significance of the igneous rocks in the northern Agrio Fold-Thrust Belt is that crosscutting relationships between the magmatic rocks and sedimentary formations indicate a minimum age of 102 Ma for the beginning of deformation in this belt, and that there were at least two deformation episodes, one during Lower to Middle Cretaceous, and a second in the middle Miocene (Zamora Valcarce et al, 2006).

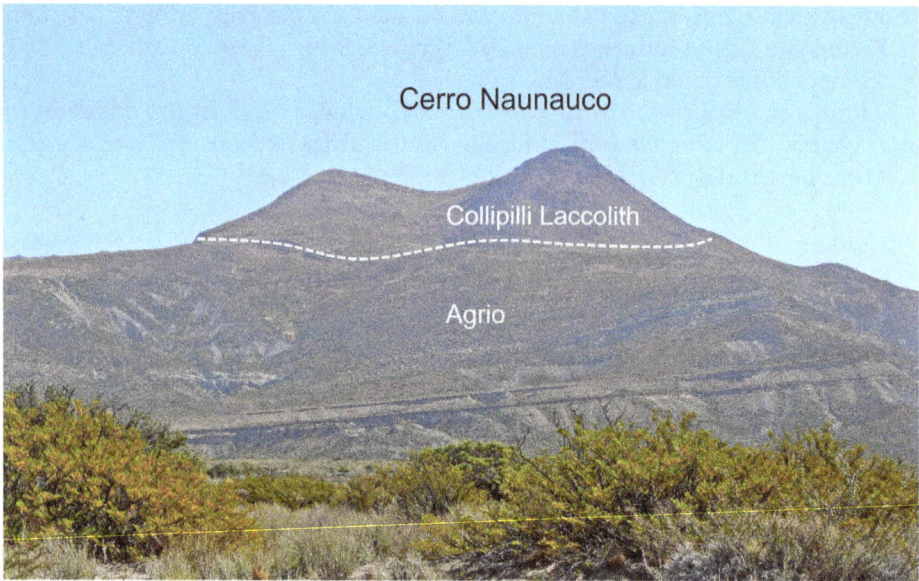

View west to Cerro Naunauco and the Collipilli Laccolith.

Collipilli Laccolith, Cerro Naunauco to Agrio Formation, Las Máquinas Anticline:
*Continue driving north on RN-40 to the southern edge of Chos Malal; bear left on RP-43 and drive through town to **Stop 16, Agrio Formation, Las Máquinas Anticline** (–37.36668, –70.29877) on the west side of Chos Malal for a total of 35.8 km (22.2 mi; 32 min).*

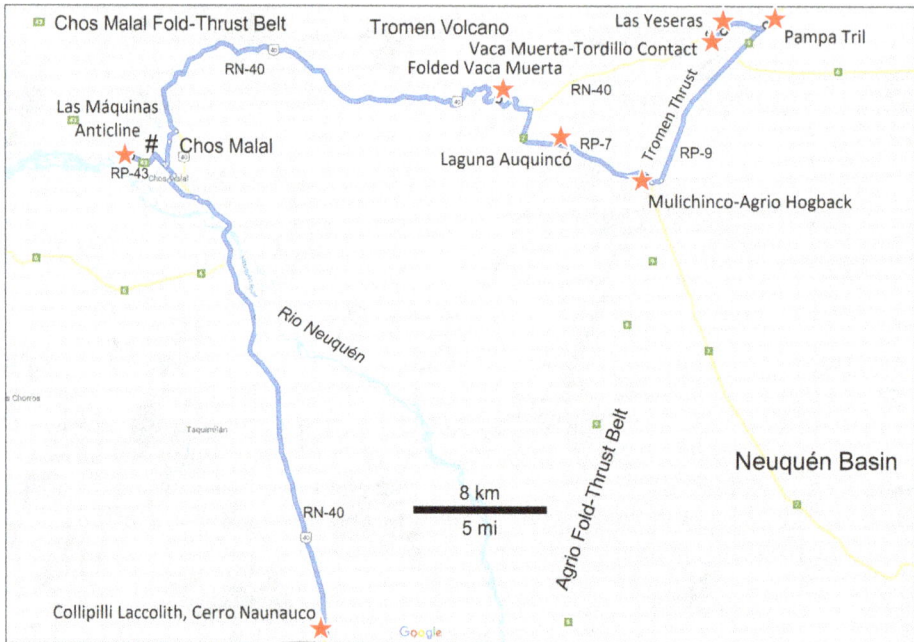

Stop 16 Agrio Formation, Las Máquinas Anticline

An extensive section of the Agrio Formation is exposed along the north side of RP-43 on the western outskirts of Chos Malal. The east-dipping beds are on the east flank of the Las Máquinas Anticline (Sánchez et al., 2018). These rocks are the lower member of the Early Cretaceous Agrio Formation, an important hydrocarbon source rock in the basin. The Pilmatué member contains deep- marine to near shore gray-green to dark gray mudstone and shale and can be up to 577 m (1890 ft) thick.

The Agrio Formation is exposed along the western outskirts of Chos Malal. Beds dip east along the east flank of the Las Máquinas Anticline. View northwest from RP-43.

To avoid Side Trip 1:

Agrio Formation, Las Máquinas Anticline to West-Verging Fold, Vaca Muerta: *Return east on RP-43 to Chos Malal; turn left (northeast) on San Martín and drive to Tucumán; turn left (north) on Tucumán and drive to Jaime de Nevares; continue straight on Jaime de Nevares to RN-40; turn left (north) on RN-40 and drive to **Stop 17, West-Verging Fold, Vaca Muerta** (−37.33486, −70.03663) for a total of 35.3 km (21.9 mi; 31 min).*

To go on Side Trip 1:

**Side Trip 1-Agrio Formation, Las Máquinas Anticline to Agrio Formation, East
Flank Chos Malal Anticline:** *Continue driving west on RP-43 for 1.3 km (0.8 mi; 1 min)
to* **Stop ST 1.1, Agrio Formation, East Flank Chos Malal Anticline** *(–37.36355,
–70.31331) and pull over on the right.*

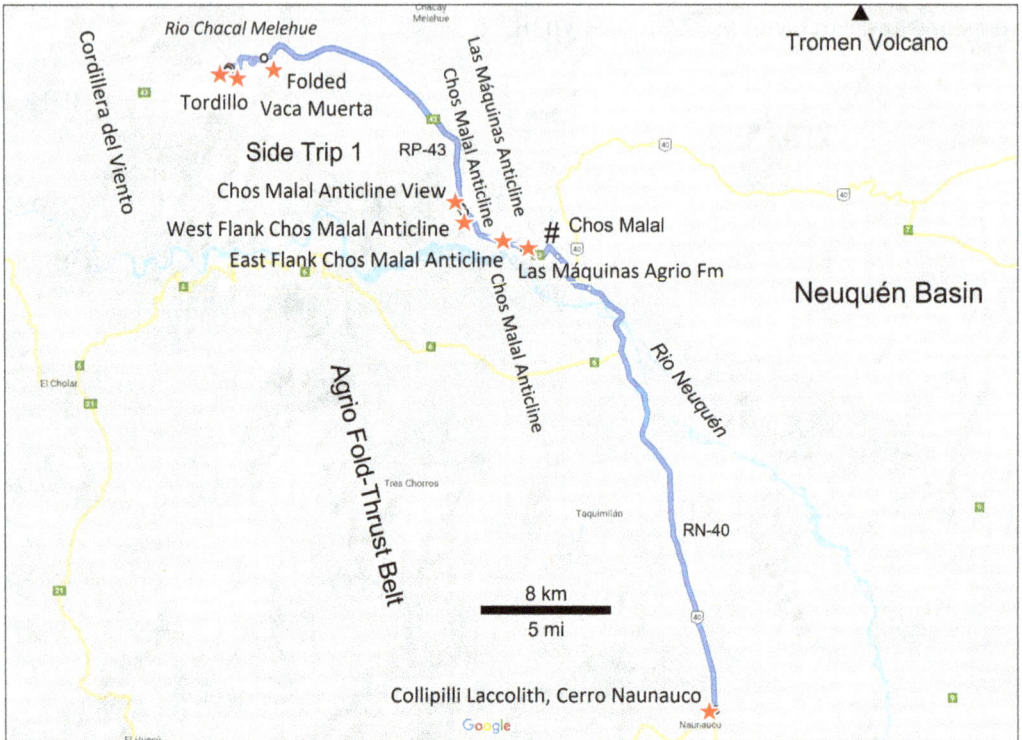

Side Trip 1 Ruta Provincial 43 along Río Chacal Melehue

Somewhere around Naunauco, the Agrio Fold-Thrust Belt merges into the Chos Malal
Fold-Thrust Belt. It is an arbitrary distinction, since the deformation is more-or-less continuous. The Chos Malal Fold-Thrust Belt is primarily thick-skinned deformation, that
is, it contains basement uplifts bounded by high-angle reverse faults formed during the
Andean Orogeny. In the shallow subsurface, many of these reverse faults merge into thin-
skinned thrust faults. Structural analysis suggests a strong link between thick and thin-
skinned structures. Major Andean faults branching from a detachment at 12 km depth
created large basement uplifts. The westernmost of these uplifts forms the Cordillera del
Viento. These basement-cored structures transferred offset and deformation to the cover
by way of the Jurassic Auquilco Formation evaporites, mainly gypsum, thus creating
the thin-skinned structures encountered in the fold-thrust belt. Shortening between the
Cordillera del Viento and the Tromen Massif is calculated between 14.1 and 16.9 km (8.5
and 10.1 mi), equivalent to 26.3 to 29.7% (Sánchez et al., 2015).

RP-43 passes through a number of folds in the Chos Malal Fold-Thrust Belt and provides excellent roadcut exposures of the structures and the Agrio and Vaca Muerta dark marine shales.

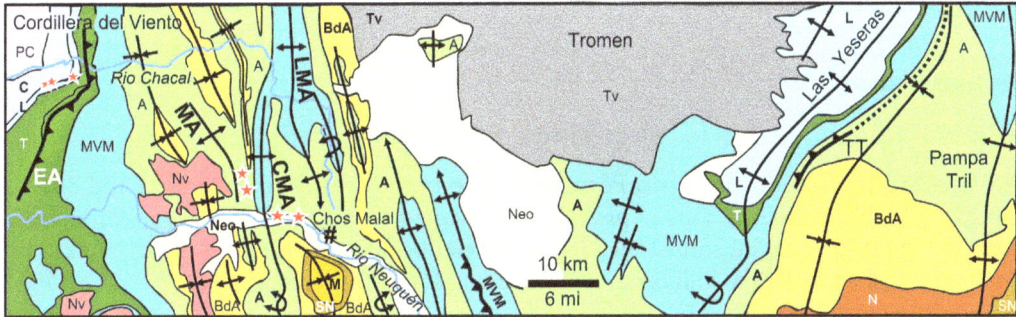

Geologic map of the Chos Malal and Rio Chacal Melehue area. PC = Pre-Cuyo; C = Cuyo Group; L = Lotena Group; T = Tordillo Formation; MVM = Mulichinco-Vaca Muerta; A = Agrio Formation; BdA = Bajada del Agrio Group; N = Neuquén Group; Nv = Naunauco volcanics; SN = Sierra Negra Formation; M = Miocene units; Tv = Tromen Volcanics; Neo = Neogene deposits; MA = Mayal Anticline; CMA = Chos Malal Anticline; LMA = Las Máquinas Anticline; EA = El Alamito Thrust; TT = Tromen Thrust. Modified after Sánchez et al., 2018.

Stop ST 1.1 Agrio Formation, East Flank Chos Malal Anticline

This RP-43 roadcut exposes the east-dipping Agrio Formation on the east side of the Chos Malal Anticline just west of Chos Malal.

Agrio Formation exposed in a roadcut on the south side of RP-43 just west of Chos Malal.

*Side Trip 1, Agrio Formation, East Flank Chos Malal Anticline to Folded Agrio, West Flank Chos Malal Anticline: Continue driving west on RP-43 for 3.5 km (2.1 mi; 3 min) to **Stop ST 1.2, Folded Agrio, West Flank Chos Malal Anticline** (−37.346448, −70.342755) and pull over on the right.*

Stop ST 1.2 Folded Agrio, West Flank Chos Malal Anticline

A west-verging fold in the Agrio Formation is exposed in the roadcut near the core of the Chos Malal Anticline. This is probably equivalent to the middle Avilé member, as it contains abundant fluvial sandstone. The Avilé can be 10 to 90 m (33 to 295 ft) thick. This appears to be a subsidiary fold on the main Chos Malal Anticline.

This outcrop on the north side of RP-43 west of Chos Malal displays a spectacular fold in the Agrio Formation interbedded thin sandstones and shales.

*Side Trip 1, Folded Agrio, West Flank Chos Malal Anticline to Folded Vaca Muerta Formation: Continue driving west on RP-43 for 20.0 km (12.4 mi; 13 min) to **Stop ST 1.3, Folded Vaca Muerta Formation** (−37.25928, −70.48940), and pull over on the right.*

Stop ST 1.3 Folded Vaca Muerta Formation

This stop has gently folded Vaca Muerta Formation black shale exposed near the core of the Mayal Anticline. The formation is choc-full of ammonites weathering out as concretions. The only place we previously have been able to touch the Vaca Muerta Formation was at Los Catutos, and there we saw the carbonate ramp member. Here you can see one of the world's great hydrocarbon source rocks.

Anticline-syncline pair in Vaca Muerta Formation black shales along RP-43. View east.

Ammonites weathering out of the Vaca Muerta as concretions.

View east toward Tromen Volcano.

Side Trip 1, Folded Vaca Muerta to Folded and Thrusted Tordillo Formation, Cordillera del Viento: *Continue driving west on RP-43 for 3.7 km (2.3 mi; 3 min) to* *Stop ST 1.4, Folded and Thrusted Tordillo Formation, Cordillera del Viento* *(−37.266039, −70.516385) and pull over on the right.*

Stop ST 1.4 Folded and Thrusted Tordillo Formation, Cordillera del Viento

See the fold-thrust belt in microcosm. Two small-offset, east-directed thrusts in probable Tordillo Formation sandstone cut these west-verging parasitic folds on the east flank of the Cordillera del Viento Uplift. We are just west of the west-directed El Alamito Thrust (Sánchez et al., 2014; Sánchez et al., 2018).

Folded and thrusted Tordillo Formation sandstone. Dotted lines indicate faults.

*Side Trip 1, Folded and Thrusted Tordillo Formation, Cordillera del Viento to Folded and Thrusted Tordillo Sandstone: Turn around and drive east on RP-43 for 120 m (384 ft; 1 min) to **Stop ST 1.5, Folded and Thrusted Tordillo Sandstone** (−37.265346, −70.515391) and pull over on the left.*

Stop ST 1.5 Folded and Thrusted Tordillo Sandstone

Probable Tordillo Sandstone with multiple folds and thrusts is exposed in a roadcut on the south side of RP-43. As with the last stop, this location is on the east flank of Cordillera del Viento Uplift and in the footwall of the El Alamito Backthrust (Sánchez et al., 2014; Sánchez et al., 2018).

Multiple thrusts in Tordillo Formation sandstone, south side RP-43. Dotted lines indicate faults.

Side Trip 1, Folded and Thrusted Tordillo Sandstone to Chos Malal Anticline:
*Continue driving east on RP-43 for 23.2 km (14.4 mi; 16 min) to **Stop ST 1.6, View into Core of the Chos Malal Anticline** (–37.343595, –70.344943).*

Stop ST 1.6 View into Core of the Chos Malal Anticline

View south to the Chos Malal Anticline southwest of Chos Malal (Sánchez et al., 2018). Agrio Formation shale is in the core, with Huitrín Formation sandstone forming the ridges above. The symmetrical structure is along trend with the northern extension of Cerro Pitrén Anticline.

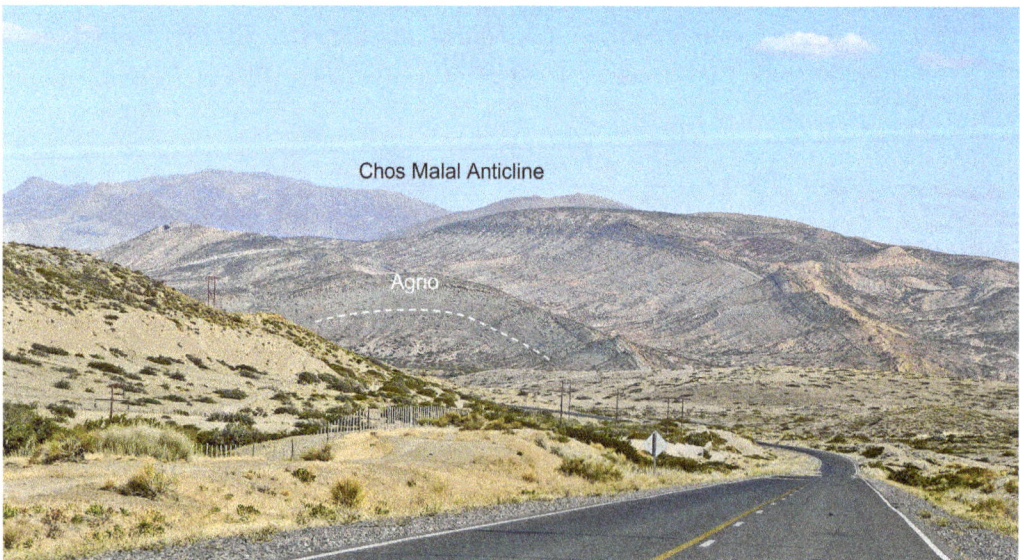

View south from RP-43 into the core of the Chos Malal Anticline.

Side Trip 1, View into Core of the Chos Malal Anticline to West-Verging Fold, Vaca Muerta: *Continue driving east on RP-43 to Chos Malal; turn left (northeast) on San Martín and drive to Tucumán; turn left (north) on Tucumán and drive to Jaime de Nevares; continue straight on Jaime de Nevares to RN-40; turn left (north) on RN-40 and drive to* **Stop 17, West-Verging Fold,Vaca Muerta** *(−37.33486, −70.03663) for a total of 40.5 km (25.1 mi; 35 min).*

Stop 17 West-Verging Fold, Vaca Muerta

The deformation we saw west of Chos Malal continues beneath the Tromen volcanic edifice. Most compression occurred before the volcanism began around 2.3 million years ago. At this stop, you see a roadcut through a west-verging fold in thin sands and shales of the Vaca Muerta Formation. The rocks here, including the volcanics, are carried on the east-directed Tromen Thrust. West-verging backthrusts have been mapped in the area (Galland et al., 2007).

West-verging fold in the Vaca Muerta Formation, north side of RN-40.

West-Verging Fold to Tromen Volcano from Laguna Auquincó: *Continue driving east on RN-40 to RP-7; at sign for Cortaderas and Neuquén turn right (south) on RP-7 (graded gravel road) and drive to* **Stop 18**, *Tromen Volcano from Laguna Auquincó (-37.35806, -69.99417) for a total of 7.2 km (4.5mi; 8 min).*

Stop 18 Tromen Volcano from Laguna Auquincó

The landscape around Chos Malal is dominated by Tromen Volcano, the second-highest peak in Patagonia (4,114 m, or 13,498 ft high). It would be reasonable to think that this impressive mountain is all volcanics, but that would be wrong. The volcanics are a relatively thin veneer, mostly less than 300 m (1,000 ft) thick sitting unconformably on a tectonically thickened section of Mesozoic sediments. The volcano sits atop the Tromen Thrust, an east-directed, deep-seated thrust punctuated by west-directed backthrusts. The Tromen Thrust comes to the surface along RN-40 east of the mountain.

The volcano is part of the Andean back-arc volcanic zone. Radiometric age dating indicates that volcanism here began around 2.3 million years ago and continues to be active: The last basalt flows erupted in historical times (Galland et al., 2007). It is commonly accepted that volcanism occurs in areas where extension provides faulted pathways for magma to work its way to the surface. Tromen, however, erupted during a compressional phase, proving that volcanism does not need extension to occur (Galland et al., 2007). The eruptions occurred through east-west fissures that were opened as a result of east-west compression.

The Chos Malal Fold-Thrust Belt terminates in the east with a series of structures known as the Las Yeseras and Pampa Tril anticlines. These east-verging folds are characterized by relatively flat crests and steeply dipping east flanks.

Tromen Volcano as seen looking north from Laguna Auquincó.

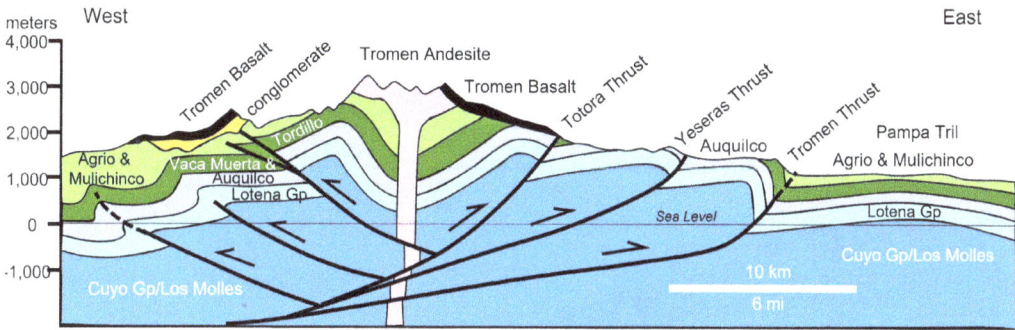

Cross-section through Tromen Volcano showing its structural underpinning. Modified after Galland et al., 2007.

> ***Tromen Volcano from Laguna Auquincó to Mulichinco-Lower Agrio Hogback:***
> *Continue driving east on RP-7 for 6.3 km (3.9 mi; 8 min) to* **Stop 19.1, Mulichinco-**
> **Lower Agrio Hogback** *(–37.377799, –69.931340).*

Stop 19 Tromen Thrust

Regional work suggests that the Tromen Thrust was active throughout Neogene (Cobbold and Rossello, 2003). Deformation of the volcanics indicates that tectonism is ongoing. The Tromen Thrust appears to be a major, deep-seated and east-directed thrust with subsidiary west-directed backthrusts. Jurassic through Holocene units are carried on the Tromen Thrust, which is detached in Jurassic evaporites of the Auquilco Formation. The amount of east-west shortening is estimated at less than 10% (Galland et al., 2007).

Stop 19.1 Mulichinco-Lower Agrio Hogback

The trace of the Tromen Thrust comes to the surface along a roughly northeast-oriented topographic depression between the Mulichinco-Agrio Hogback on the west and the Pampa Tril Anticline to the east. The Mulichinco-Lower Agrio section exposed along this hogback is carried on the leading edge of the Tromen Thrust and is inclined to the west.

> ***Mulichinco-Lower Agrio Hogback to Tromen Thrust at Pampa Tril:*** *Drive 470 m*
> *(1,540 ft) east on RP-7 to the intersection with RP-9; turn left (north) on RP-9 and drive*
> *to* **Stop 19.2, Tromen Thrust at Pampa Tril** *(–37.294254, –69.847291), for a total of*
> *12.8 km (8.0 mi; 19 min).*

Mulichinco and lower Agrio formations carried on the Tromen Thrust. View north from the intersection of RP-7 and RP-9.

Stop 19.2 Tromen Thrust at Pampa Tril

The east-directed Tromen Thrust is located at the break in slope just east of outcrops of the Lower Cretaceous Mulichinco Formation sandstone. The thrust carries Jurassic and Lower Cretaceous units. The detachment is in or near the Jurassic Auquilco Formation gypsum.

View north to the Tromen Thrust from near the intersection of RP-9 and RN-40.

Tromen Thrust at Pampa Tril to Las Yeseras Anticline, Auquilco Gypsum, and Tromen Volcano: *Drive 315 m (1,030 ft) north on RP-9 to RN-40; turn left on RN-40 and drive to* **Stop 20, Las Yeseras Anticline, Auquilco Gypsum, and Tromen Volcano** *(−37.297499, −69.877107) for a total of 3.6 km (2.2 mi; 4 min) and pull over on the right.*

Stop 20 Las Yeseras Anticline, Auquilco Gypsum, and Tromen Volcano

The Las Yeseras Anticline formed along the leading edge of the Tromen Thrust. Highly eroded Jurassic Auquilco Formation gypsum (the light-colored unit) is exposed in the core of the structure. The gypsum may have been structurally thickened in the core of the fold, or may indicate a splay of the thrust fault.

View looking northwest from RN-40 over light-colored gypsum beds to Tromen Volcano.

View north up the east flank of Las Yeseras Anticline.

Las Yeseras Anticline to Vaca Muerta – Tordillo Contact: Continue driving south on RN-40 for 1 km (0.6 mi; 1 min) to **Stop 21, Vaca Muerta – Tordillo Contact** (−37.299287, −69.886618) and pull over on the right just after the guardrail.

Stop 21 Vaca Muerta – Tordillo Contact

The abrupt change from the light-colored Tordillo Formation marine sandstone at the base to dark and organic-rich Vaca Muerta Formation shale at the top of this roadcut indicates a sudden deepening of the foreland basin in latest Jurassic time.

Contact between the Tordillo Formation light-colored sandstone (below) and the dark Vaca Muerta Formation shale (above). View south along RN-40.

Vaca Muerta –Tordillo Contact to Mulichinco Formation Flatirons: Make a U-turn and drive north on RN-40 for 15.6 km (9.7 mi; 10 min) to **Stop 22, Mulichinco Formation Flatirons** (−37.20719, −69.78916) and pull over on the right.

Stop 22 Mulichinco Formation Flatirons

RN-40 closely follows the Tromen Thrust front. The thrust trace lies between near-vertical sediments on the west and gently east-dipping units to the east (Galland and Sassier, 2016). The east-dipping units are probably carried on the thrust. The trace of the thrust is just east of the highway. Folding at the thrust front caused these spectacular and colorful flatirons.

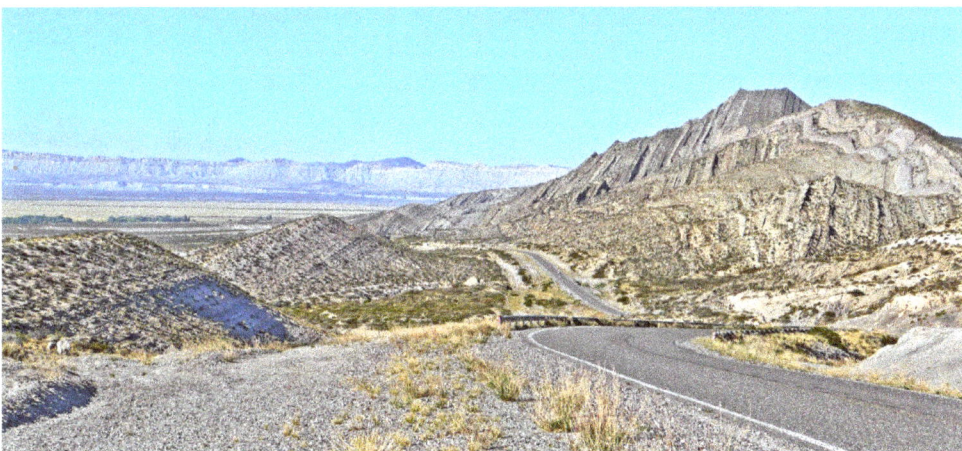

Mulichinco Sandstone is carried on the Tromen Thrust. View south along RN-40. The Las Yeseras Anticline is to the west.

Spectacular flatirons in the Lower Cretaceous Mulichinco Sandstone at leading edge of the Tromen Thrust. The trace of the thrust is under recent cover near the base of the outcrops. View northwest from RN-40.

> ***Mulichinco Formation Flatirons to Tromen Volcano:*** *Continue driving north on RN-40 for 5.9 km (3.7 mi; 4 min) to **Stop 23, Tromen Volcano** (−37.15555, −69.79420).*

Stop 23 Tromen Volcano

This scenery stop affords a particularly nice view of Tromen Volcano and its volcanic plateau.

View west to Tromen Volcano from RN-40. More recent basalt flows make up the dark patch on the north (right) side of the peak.

*Tromen Volcano to Domuyo View: Continue driving north on RN-40 for 59.6 km (37.0 mi; 41 min) to **Stop 24, Domuyo View** (−36.789007, −69.858320).*

Stop 24 Domuyo View

From Buta Ranquil north to Barrancas, RN-40 takes us between the Tromen volcanic plateau and the west side of the Sierra de Reyes. The Sierra de Reyes is another broad, leading-edge anticline that exposes Jurassic gypsum in its core, similar to the Las Yeseras Anticline to the south. The Sierra de Reyes anticline, however, is carried on a deep, east-directed thrust and a shallow, west-directed backthrust. East of this structure, the Agua de Reyes Thrust and associated east-verging La Salinita Anticline produces hydrocarbons from Upper Jurassic-Lower Cretaceous reservoirs. East-west shortening here is about 4.5 km or 20% (Sagripanti et al., 2012).

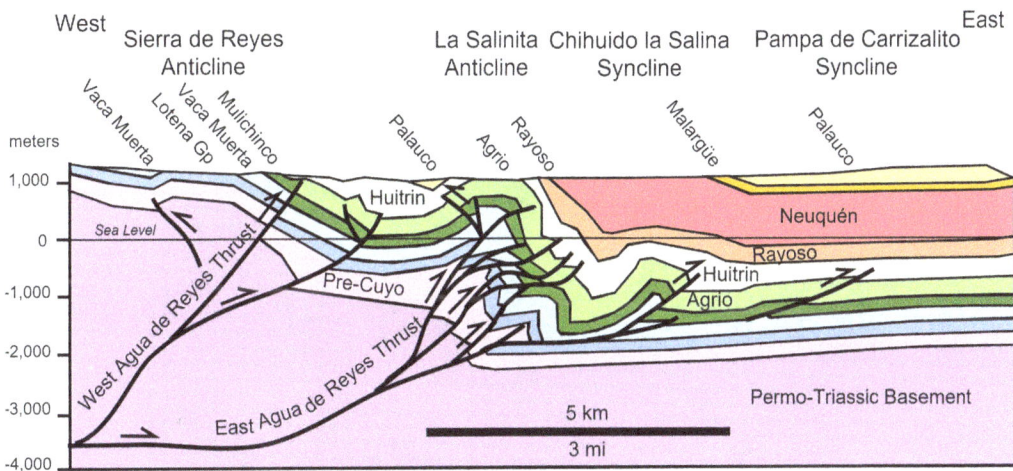

Cross-section through the Sierra de Reyes and oil fields on the La Salinita Anticline. Modified from Sagripanti et al., 2012.

Domuyo, the "Roof of Patagonia," is a massive snowcapped landmark 4,709 m (15,450 ft) high in the eastern Andes. The highest peak in Patagonia, on many maps it is shown as Domuyo Volcano. It is *not* a volcano. It is a Triassic-Jurassic sedimentary rock-cored basement uplift that could be considered the northern extension of the Cordillera del Viento. The Cordillera is an uplift west and northwest of Chos Malal that exposes the pre-Jurassic basement within the Chos Malal Fold-Thrust Belt (Galland and Sassier, 2016; Rojas Vera et al., 2016).

View west to Domuyo, a Jurassic-cored uplift along the crest of the Andes. Badlands in the foreground are developed in shales of the Late Cretaceous Neuquén Group.

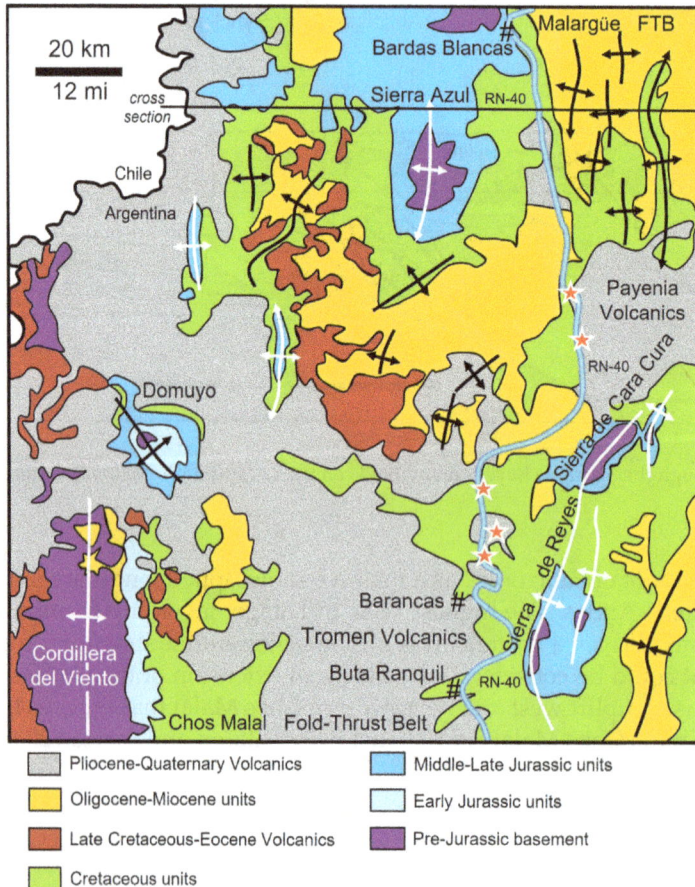

Geologic map of the Buta Ranquil to Bardas Blancas area. Information drawn from Pons et al., 2010; Rojas Vera et al., 2016. Cross-section is shown in the next figure.

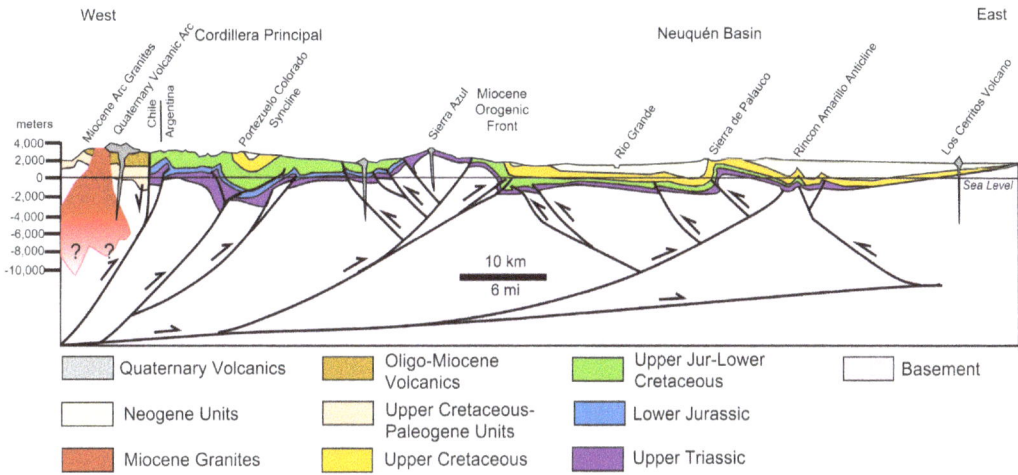

Cross-section through the Sierra Azul south of Bardas Blancas. Modified after Ramos et al., 2014 and Gianni et al., 2018.

*Domuyo View to Cochiquito Volcano: Continue driving north on RN-40 for 5.7 km (3.5 mi; 4 min) to **Stop 25, Cochiquito Volcano** (−36.741945, −69.861419).*

Stop 25 Cochiquito Volcano

Pleistocene Cochiquito Volcano sits halfway between Mount Tromen and Cerro Payún Volcano. The rhyolitic cinder cone sits above an associated basalt flow. Such bimodal volcanism appears to be common in this back-arc setting. The lack of vegetation is indicative of its young age.

View east to the large Cochiquito cinder cone and basalt flow from RN-40.

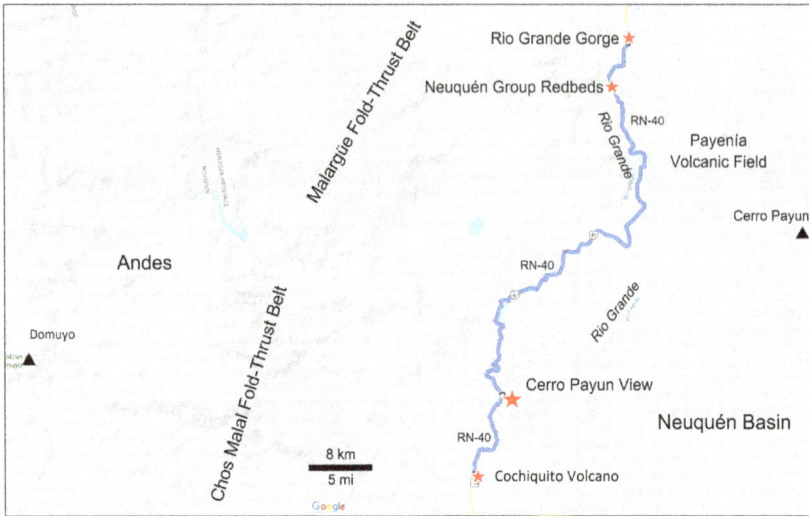

Cochiquito Volcano to Cerro Payún View: *Continue driving north on RN-40 for 11.8 km (7.3 mi; 9 min) to* **Stop 26, Cerro Payún View and the Payunia Volcanic Field** *(−36.66399, −69.82636) and pull over on the right.*

Stop 26 Cerro Payún View and the Payunia Volcanic Field

This stop provides a view of the magnificent and classic composite volcanic cone of Cerro Payún about 51 km (32 mi) to the east-northeast. It is one of several large volcanoes in the Payunia Volcanic Field.

Cerro Payún looking east from RN-40. The volcanic cone towers above surrounding landscape.

The extensive Payunia Volcanic Field is part of a larger back-arc volcanic province that includes the Tromen and Llancanelo volcanic fields. It contains several prominent composite volcanic cones including Cerro Payún (3,680 m; 12,074 ft), Payún Matrú caldera (3,715 m; 12,188 ft), and Cerro Diamante (2,354 m; 7,723 ft). The volcanic field also claims the longest single Quaternary lava flow on earth, a 181 km (112 mi) long tongue of basalt. Basalt, as we know, is one of the more runny magmas and tends to flow long distances. The Payún Matrú and Llancanelo fields combined cover over 15,900 km^2 (6,140 mi^2) with basalt flows and over 800 cinder cones (Risso et al., 2009; Galland and Sassier, 2016).

The province erupted in two main phases: The first, an earlier Miocene-Pliocene event erupted mainly mafic (dark) andesites. The second stage erupted Pliocene to Recent basalt flows and composite cones (Kay and Ramos, 2006).

Payún Matrú is a Hawaiian-type eruption in that lavas flow from vents. It is a shield volcano with a 9 by 7 km (5.4 by 4.2 mi) caldera. Much of the basaltic flows originate from east-west faults in the Payunia Volcanic Field; eruptions in the Llancanelo field appear to be controlled by northwest-southeast faults. The Llancanelo field was primarily Hawaiian and Strombolian (relatively mild blasts with abundant cinders, spatter cones, and lava bombs) type eruptions.

Geochemical studies indicate that these basalts have a common mantle source characteristic of a back-arc environment. The volcanics have an alkaline "intra-plate" chemical signature considered related to extensional tectonics (Kay and Ramos, 2006; Risso et al., 2009), a conclusion opposite that of Galland et al. (2007) for the Tromen Volcanic Field to the south.

Google Maps image showing the Payunia Volcanic Field and stops. Imagery ©2023 TerraMetrics.

Cerro Payun View to Neuquén Group Redbeds: Continue driving north on RN-40 for 57.2 km (35.5 mi; 41 min) to Stop 27, Neuquén Group Redbeds (–36.357306, –69.685487) and pull over on the right.

Stop 27 Neuquén Group Redbeds

On the west side of the Río Grande there are massive exposures of brightly colored syn-tectonic claystones and siltstones of the Upper Cretaceous Neuquén Group, perhaps the Lisandro Formation. Alluvial, river, swamp, and wind-blown deposits were eroded off actively growing structures and deposited in a non-marine basin. These units create a highly scenic landscape in an otherwise volcanic-dominated terrain.

Redbeds in the Neuquén Group looking southwest from RN-40.

The east side of the Río Grande consists largely of basalt flows from the Quaternary Payunia Volcanic Field.

Cinder cones and basalt flows of the Payunia Volcanic Vield. View to the east from RN-40.

> **Neuquén Group Redbeds to Río Grande Gorge:** *Continue driving north on RN-40 for 6.4 km (4 mi; 4 min) to* **Stop 28, Río Grande Gorge** *(−36.312744, −69.666700) and pull over on the left just before the bridge.*

Stop 28 Río Grande Gorge

This stop at the RN-40 bridge over the Río Grande Gorge shows the power of water to cut through hard rock. These aa (rough, rubbly) and pahoehoe (ropy) lava flows originate from fissures at or near Payún Matru some 36 km (23 mi). distant. The Río Grande Valley is filled with up to 100 m (328 ft) of lava (Risso et al., 2009). The river has cut this 14 m (46 ft) wide gorge through about 15 m (49 ft) of basalt.

View south from RN-40 bridge over the Río Grande.

*Río Grande Gorge to Mirador Río Grande: Continue driving north on RN-40; just past the town of Bardas Blancas is a sign on the left for the Reserva Natural Caverna de las Brujas; turn left (north) onto a graded gravel road and drive to **Stop 29.1, Mirador Río Grande** (−35.833116, −69.799173) for a total of 63.2 km (39.3 mi; 50 min).*

Stop 29 Bardas Blancas

Somewhere between Chos Malal and Malargüe, the Chos Malal Fold-Thrust Belt transitions to the Malargüe Fold-Thrust Belt. The stratigraphy is largely unchanged, and the distinction is somewhat arbitrary. The structure is somewhat simpler: From west to east there is the Río Grande Basement High (a basement-involved east-verging thrust-cored anticline), a large north-south backthrust extending from Bardas Blancas to Castillo de Pincheira, the piggy-back Malargüe Syncline just to the east, and the basement-involved east-verging thrust-cored Malargüe Anticline. Then the bottom drops out and you are in the deep basin.

On the east flank of the east-verging Río Grande basement high is the east-dipping Upper Jurassic La Manga Formation limestone. A large cavern system known as Caverna de las Brujas has developed in the La Manga Formation.

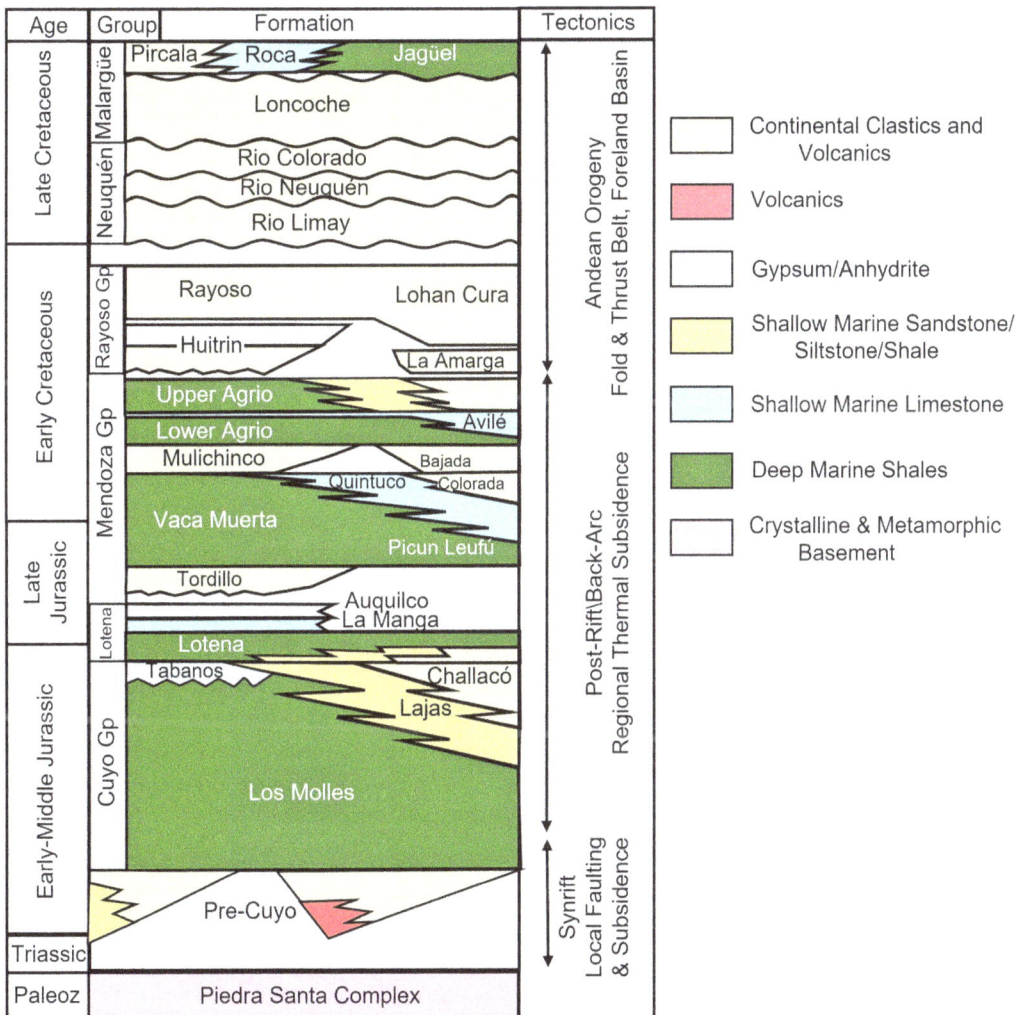

Stratigraphy and tectonic setting of the Bardas Blancas-Malargüe area. Information derived from Howell et al., 2005; Pazos et al., 2007; Rocha-Campos et al., 2011, Pazos and Krapovicas, 2014.

Stop 29.1 Mirador Río Grande

This stop provides a view south over Bardas Blancas. The Río Grande Basement High is to the southwest, with Jurassic units at the surface. This east-verging anticlinal feature encounters the Bardas Blancas Backthrust before the section plunges into the Malargüe Syncline. From the basement high on the west to the synclinal axis on the east, pretty much the entire Jurassic through Neogene section is exposed.

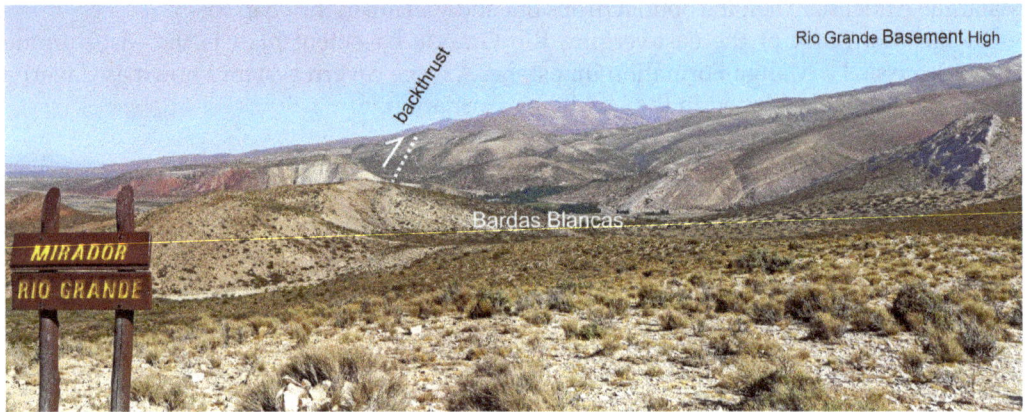

View south from Mirador Rio Grande. The town of Bardas Blancas is in the valley, center.

> ***Mirador Río Grande to Caverna de las Brujas:*** *Continue driving north on the gravel road for 4.6 km (2.8 mi; 8 min) to **Stop 29.2, Caverna de las Brujas** (–35.801887, –69.819277).*

Stop 29.2 Caverna de las Brujas (Witches' Cave)

Caverna de las Brujas is about 65 km (39 mi) south of Malargüe and 8 km (5 mi) north of Bardas Blancas. The cavern is developed in the Upper Jurassic La Manga Formation. The La Manga contains ammonites, corals, sponges, gastropods, and other fossils. Near vertical fractures assisted with dissolving the limestone, and the flowing water during the last ice age, when precipitation was much higher than now, completed development of the caves seen today. The caverns are known for impressive stalactites and stalagmites, large galleries, and vaulted rooms (Las Flores, Sala de la Virgen). Over 6 km (3.6 mi) of passages have been discovered, but visitors are able to visit only the first 200 m (650 ft) during the 2-hour guided tour.

The galleries near the entrance provide a refuge for bats and other outside creatures. The caves also contain fauna that evolved to live underground, some translucent, and some without eyes.

At least one of the galleries (Sala de la Virgen) has artifacts that indicate the caves have been used intermittently by humans since 3,000 BC. The caverns became a Mendoza

Provincial Nature Reserve in 1990 and are part of a system of protected eco-parks. It is also a National Natural Monument.

Visit

There is usually only one tour per day. Starting time varies. Groups must arrange tours in advance and have a guide provided by the Dirección de Recursos Naturales Renovables in Malargüe.

Tickets can be obtained from the Centro Turismo, Malargüe. Cost is 502 pesos (2019) for a 2-hour excursion. Bring appropriate clothing: The caves are a constant 8–11° C (48–52° F) with 90% humidity.

Phone: (0261) 4258751 or (02627) 470215.

Email: informes-anp@mendoza.gov.ar.

Wwebsite: http://www.ambiente.mendoza.gov.ar/visitas-educativas-anp/.

For educational visits to the Natural Reserve Caverna de Las Brujas (as well as La Payunia and Castles de Pincheira), contact the tourist office in Malargüe by email at infoturismo@malargue.gov.ar or infoturismomalargue@gmail.com. Indicate the day of the trip and the nature reserve you wish to visit. You will receive a list of authorized guides for hire.

Entrance to Caverna de las Brujas.

This strangely phallic stalagmite is in Caverna de las Brujas. Photo courtesy of Pablo-Flores, https://commons .wikimedia.org/wiki/File:Witches%27_Cave_stalagmite.jpg.

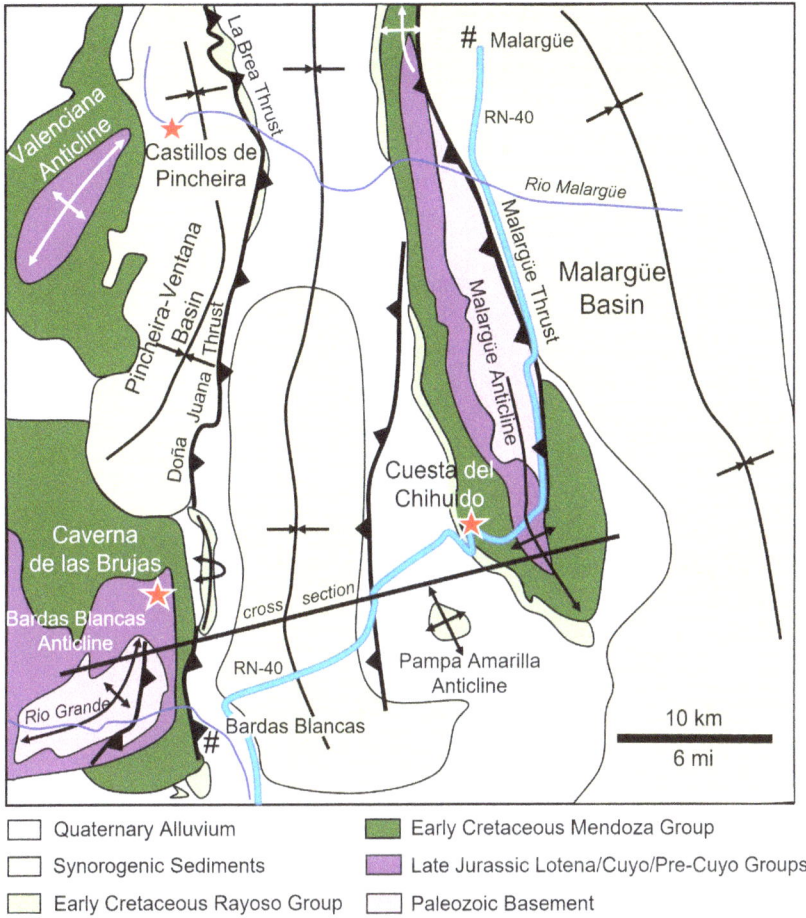

Geologic map of the Bardas Blancas to Malaragüe area. Modified after Silvestro et al., 2005.

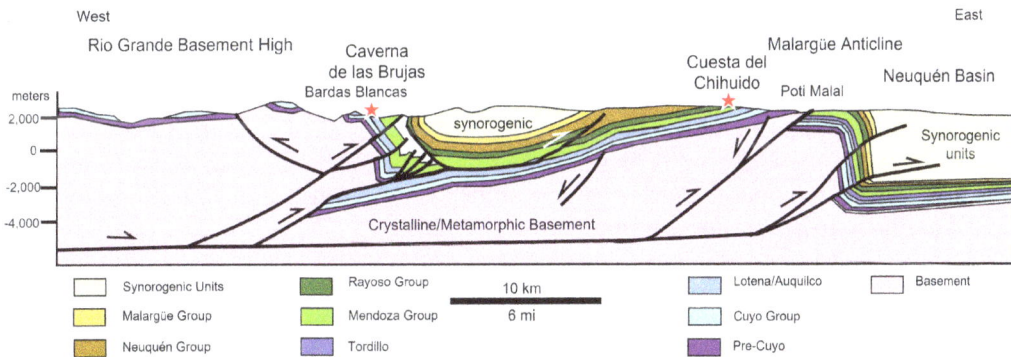

Cross-section from Bardas Blancas through Cuesta del Chihuido to the Malargüe Anticline. Modified after Branellec et al., 2016.

*Caverna de las Brujas to Agrio Formation, Dike, and Sill: Return to RN-40 and turn left (east, then north); continue driving north on RN-40 to **Stop 30.1, Agrio Formation, Dike, and Sill** (–35.74886, –69.59025) for a total of 32.5 km (20.2 mi; 35 min).*

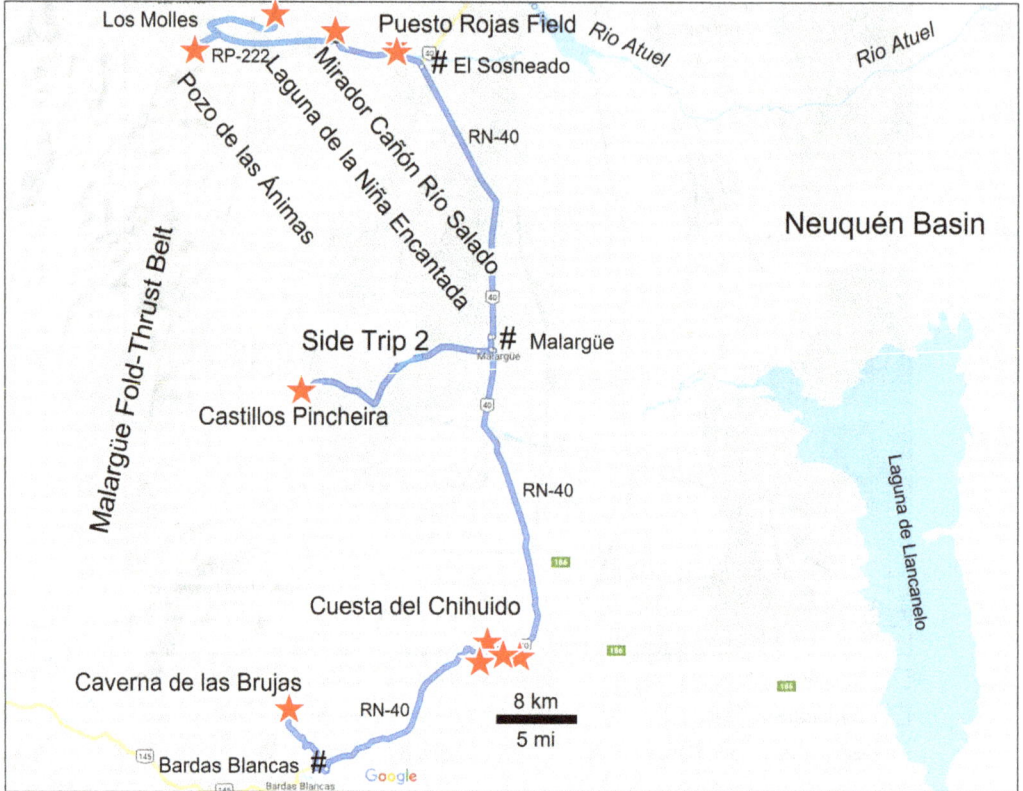

Stop 30 Cuesta del Chihuido

We are on the gently west-dipping west flank of the Malargüe Anticline. A cuesta is a ridge with a gentle dip slope on one side and a steep erosional slope on the other. The Cuesta del Chihuido is a 300-meter- high (1,000 ft) topographic escarpment. The road cuts, from top to base, through the Cretaceous Neuquén Formation, and the Jurassic Agrio Formation, Chachao Formation, and Vaca Muerta Formation. The Chachao and Vaca Muerta formations are rich in fossils, reflecting their marine origins.

The Agrio Formation here was deposited on a gentle marine slope. It comprises rhythmic carbonate-shale beds (Spacapan et al., 2016).

The Chachao Formation is a thin oyster-bed limestone that formed on a carbonate ramp. The Chachao is characterized by yellowish, fossil-rich limestone composed largely of oysters and scallops. The fossil-rich layers are interbedded with massive marl and shale beds (Venari et al., 2012). It is a good petroleum reservoir in the adjacent basin.

The Vaca Muerta is an organic-rich hydrocarbon source rock for the basin. In southern Mendoza Province, the Vaca Muerta grades up into the Chachao Formation.

An interesting feature of these outcrops is the intrusion of andesitic dikes and sills.

Recent work suggests that these highly-altered sills were emplaced during late Miocene (between 10.5 and 7 Ma). They may be related to nearby hydrocarbon-producing sills intruded into the Vaca Muerta Formation during Miocene (Spacapan et al, 2016). A volcanic rock can produce oil if it was fractured before the oil migrated into the rock.

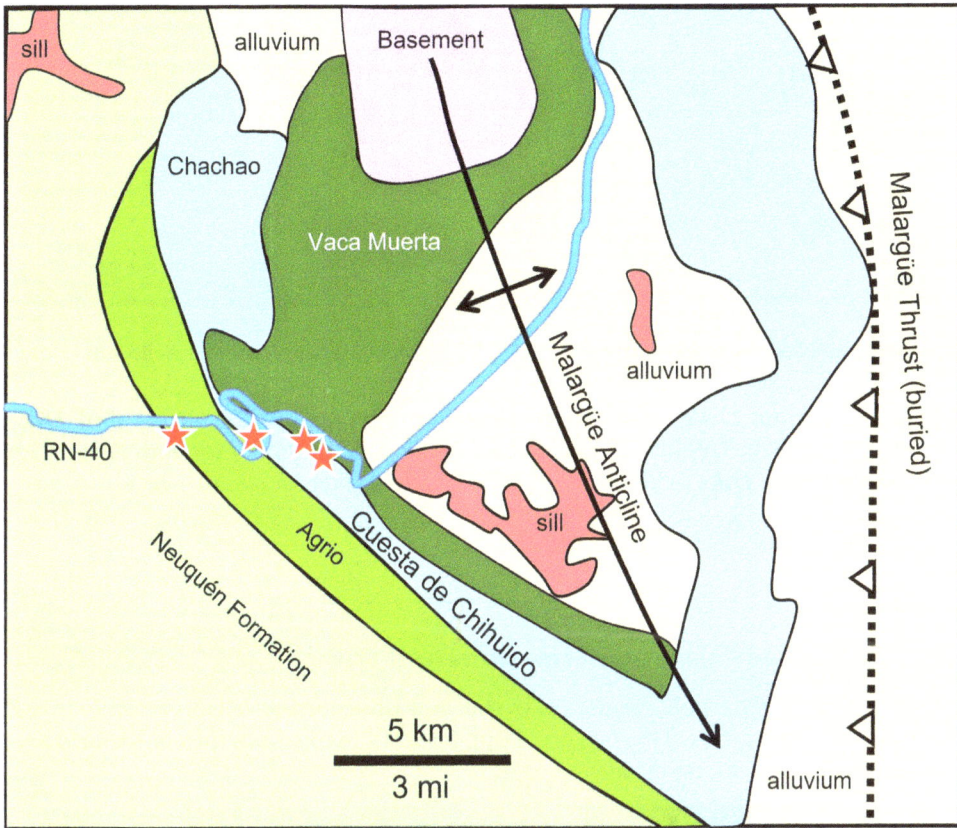

Geology of the Cuesta del Chihuido area. Modified after Spacapan, et al., 2016.

Stop 30.1 Agrio Formation, Dike, and Sill

At the top of the cuesta, RN-40 cuts through buff-colored siltstones and shales of the Late Jurassic-Early Cretaceous Agrio Formation. Andesitic dikes can be seen cutting through the shale. In at least one outcrop the dike cuts the sills, so is younger. The sills were intruded during the late Miocene Huincán eruptive episode (10.5 to 7 Ma; Spacapan et al., 2016). The bedding in this area is all dipping southwest, away from the axis of the Malargüe Anticline.

West-dipping Agrio Formation and near-vertical dikes, Cuesta del Chihuido. View east on RN-40.

> ***Agrio Formation, Dike, and Sill to Chachao Formation Limestone and Dike:***
> *Continue north on RN-40 for 1.8 km (1.1 mi; 1 min) to **Stop 30.2, Chachao Formation***
> ***Limestone and Dike** (–35.748370, –69.579999) and pull over on the right.*

Stop 30.2 Chachao Formation Limestone and Dike

The Lower Jurassic Chachao Formation oyster-bed limestone at this location is intruded by andesitic dikes and sills. This is a good opportunity to examine the characteristics of a petroleum reservoir rock up close.

Chachao Formation, sill, and feeder dike. Google Street View south from RN-40. The dike clearly crosscuts and offsets the sill, indicating that it is both younger and probably intruded along a fault.

Chachao Formation Limestone and Dike to Dike and Sill: *Continue north on RN-40 for 1.5 km (0.9 mi; 1 min) to **Stop 30.3, Dike and Sill** (−35.748296, −69.574140) and carefully pull over on the right.*

Stop 30.3 Dike and Sill

The Lower Jurassic Vaca Muerta Formation organic-rich shale and major source rock is exposed at this roadcut. It is intruded by andesitic dikes and sills. Bedding abruptly changes dip at the dike, suggesting that it was intruded along a fault. The sill and underlying bedding appear to roll into the dike/fault, suggesting they are on the upthrown side of a thrust fault. Notice the large concretions in the Vaca Muerta shale.

Vaca Muerta Formation with sill and near-vertical feeder dike. Google Street View south from RN-40.

Dike and Sill to Vaca Muerta: *Continue north on RN-40 for 160 m (0.1 mi; 1 min) to **Stop 30.4, Vaca Muerta** (−35.748963, −69.572623) and carefully pull over on the right.*

Stop 30.4 Vaca Muerta

The Vaca Muerta Formation shale is characterized here by thin limestone beds and large concretions. Concretions tend to form after sediment burial but before the sediment is fully lithified. Typically, they form when a mineral precipitates around a seed point, which is often organic, such as a leaf, tooth, shell fragment, or other fossil.

Vaca Muerta in RN-40 roadcut. Some concretions are indicated by arrows. Google Street View south.

To avoid Side Trip 2:

> ***Vaca Muerta to Sosneado Thrust View and El Sosneado Field:*** *Continue north on RN-40 to El Sosneado; turn left (north) at the Service Station and sign for RN-40 and drive 2 km (1.3 mi) on unnamed graded gravel road to **Stop 31, Sosneado Thrust View and El Sosneado Field** (−35.069857, −69.565521) for a total of 86.2 km (53.6 mi; 1 hr 5 min).*

To go on Side Trip 2:

> ***Side Trip 2, Vaca Muerta to Castillos Pincheira:*** *Continue north on RN-40 to Malargüe; turn left (west) onto Fortin Malargüe Oeste; continue straight (west) on A Castillos Pincheira to **Stop ST 2, Castillos Pincheira** (−35.51437, −69.79578) for a total of 57.5 km (35.8 mi; 1 hr 4 min).*

To go directly to Side Trip 3:

> ***Vaca Muerta to Side Trip 3.1, Puesto Rojas Field and Cañada Ancha Anticline:*** *Continue north on RN-40 to sign for Los Molles/Las Leñas; continue straight (north) on RP-222 to **Stop ST 3.1, Puesto Rojas Field and Cañada Ancha Anticline** (−35.199229, −69.762482) for a total of 74.3 km (46.2 mi; 58 min).*

Side Trip 2 Castillos Pincheira

As you drive west along the road "A Castillos de Pincheira," you pass through the north-south-oriented Pincheira Syncline, a piggyback structure carried on the east-directed Malargüe Thrust. The syncline is filled with synorogenic deposits (deposits shed off the growing Andes) of the Late Cretaceous-Oligocene Malargüe Group. After crossing the La Brea Backthrust you enter younger, dominantly Miocene synorogenic deposits in the Western Basin Syncline (Branellec et al., 2016).

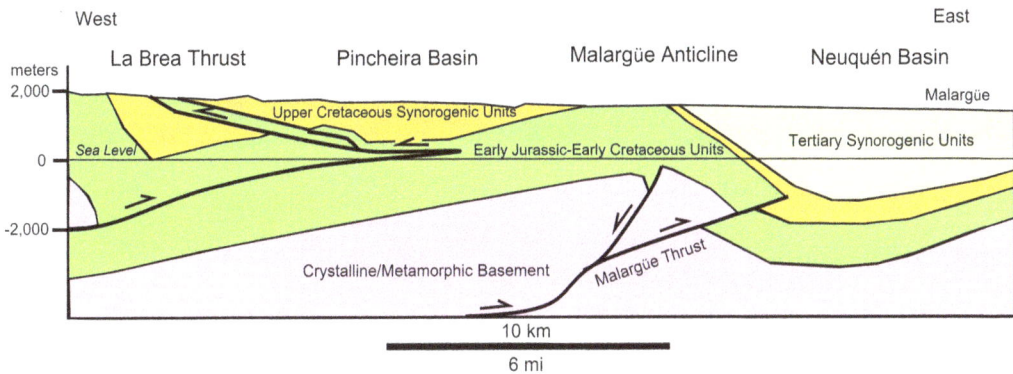

Cross-section from Malargüe west to the La Brea Thrust. Modified after Branellec et al., 2016.

Castillos de Pincheira is a Mendoza Provincial Reserve and protected natural area. Located about 27 km (16 mi) west of the city, it consists of Miocene sandstone that has been eroded into castle-like palisades and abutments by the action of glaciers and the Malargüe River.

Prior to the arrival of Europeans, this area was inhabited by the Huarpe people. They were an agrarian group who grew corn, beans, squash, and quinoa. They occupied the current Argentinian provinces of San Luis, Mendoza, San Juan, and northern Neuquén. They were influenced by Inca culture and adopted llama ranching and the Quechua language. Huarpe arrowheads and pot shards are still found throughout the area.

After the arrival of Europeans, the indigenous peoples were replaced by the ranching culture of Argentina.

The name derives from the Pincheira brothers, who ran a legendary outlaw gang in the area between 1811 and 1833.

The 650 ha (1,600 ac) park was created in 1999 to protect this unique landscape.

Visit

As with Caverna de las Brujas, contact the tourist office in Malargüe by email to arrange a stay in the park.

Email: infoturismo@malargue.gov.ar.

Phone: +54 260 460-0859.

Entry fees: see website for current fees.

Website: www.castillosdepincheira.com.

View southwest to the Castillos Pincheira.

To avoid Side Trip 3:

> **Castillos Pincheira to Sosneado Thrust View and El Sosneado Field:** *Return east on A Castillos Pincheira to Malargüe; turn left (north) on RN-40 and drive to El Sosneado; turn left (north) at the service station and sign for RN-40 and drive 2 km (1.3 mi) on unnamed graded gravel road to* **Stop 31, Sosneado Thrust View and El Sosneado Field** *(−35.069857, −69.565521) for a total of 75.6 km (47.0 mi; 1 hr 19 min).*

To go on Side Trip 3:

> **Side Trip 3, Castillos Pincheira to Puesto Rojas Field and Cañada Ancha Anticline:** *Return east on A Castillos Pincheira to Malargue; turn left (north) on RN-40 and drive to sign for Los Molles/Las Leñas; continue straight (north) on RP-222 to* **Stop ST 3.1, Puesto Rojas Field and Cañada Ancha Anticline** *(−35.199229, −69.762482) for a total of 63.7 km (39.6 mi; 1 hr 11 min).*

Side Trip 3 Río Salado Canyon

A quick drive up the Río Salado Canyon allows us to see some lesser-known local geologic points of interest.

Google Earth oblique view north over the Río Salado and El Sosneado. Imagery ©2023 CNES/Airbus, Landsat/ Copernicus, Maxar Technologies.

Stop ST 3.1 Puesto Rojas Field and Cañada Ancha Anticline

Just after leaving RN-40, driving west up the Río Salado, we pass through the Puesto Rojas Oil Field. Discovered in 1974 by YPF (Yacimientos Petroliferos Fiscales), this thrust-belt field, along with the adjacent Cerro Mollar Oil Field, contains 5.56 million m³ (35 million barrels; Fuentes et al., 2016). Puesto Rojas produced 3.66 million m³ (23 million barrels) of 22° API oil through 2015, mainly from the Early Cretaceous Chachao Formation limestone at a depth of 1,930 m (6,230 ft). The field is trapped in a shallow, north-south trending anticline detached at the top Jurassic at a depth of about 2,000 m (6,560 ft; Horton et al., 2016).

Growth strata in the Oligocene Agua de la Piedra Formation, Malargüe Group, suggests that the last Andean compressional phase started between 20 and 15 Ma in this area. The main contractional event occurred during Miocene (Martos et al., 2020).

The Cañada Ancha Anticline is an east-verging fold carried on a thrust detached in evaporites either at the top of the Jurassic Cuyo Group or Lotena Group. It is part of a shallow duplex system that includes the Puesto Rojas Anticline immediately to the east. One possible interpretation is given in the following cross-section by Horton et al., 2016.

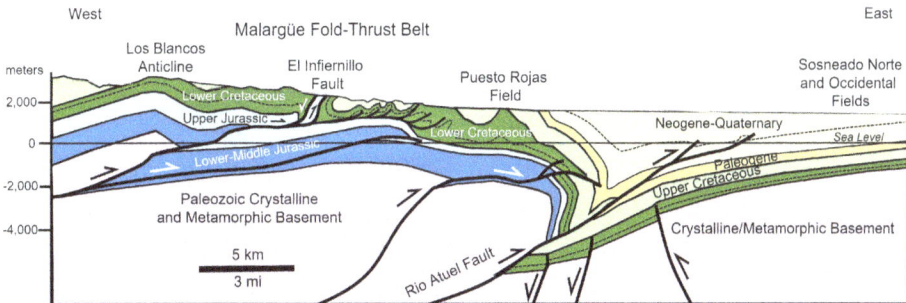

West-east cross-section through the Puesto Rojas Field. Section location is shown on the following figure. Modified after Horton et al., 2016.

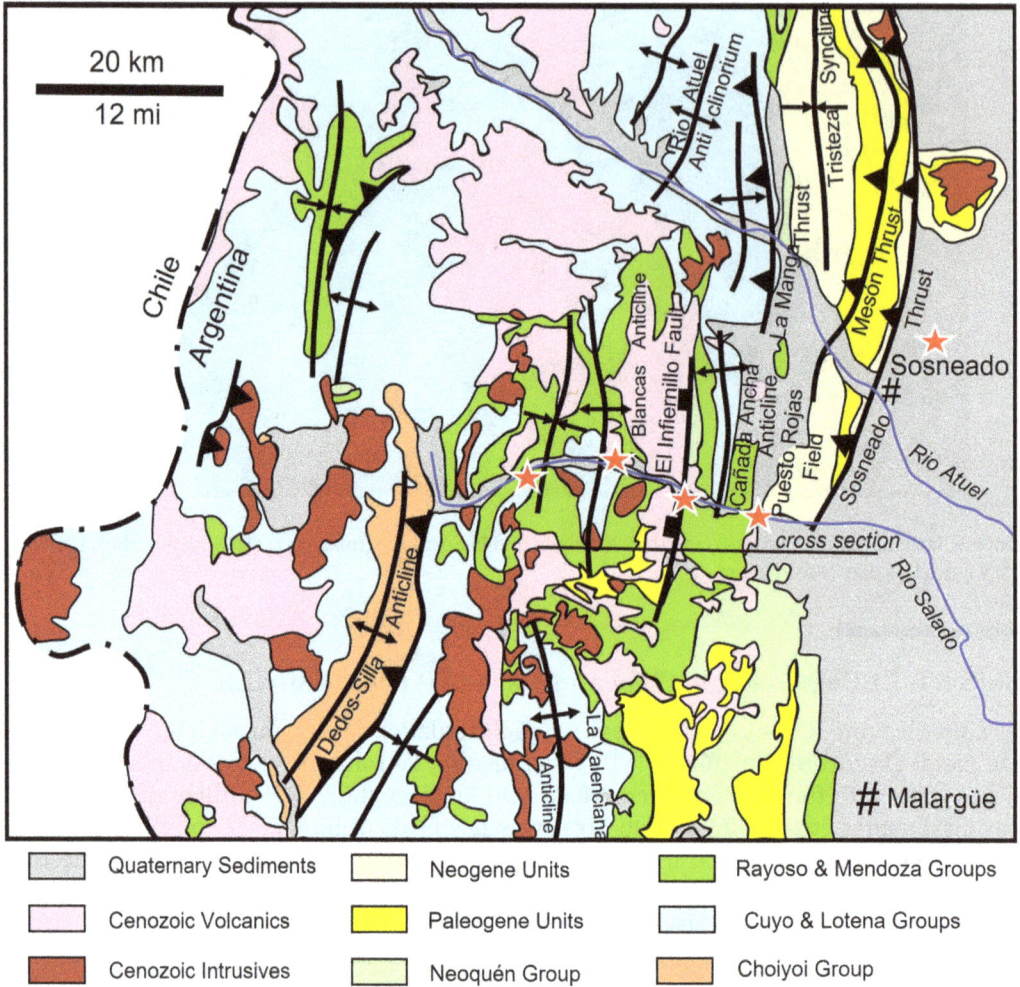

Geologic map of the Río Salado area. Modified after Horton et al., 2016; Fuentes et al., 2016.

Side Trip 3, Puesto Rojas Field and Cañada Ancha Anticline to Infiernillo Basalt Flow, Mirador Cañón Río Salado: *Continue driving west on RP-222 for 2.6 km (1.6 mi; 3 min) to* **Stop ST 3.2, Mirador Cañón Río Salado** *(−35.197235, −69.782665).*

Stop ST 3.2 Infiernillo Basalt Flow, Mirador Cañón Río Salado

From the Río Salado Canyon Overlook, you can see the very recent Infiernillo (Little Hell) basalt flow across the canyon to the north. The flow, actually a late Pleistocene-Holocene basaltic-andesite, last erupted around 8,840 years ago from the Volcán Hoyo Colorado vent (Volcano Discovery, online; Smithsonian, online). Extensional tectonics resulted in fissure eruptions from multiple vents in the area. The Hoyo Colorado cone produced a flow that for a time dammed the Rio Salado.

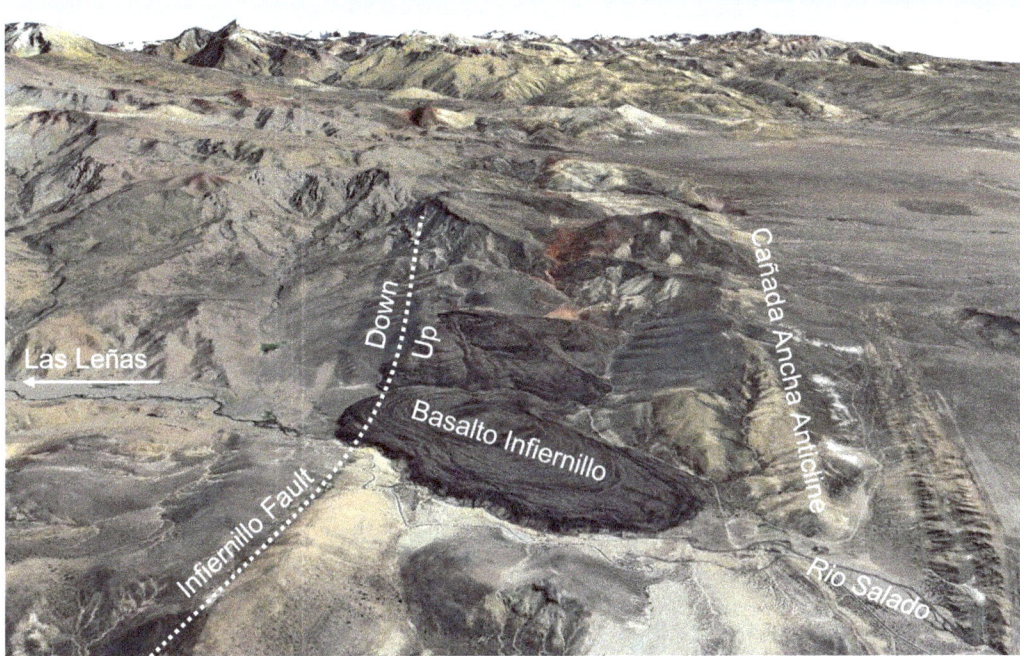

Google Earth oblique view north at Infiernillo Basalt flow. Imagery ©2023 Airbus Landsat/Copernicus. CNES/ Airbus.

> ***Side Trip 3, Infiernillo Basalt Flow, Mirador Cañón Río Salado to Laguna de la Niña Encantada:*** *Continue driving west on RP-222 to Los Molles; turn right (north) onto an unnamed graded gravel road and drive 1.2 km (0.8 mi) across the Río Salado to the first intersection where you can turn right onto Acceso a Laguna de la Niña Encantada; drive east on the north side of Rio Salado for another 6.9 km (4.3 mi) to **Stop ST 3.3, Laguna de la Niña Encantada** (−35.160093, −69.869271) on the left for a total of 24.8 km (15.4 mi; 32 min).*

Stop ST 3.3 Laguna de la Niña Encantada

A phreatic explosion (a steam blast that occurs when lava encounters near-surface groundwater) formed the Niña Encantada lagoon. A local legend explains how this lake got its name.

A long time ago there was a most beautiful Indigenous princess named "Elcha." Her beauty was remarkable, and she was the pride of her tribe. Elcha had a childhood companion, a boy of no nobility, and over time a genuine love grew between them.

At some point a confrontation developed with a neighboring tribe. The other tribe's shaman, an evil woman, convinced the princess's father that the best way to placate her tribe was to arrange a marriage between their prince and the princess. An agreement was reached between chiefs, and they fixed the date at the next new moon.

Princess Elcha was only informed of the decision the night before the wedding. Desperate, she and her lover quickly escaped to the north. When the neighboring tribe arrived for the wedding, they learned of the flight of the two youths and immediately set out to bring back the princess.

It was a dark night and soon began to rain. The youths, who had left the trail, came to a precipice that dropped to a lagoon far below. Behind them, they saw the torches of their pursuers getting ever closer. Without a thought to their safety, Elcha and the youth jumped into the icy water and died.

The first to arrive and see the lagoon was the witch, and the moment she did a powerful beam of light hit her, leaving her petrified. The others, frightened, saw the image of Elcha reflected from the pond's surface. Since then, it has been known as the lagoon of the enchanted girl, Elcha, which in the indigenous language means "mirror."

Laguna De La Niña Encantada (Enchanted Girl Lagoon). Photo courtesy of Cristina López Palacios, https://commons.wikimedia.org/wiki/File:Laguna_de_la_Ni%C3%B1a_Encantada.JPG.

**Side Trip 3, *Laguna de la Niña Encantada to Pozo de las Ánimas:* Return *west to Los Molles and RP-222; turn right (west) on RP-222 and drive a total of 14.8 km (9.2 mi; 21 min) to* Stop ST 3.4, *Pozo de las Ánimas (−35.189533, −70.001814) on the left.*

Stop ST 3.4 Pozo de las Ánimas

A pair of sinkholes, or dolines, likely developed as a karst feature (cavern collapse) over Jurassic gypsum in the Auquilco Formation or perhaps Cretaceous Mendoza Group limestones (Chauchau Formation). The "wells of the souls" are 26 to 30 m (85 to 100 ft) deep in coarse alluvial material at the surface.

Pozo de las Ánimas. Photo courtesy of Fabio from Mendoza, https://commons.wikimedia.org/wiki/File:Pozo _de_Las_Animas_Mendoza_Argentina.jpg.

Side Trip 3, Pozo de las Ánimas to Sosneado Thrust View and El Sosneado Field:
Return east on RP-222 to RN-40; turn left (north) on RN-40 and drive to El Sosneado; turn left (north) at the Service Station and sign for RN-40 and drive 2 km (1.3 mi) on the unnamed graded gravel road to **Stop 31, Sosneado Thrust View and El Sosneado Field** *(−35.069857, −69.565521) for a total of 57.0 km (35.4 mi; 49 min).*

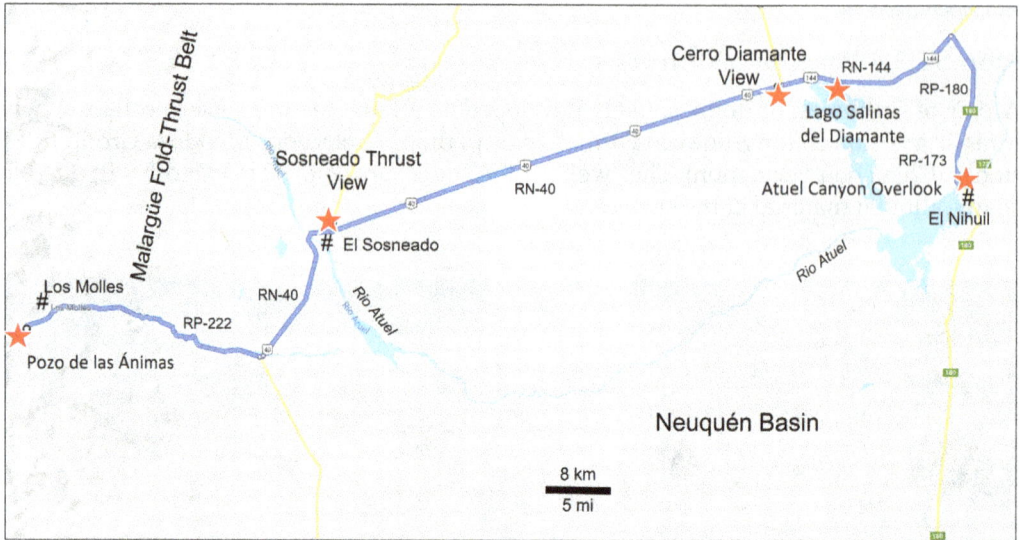

Stop 31 Sosneado Thrust View and El Sosneado Field

Looking northwest from this location, you can see the topographic front that results from the Sosneado Thrust bringing the Neogene Agua de la Piedra Formation to the surface. A surface anticline formed by the thrust lies over a deeper anticline that produces oil from Early Cretaceous Rayoso and Huitrín formation reservoirs in the north and west Sosneado Field, and from the Upper Cretaceous Loncoche Formation in the eastern Sosneado Field (Giampaoli et al., 2005; Horton et al., 2016). Oil is also trapped in updip erosional truncations and updip sandstone pinchouts. This foreland field (in Canada it would be called a "Foothills Belt" field) and the contiguous Loma de la Mina Field together contain 3.02 million m³ (19 million barrels) of oil (Fuentes et al., 2016).

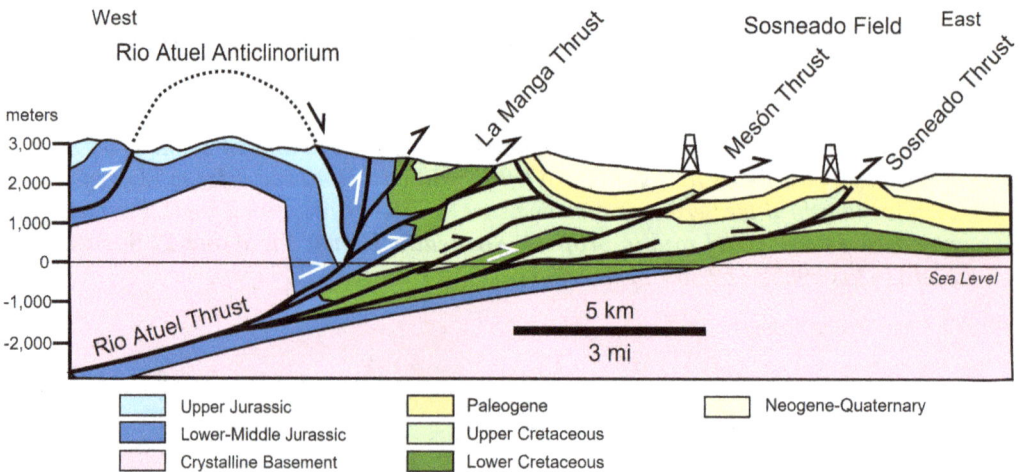

West-east cross-section through the Sosneado Thrust. Modified after Horton et al., 2016.

> **Sosneado Thrust View and El Sosneado Field to Cerro Diamante View:** *Return to RN-40; turn left (east) on RN-40 and drive to* **Stop 32, Cerro Diamante View** *(-34.91538, -68.93776) for a total of 63.6 km (39.5 mi; 43 min).*

Stop 32 Cerro Diamante View

Driving east from the Sosneado Thrust we enter the essentially undeformed Neuquén foreland basin. Cerro Diamante (Diamond Hill) is a solitary composite cone that rises 800 m (2,600 ft) above the flat pampas to an elevation of 2,362 m (7,749 ft). This isolated volcano has been linked to the extensive late Pliocene to Quaternary Payunia back-arc volcanic province (Ramos and Kay, 2006).

View northwest to Cerro Diamante from RN-144. It is so perfect you can almost imagine a puff of steam coming from the top.

> **Cerro Diamante View to Lago Salinas del Diamante:** *Continue east on RN-40 to RN-144; continue straight on RN-144 to* **Stop 33, Lago Salinas del Diamante:** *(−34.915444, −68.855118) on the right for a total of 8.4 km (5.2 mi; 6 min).*

Stop 33 Lago Salinas del Diamante

Lago Salinas del Diamante is a playa lake, a shallow lake that fills during the rainy season and is dry the rest of the year. A salt layer at the surface dissolves when the rain comes and floods the depression. When the water evaporates, the salt precipitates at the surface and is mined.

The salt flats have been privately held by the Remaggi family since 1916, and the salt is actively mined today. Nevertheless, the lake was declared a National Protected Area due to its unique ecosystem, and the family opened it to the public in 2015. There is a small museum dedicated to educating the public about the salt found here.

Lago Salinas del Diamante.

Mining salt, Lago Salinas del Diamante. View south from RN-144.

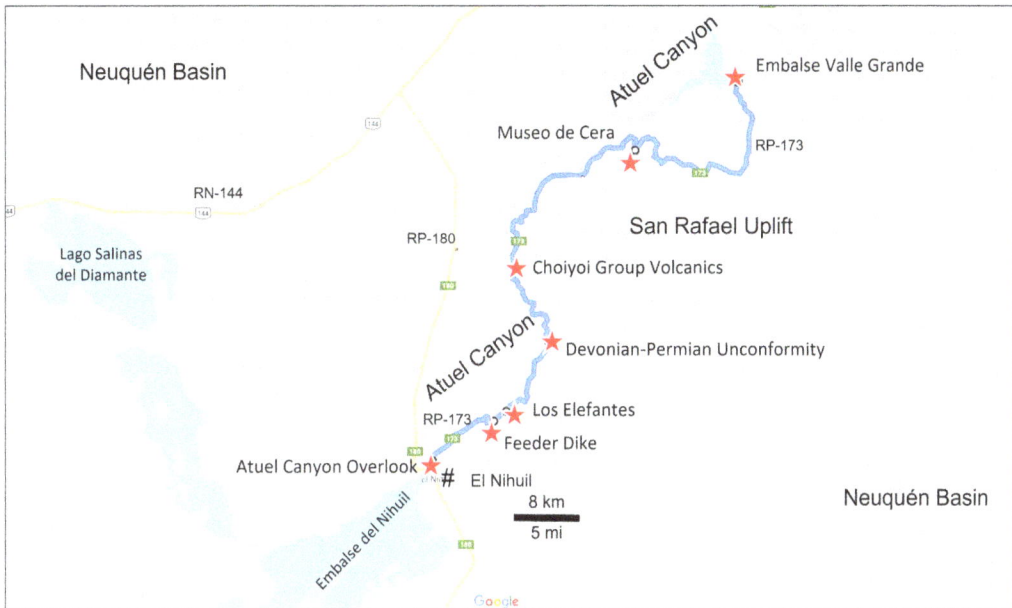

*Lago Salinas del Diamante to Atuel Canyon Overlook, Cochicó Group: Continue east on RN-144 to RP-180; turn right (south) on RP-80 and drive to El Nihuil Dam and RP-173; turn left (east) on RP-173 and drive to **Stop 34.1, Atuel Canyon Overlook, Cochicó Group** (−35.02043, −68.67518) for a total of 37.8 km (23.5 mi; 27 min).*

Stop 34 Atuel Canyon

So far on this trip, we have seen Jurassic to Recent strata. Here we will see the oldest rocks on the trip.

Atuel Canyon formed where the Río Atuel cuts through the San Rafael Block. The San Rafael Block is an uplift within the Cuyania Terrane, a terrane formed opposite the Ouachita Mountains when the western margin of Gondwana collided with proto-North America. Paleozoic metasediments in the region overlie a Precambrian (Grenvillian) basement. The Upper Carboniferous -Lower Permian El Imperial Formation unconformably overlies metamorphosed middle Paleozoic units.

Structurally, the San Rafael Block is an east-vergent asymmetric basement uplift with a steeper eastern flank and gentle west flank. The east flank is uplifted along a series of high-angle reverse faults. These reverse faults were originally normal faults that defined the western margin of the Triassic Alvear extensional basin. This basement-involved uplift resulted from inversion of the normal faults during the Andean Orogeny (Folguera et al, 2009).

It is unclear whether the Atuel River was superposed on the block (eroded down onto the structure) or is antecedent (the structure rose as the river maintained its elevation).

The canyon is thought to have been carved since the start of the last Ice Age (Pazos and Krapovikas, 2014).

The El Imperial Formation consists of unmetamorphosed and gently folded deltaic, marine, and glacio-marine sediments. The folding occurred during the Early Permian, an event known locally as the San Rafael Tectonic Stage. San Rafael compression was associated with abundant rhyolitic volcanism of the Early Permian Cochicó Group (Arroyo Punta del Agua and Yacimiento Los Reyunos formations). The Yacimiento Los Reyunos Formation is composed of several members: A lower siliciclastic (sandy and silty) member with ignimbrites is the first non-marine unit in the region. An aeolian middle member, the Areniscas Atigradas, contains uranium mineralization. The upper member comprises several ignimbrite flows.

After Cochicó Group volcanism came the Permo-Triassic Choiyoi Group volcanics. These were deposited during an extensional event that is responsible for several rift basins. One of them contains the Triassic Puesto Viejo and Cerro Carrizalito formations consisting of continental sediments, rhyolitic ignimbrites, and basalts. It is exposed in the Atuel Canyon area. Finally, the Miocene Aisol Formation sandstone is covered by basalt flows. The Aisol Formation is exposed very close to Atuel Canyon, and the basalt flows can be seen in some parts of the canyon.

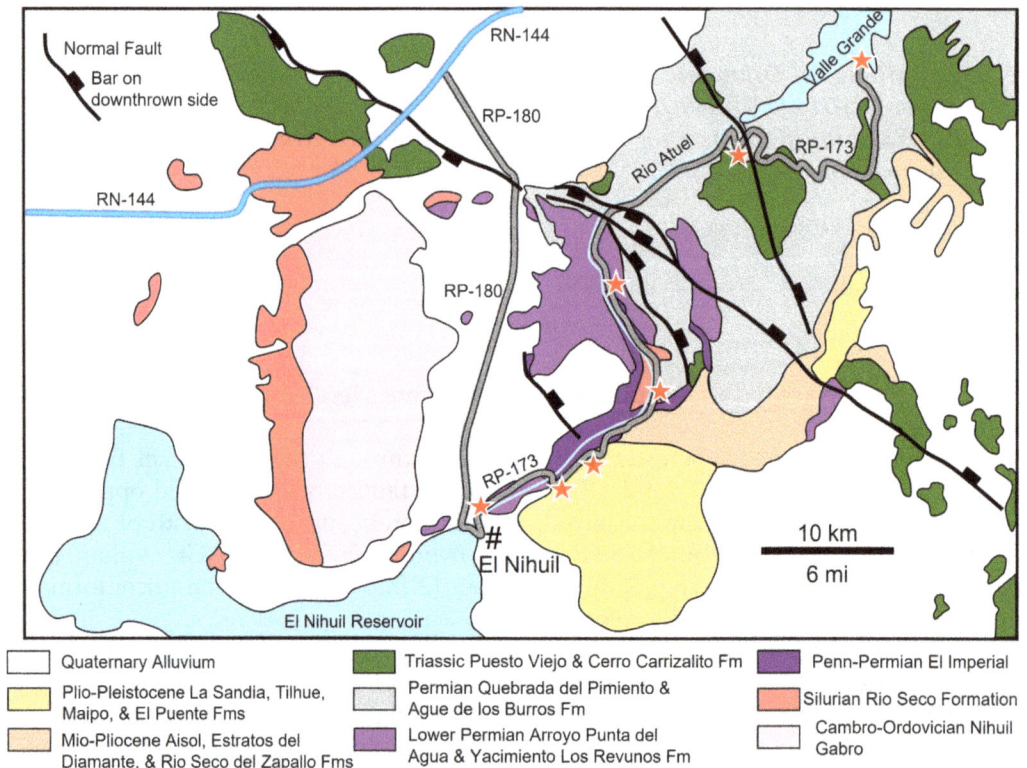

Geologic map of the Atuel Canyon area. Modified after Pazos et al., 2007; Rocha-Campos et al., 2011; and Pazos and Krapovickas, 2014.

Age	Group	Formation	Lithology
Pliocene-Pleistocene		La Sandia/ Tilhué/ Maipo/ El Puente	Conglomerates, Tuffs, Basalts, Andesites
Miocene-Pliocene		Aisol/ Estratos del Diamante/ Rio Seco de Zapallo	Basalts, Andesites, Conglomerates, Tuffaceous Sandstones
Triassic	Choiyoi	Puesto Viejo/ Cerro Carrizalito	Sandstones, Siltstones, Shales, Rhyolitic Ignimbrites, Rhyolites, Basalts
Upper Permian	Upper	Quebrada del Pimiento/ Agua de los Burros	Volcanics (Andesites, Rhyolites)
Lower Permian	Cochicó Group Lower Choiyoi	Arroyo Punta del Agua/ Yacimientos Los Reyunos	Volcanics (Andesites, Dacites, Rhyolites)
Penn-Perm		El Imperial	Deltaic Sandstones; Marine Sediments
Devon.		La Horqueta	Granitic Stock
Cambro-Ordovic		El Nihuil	Gabbro/Dolerite

Stratigraphy, Atuel Canyon. Information derived from Pazos et al., 2007; Rocha-Campos et al., 2011, Pazos and Krapovicas, 2014.

Stop 34.1 Atuel Canyon Overlook, Cochicó Group

The stunning view from the entrance to Atuel Canyon looks out over the Early Permian Cochicó Group. These light-colored rhyolitic volcanics are associated with Early Permian San Rafael tectonic compression prior to the breakup of Gondwana.

Entrance to Atuel Canyon near El Nihuil, view northeast from RP-173. Early Permian rhyolitic volcanics of the Cochicó Group (Arroyo Punta del Agua and Yacimiento Los Reyunos formations).

> ***Atuel Canyon Overlook, Cochicó Group to Feeder Dike Intruding El Imperial Formation:*** *Continue east on RP-173 for 4.5 km (2.8 mi; 7 min) to **Stop 34.2, Feeder Dike Intruding El Imperial Formation** (−35.00332, −68.64093).*

Stop 34.2 Feeder Dike Intruding El Imperial Formation

A basaltic feeder plug can be seen intruding the Upper Carboniferous-Lower Permian El Imperial Formation. The El Imperial consists of deltaic sandstones, marine, and glacio-marine sediments. This feeder dike was probably one source of the Oligocene-Pliocene basalt flows at the surface.

Basalt feeder dike (arrow) intruded into the El Imperial Formation.

Feeder Dike Intruding El Imperial Formation to Los Elefantes and El Imperial Formation: *Continue east on RP-173 for 0.9 km (0.6 mi; 2 min) to* **Stop 34.3, Los Elefantes and El Imperial Formation** *(−34.998857, −68.633837) and pull over on the right.*

Stop 34.3 Los Elefantes and El Imperial Formation

The location is named Los Elefantes (The Elephants) due to the odd resemblance of the white El Imperial Formation to these animals. The El Imperial Formation in the background is quite a bit higher: It is separated from the closer El Imperial in front of you by a fault, down toward the viewer.

Los Elefantes. These light-colored sandstones, in some eyes evocative of elephants, are the Upper Carboniferous-Lower Permian El Imperial Formation.

Los Elefantes and El Imperial Formation to Devonian-Permian Unconformity: *Continue north on RP-173 for 4.9 km (3.1 mi; 8 min) to* **Stop 34.4, Devonian-Permian Unconformity** *(−34.96773, −68.61052) and pull over on the left.*

Stop 34.4 Devonian-Permian Unconformity

Horizontally-layered Permo-Triassic Cochicó/Choiyoi groups sit unconformably over near-vertical Precambrian to Devonian sediments and metasediments. Both of these Permo-Triassic groups are mostly rhyolite and ignimbrites (tuff, ash, and volcanics). They erupted and were deposited during a compressional pulse just prior to the breakup of Gondwana.

Unconformity (dashed line) between near-vertical Precambrian-Devonian metasediments (below) and Permo-Triassic Cochicó/Choiyoi Groups mostly rhyolitic volcanics (above).

> *Devonian-Permian Unconformity to Choiyoi Group Volcanics: Continue north on RP-173 for 5.2 km (3.2 mi; 9 min) to **Stop 34.5, Choiyoi Group Volcanics** (−34.93491, −68.6296) and pull over on the right.*

Stop 34.5 Choiyoi Group Volcanics

This canyon just keeps getting more scenic. Ahead and above are brilliant and multi-hued volcanic ash/tuff layers of the Permian Choiyoi Group.

Vivid colors of the Choiyoi Group rhyolitic volcanic strata.

> ***Choiyoi Group Volcanics to Museo de Cera:*** *Continue northeast on RP-173 for 12.6 km (7.8 mi; 19 min) to* **Stop 34.6, Museo de Cera** *(–34.88332, –68.56243) and pull over on the right.*

Stop 34.6 Museo de Cera

The Museo de Cera (Wax Museum) stop presents another fine view of the multicolored Choiyoi Group volcaniclastics.

Museo de Cera, fantastic erosional forms in the Choiyoi Group.

> ***Museo de Cera to Embalse Valle Grande:*** *Continue east and north on RP-173 for 13.6 km (8.5 mi; 17 min) to* **Stop 34.7, Embalse Valle Grande** *(–34.852767, –68.506165).*

Stop 34.7 Embalse Valle Grande

Embalse Valle Grande (Big Valley Reservoir) is a man-made lake at the lower end of Atuel Canyon. Most of the rocks you see are Permo-Triassic Choiyoi Group volcanic sediments, flows, and air-fall tuffs.

This lake is the lowest of a group of four reservoirs along the Río Atuel in the Atuel Canyon. All of the reservoirs are used to generate hydroelectric power; this and the Embalse de Nihuil at the head of the canyon are also used for recreation (swimming, boating, fishing). The dams also provide flood control and siphon water for irrigation of nearby farms. Built in 1964, the dam has a maximum height of 115 m (377 ft).

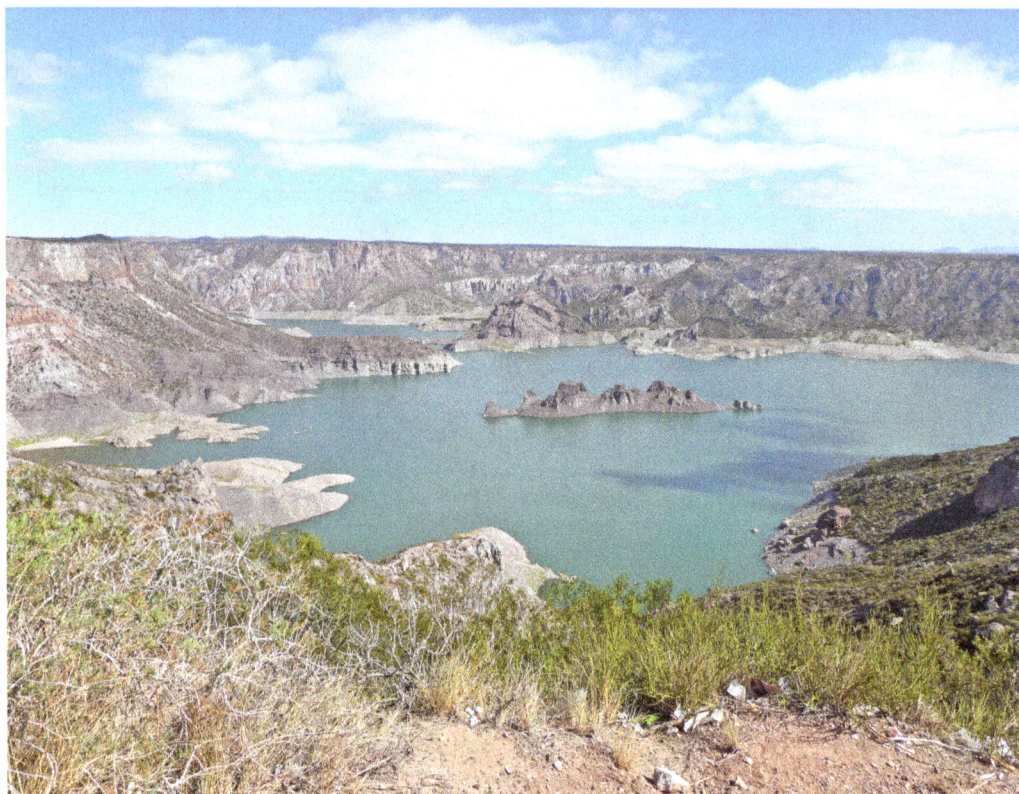

Emerald-green waters of the Embalse Valle Grande.

Embalse Valle Grande to Pareditas, Valle de Uco, and the Terroir of Mendoza: *Continue east and north on RP-173 to RN-143; turn left (northwest) on RN-143 and drive to Los Filtros/RP-150; turn left (northwest) on Los Filtros/RP-150 and drive to RN-143; continue straight (northwest) on RN-143 to RN-40; continue straight (north) on RN-40 to* **Stop 35, Pareditas, Valle de Uco, and the Terroir of Mendoza** *(−33.929687, −69.082431) and pull over on the right for a total of 147 km (91.4 mi; 1 hr 58 min).*

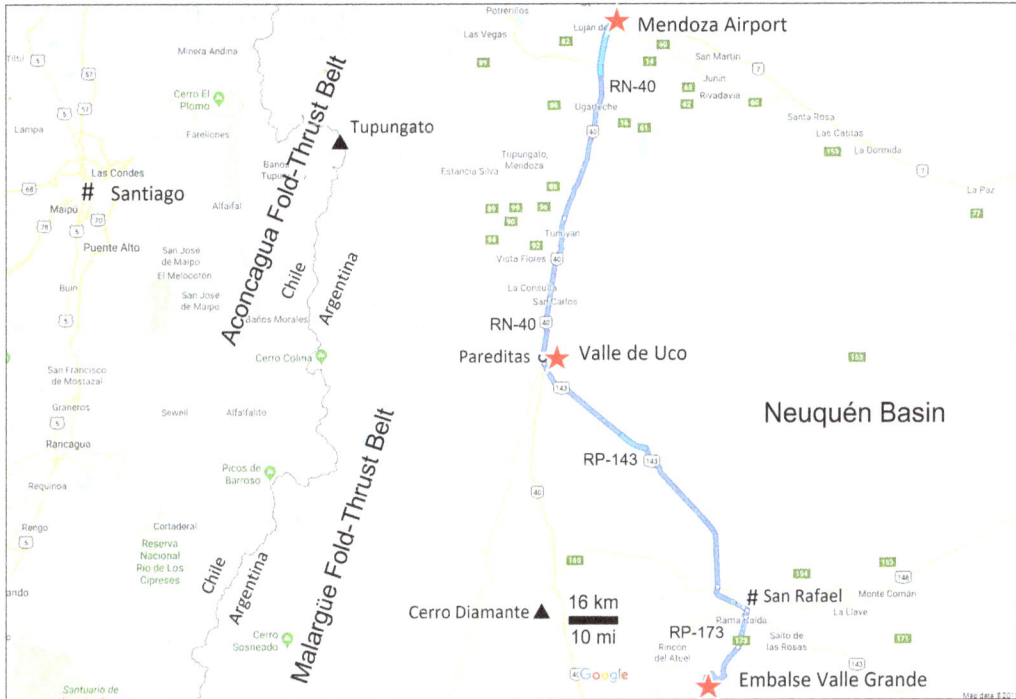

Stop 35 Pareditas, Valle de Uco, and the Terroir of Mendoza

Argentina is the 5th largest wine producer in the world, and the Cuyo Region is the best-known and highest-producing wine region in Argentina. Cuyo means "desert country" in the Huarpe Millcayac language of the native group that lived here before colonization. Mendoza is the wine capital of the Cuyo Region, producing 70% of Argentina's wine. Mendoza is broken up into 5 subregions: 1) Uco Valley, including Tunuyán, Tupungato, and Saint Charles; 2) Luján de Cuyo and Maipú; 3) Lavalle and Las Heras, 4) San Martín, Rivadavia, Junín, Santa Rosa, and La Paz), and 5) San Rafael, Malargüe, and General Alvear. The roughly 300,000 ha (741,000 ac) of Mendoza vineyards are at altitudes ranging from 430–2,000 m (1,400–6,500 ft).

Terroir

Terroir describes the sum of all environmental factors involved in producing wine and imparting flavor to the wine. These include, but are not limited to climate, soil, and topography.

Nestled at the foot of the Andes, with snow-capped peaks as a backdrop, the region's arid climate, well-drained soils, and glacial water conspire to make great wine. The low humidity means few insects and makes vines less susceptible to grape diseases, fungus, and mold. This means there is very little use of insecticides and fungicides, and some vineyards are totally organic. The vines are affected by four distinct seasons, including winter dormancy. In this desert-like climate, annual rainfall rarely exceeds 250 mm (10 in) per year and summer daytime temperatures can be upwards of 40°C (104°F), with nighttime temperatures as low as 10°C (50°F). Winter temperatures can drop below 0°C (32°F), but frost is rare. Warmer regions have an average of 320 sunny days per year.

The vines are planted in well-drained sandy alluvial soils with some gravel and limestone and a clay substrate. The very lack of nutrients stresses the vines, concentrating the flavors. Irrigation water is said to have a high dissolved mineral content, imparting some flavor to the wines.

Several grape varieties are grown in the Mendoza Region. Dominant grapes include Bonarda, Cabernet Sauvignon, Chardonnay, Syrah, Tempranillo, and Malbec. The high-altitude vineyards of Tupungato, in the Uco Valley southwest of Mendoza, are gaining popularity for their Chardonnays. Cabernet Sauvignon is receiving attention in the cooler climate of the Maipú region. Malbec, however, is the region's, and the country's flagship wine.

Whereas the Malbec grape has been all but abandoned in Europe, it is the premier grape in Mendoza, with 20% of the total area of Malbec planted in the country. In Mendoza, this grape thrives on the long hours of sun, warm days, and cool nights, the well-drained alluvial soil, and the glacial meltwater with its high dissolved mineral content. After harvesting the grape, wine growers convert the tart malic acid, naturally present in grape must (freshly crushed juice with skins, stems, and seeds), to lactic acid using a process called "malolactic fermentation" (MLF). MLF reduces the acidity and makes the flavor smoother and creamier. This is discussed further in Tour 4, Stop 1, Luján de Cuyo.

History of Wine in Argentina and Mendoza

There are several stories about how Malbec got its name. In one, "mal bec" means "bad mouth" in French. Another claims that Malbec is named after a Hungarian peasant who first spread the grape variety throughout France. One thing is indisputable: It came from France, probably around Cahors in southwest France. In France, the grape is often called côt, but also goes by cor, cos, cau, Cahors, Pressac, Gaillac, Auxerrois, or Côt Noir. It was first recorded as malbec in the 1780s, supposedly planted in Bordeaux by a Monsieur Malbeck.

In 1556, Father Juan Cedrón established the first vineyard in the Mendoza region with cuttings from the Chilean Central Valley. Between 1569 and 1589, missionaries and settlers built extensive irrigation ditches to irrigate the vineyards with water from melting glaciers in the Andes. In the 1860s, Domingo Sarmiento, provincial governor, asked French agronomist Miguel Aimé Pouget to bring grapevines from France. Among the varieties he brought were the first Malbec vines.

Argentina's most highly rated Malbec wines come from Mendoza's high-altitude regions of Luján de Cuyo and Valle de Uco. These districts are between 800 m and 1500 m elevation (2,800 to 5,000 feet). Following the example of neighboring Chile, in the late 1900s the Argentine wine industry started to more aggressively focus on the export market – particularly the lucrative British and American markets. Winemakers from France, California, and Australia brought modern technical know-how for viticultural and winemaking techniques such as yield control, temperature control fermentation and the use of new oak barrels.

Argentine wines started being exported during the 1990s, and continue to grow in popularity, making Argentina the largest wine exporter in South America. By the end of the 1990s, Argentina was exporting more 3.3 million gallons (12.5 million liters) to the United

States alone. The 2002 devaluation of the Argentine peso kick-started enotourism. Wine tourism continues to grow in Argentina.

About an hour south of Mendoza, the Valle de Uco (Uco Valley) is the southernmost of the Mendoza winegrowing areas. It also has the most stunning views of the Andes. Wine tasting can be difficult, as many wineries require reservations and wineries tend to be far apart. Sampling the wines here requires more logistical work than in the Lujan de Cuyo region closer to Mendoza, where you can almost walk from one vineyard to the next.

Valle de Uco vineyard with the Andes in the background. View east from RN-40.

> *Pareditas, Valle de Uco, and the Terroir of Mendoza to Mendoza Airport:* Continue north on RN-40 to Mendoza; at 119 km (71.4 mi) stay left and follow the signs for Mendoza/Las Heras/San Juan and merge onto Accesso Este going west; use the second-from-the-right lane to turn right (north) onto Av. Ricardo Videla; at the roundabout "Rotunda Del Avion" take the second exit to continue straight on R-40/Av Base Aerea; use any lane to bear left onto RN-40/Accesso Norte; follow signs to Aeropuerto for a total of 131 km (81.7 mi; 1 hr 54 min).

We end this tour in the heart of the Argentine wine region. We have traversed dinosaur badlands, ancient seafloors, a wilderness replete with volcanoes and oil wells, premium vineyards, sheer canyons, dazzling peaks, and folded mountains rich in human and geologic history. Neuquén-Mendoza country has it all.

References

Aguirre-Urreta, B., D.G. Lazo, and P.F. Rawson. 2012. Decapod Crustacea from the Agrio Formation (Lower Cretaceous) of the Neuquén Basin, Argentina. *Palaeontology* v. 55, Pt 5, p. 1091–1103.

Arcuri, M., and C. Zavala. 2008. Hyperpycnal Shelfal lobes – Some examples of the Lotena and Lajas Formations, Neuquén Basin, Argentina. *American Association Petroleum Geologists Search and Discovery*, Article #50066, 4 p.

Barros, P., H.J. Campos, G. Pedersen, P. Plink-Bjorklund, A. Moscariello, and E. Morettini. 2016. New insights into the Cretaceous Rayoso formation: A regional overview of a large Fluvial Fan and implications for reservoir prediction (abs). American Association Petroleum Geologists Search and Discovery Article #90259, AAPG Annual Convention and Exhibition, Calgary, Alberta, Canada, June 19–22, 2016.

Borghi, P., L. Fennell, R. Gómez Omil, M. Naipauer, E. Acevedo, and A. Folguera. 2019. The Neuquén group: The reconstruction of a Late Cretaceous foreland basin in the southern Central Andes (35–37°S). *Tectonophysics* v. 767, 228177, 24 p.

Branellec, M., B. Nivière, J.-P. Callot, and J.-C. Ringenbach. 2016. Mechanisms of basin contraction and reactivation in the basement-involved Malargüe fold-and-thrust belt, Central Andes (34–36° S). *Geological Magazine* v. 153 no. 5/6, p. 926–944.

Canale, N., J.J. Ponce, N.B. Carmona, and D.I. Drittanti. 2016. Ichnology of deltaic mouth-bar systems of the Lajas Formation (Middle Jurassic) in the Sierra de La Vaca Muerta, Neuquén Basin, Argentina. *Ameghiniana* v. 53 no. 2, p. 170–183.

Cobbold, P.R., and E.A. Rossello. 2003. Aptian to recent compressional deformation, foothills of the Neuquén Basin, Argentina. *Marine and Petroleum Geology* v. 20, p. 429–443.

Cruset, D., J. Vergés, N. Rodrigues, J. Belenguer, E. Pascual-Cebrian, Y. Almar, I. Pérez-Cáceres, C. Macchiavelli, A. Travé, A. Beranoaguirre, R. Albert, A. Gerdes, and G. Messager. 2021. U–Pb dating of carbonate veins constraining timing of beef growth and oil generation within Vaca Muerta Formation and compression history in the Neuquén Basin along the Andean fold and thrust belt. *Marine and Petroleum Geology* v. 132, 24 p.

Fleming, C. 2018. Vaca Muerta slowly comes to life. World Oil, December 2018, p. 70.

Folguera, A., J.A. Naranjo, Y. Orihashi, H. Sumino, K. Nagao, E. Polanco, and V.A. Ramos. 2009. Retroarc volcanism in the northern San Rafael Block (34°–35°30'S), southern Central Andes: Occurrence, age, and tectonic setting. *Journal of Volcanology and Geothermal Research* v. 186, p. 169–185.

Fuentes, F., B.K. Horton, D. Starck, and A. Boll. 2016. Structure and tectonic evolution of hybrid thick- and thin-skinned systems in the Malargüe fold–thrust belt, Neuquén basin, Argentina. *Geological Magazine* v. 153 no. 5–6, September, p. 1066–1084.

Galland, O., and C. Sassier. 2016. Neuquén Foothills: Geology, Dinosaurs and… Asado. *GEO ExPRO* v. 13 no. 4, 18 p.

Galland, O., E. Hallot, P.R. Cobbold, G. Ruffet, and J. de Bremond d'Ars. 2007. Volcanism in a compressional Andean setting: A structural and geochronological study of Tromen volcano (Neuquén province, Argentina). *Tectonics* v. 26, TC4010, 24 p.

Galland, O., J.B. Spacapan, O. Rabbel, K. Mair, F. González Soto, T. Eiken, M. Schiuma, and H.A. Leanza. 2019. Structure, emplacement mechanism and magma-flow significance of igneous fingers – Implications for sill emplacement in sedimentary basins. *Journal of Structural Geology*, 16 p.

Giampaoli, P., J.L. Ramirez, and M.A. Gait. 2005. Estilos de entrampamiento en la faja plegad y fallada de Malargüe. VI Congreso de Exploración y Desarrollo del Hidrocarburos, Mar del Plata, Argentina. Volume: Las Trampas de Hidrocarburos de las Cuencas Productivas de Argentina, 20 p.

Gianni, G.M., A. Echaurren, L. Fennell, C. Navarrete, P. Quezada, J. Tobal, M. Giménez, F.M, Dávila, and A. Folguera. 2018. Cretaceous Orogeny and marine transgression in the Southern Central and Northern Patagonian Andes: Aftermath of a large-scale flat-subduction event? *In* Folguera, A. et al. (Eds.), *The Evolution of the Chilean-Argentinean Andes*, Chapter 12. Springer Earth System Sciences, Switzerland, p. 279–316.

Grimaldi, G.O., and S.L. Dorobek. 2011. Fault framework and kinematic evolution of inversion structures: Natural examples from the Neuquén Basin, Argentina. *American Association of Petroleum Geologists Bulletin* v. 95 no. 1, p. 27–60.

Horton, B.K., F. Fuentes, A. Boll, D. Starck, S.G. Ramirez, and D.F. Stockli. 2016. Andean stratigraphic record of the transition from backarc extension to orogenic shortening: A case study from the northern Neuquén Basin, Argentina. *Journal of South American Earth Sciences* v. 71, p. 17–40.

Howell, J.A., E. Schwarz, L.A. Spalletti, and G.D. Veiga. 2005. The Neuquén Basin: An overview. *In* Veiga, G.D., L.A. Spalletti, J.A. Howell, and E. Schwarz (Eds.), *The Neuquén Basin, Argentina: A Case Study in Sequence Stratigraphy and Basin Dynamics*. Geological Society, London, Special Publications, v. 252, p. 1–14.

Kay, S.M., and V.A. Ramos. 2006. Evolution of an Andean margin: A tectonic and magmatic view from the Andes to the Neuquén Basin (35–39°S lat). Geological Society of America Special Paper 407, Boulder.

Kurcinka, C., R.W. Dalrymple, and M. Gugliotta. 2018. Facies and architecture of river-dominated to tide-influenced mouth bars in the lower Lajas Formation (Jurassic), Argentina. *American Association of Petroleum Geologists Bulletin* v. 102 no. 5, p. 885–912.

Leanza, H.A. 2012. The Vaca Muerta Formation (Late Jurassic - Early Cretaceous): History, stratigraphic context and events of this emblematic unit of the Neuquén basin, Argentina. Conference: Vaca Muerta: The leading shale play in Latin America. Buenos Aires, Argentina. American Association of Petroleum Geologists, Geosciences Technology Workshops, 40 p.

Leanza, H.A., and C. Hugo. 2001. Hoja Geológica 3969-I - Zapala, provincia del Neuquén. Servicio Geológico Minero Argentino, Instituto de Geología y Recursos Minerales, 1:250,000.

Leanza, H.A., and C. Hugo. 2005. Hoja Geológica 3969-I - Zapala, provincia del Neuquén. Servicio Geológico Minero Argentino, Instituto de Geología y Recursos Minerales, 1:250,000.

Lebinson, F., M. Turienzo, N. Sánchez, V. Araujo, M.C. D'Annunzio, and L. Dimieri. 2018. The structure of the northern Agrio fold and thrust belt (37°30′ S), Neuquén Basin, Argentina. *Andean Geology* v. 45 no. 2, p. 249–273.

Legarreta, L., C.E. Cruz, G. Vergani, G.A. Laffitte, and H.J. Villar. 2005. Petroleum mass-balance of the Neuquén Basin, Argentina: A comparative assessment of the productive districts and non-productive trends. American Association of Petroleum Geologists, Search and Discovery Article #10080, 7 p.

Martos, F.E., L.M. Fennell, S. Brisson, G. Palmieri, M. Naipauer, and A. Folguera. 2020. Tectonic evolution of the northern Malargüe Fold and Thrust Belt,Mendoza province, Argentina. *Journal of South American Earth Sciences* v. 103, 9 p.

Mcilroy, D., S. Flint, J.A. Howell, and N. Timms. 2005. Sedimentology of the tide-dominated Jurassic Lajas Formation, Neuquén Basin, Argentina. *In* Veiga, G.D., L.A. Spalletti, J.A. Howell, and E. Schwarz (Eds.), *The Neuquén Basin, Argentina: A Case Study in Sequence Stratigraphy and Basin Dynamics*. Geological Society, London, Special Publications, v. 252, p. 83–107.

Naipauer, M., and V.A. Ramos. 2016. Changes in source areas at Neuquén Basin: Mesozoic evolution and tectonic setting based on U–Pb ages on zircons. *In* Folguera, A., et al. (eds.), *Growth of the Southern Andes*. Springer Earth System Sciences, Switzerland, p. 33–61.

Ostera, H.A., R. Garcia, D. Malizia, P. Kokot, L. Wainstein, and M. Ricciutti. 2016. Shale gas plays, Neuquén Basin, Argentina: Chemostratigraphy and mud gas carbon isotopes insights. *Brazilian Journal of Geology* v. 46, supl. 1, 22 p.

Pazos, P.J., and V. Krapovickas. 2014. Ichnology throughout the Palaeozoic in the Atuel Canyon. 4th International Palaeontological Congress, Mendoza. Mid-Congress Field Trip. 16 p.

Pazos, P.J., M. diPasquo, and C.R. Amenabar. 2007. Trace fossils of the glacial to postglacial transition in the El Imperial Formation (Upper Carboniferous), San Rafael Basin, Argentina. Society for Sedimentary Geology Special Publication no. 88, p. 137–147.

Pons, J., M. Franchini, L. Meinert, L. López-Escobar, and L. Maydagán. 2010. Geology, petrography and geochemistry of igneous rocks related to mineralized skarns in the NW Neuquén basin, Argentina: Implications for Cordilleran skarn exploration. *Ore Geology Reviews* v. 38, p. 37–58.

Ramos, V.A., and A. Folguera. 2005. Tectonic evolution of the Andes of Neuqu6n: Constraints derived from the magmatic arc and foreland deformation. *In* Veiga, G.D., L.A. Spalletti, J.A. Howell, and E. Schwarz (Eds.), *The Neuquén Basin, Argentina: A Case Study in Sequence Stratigraphy and Basin Dynamics*. Geological Society, London, Special Publications, v. 252, p. 15–35.

Ramos, V.A., and M.H. Kay. 2006. Overview of the tectonic evolution of the southern Central Andes of Mendoza and Neuquén (35–39°S latitude). *Geological Society of America Special Paper* 407, p. 1–18.

Ramos, V.A., V.D. Litvak, A. Folguera, and M. Spagnuolo. 2014. An Andean tectonic cycle: From crustal thickening to extension in a thin crust (34°–37° SL). *Geoscience Frontiers* v. 5, p. 351–367.

Risso, C., K. Németh, and F. Nullo. 2009. Field guide to Payún Matru and Llancanelo volcanic fields, Malargüe – Mendoza. 3IMC. 3rd International Maar Conference, April 14–17, 2009, Malargüe, Argentina, 28 p.

Rocha-Campos, A.C., M.A. Basei, A.P. Nutman, L.E. Kleiman, R. Varela, E. Llambias, F.M. Canile, and O. de C.R. da Rosa. 2011. 30 million years of Permian volcanism recorded in the Choiyoi igneous province (W Argentina) and their source for younger ash fall deposits in the Paraná Basin: SHRIMP U–Pb zircon geochronology evidence. *Gondwana Research* v. 19, p. 509–523.

Rojas Vera, E.A., A. Folguera, G.Z. Valcarce, M. Giménez, F. Ruiz, P. Martínez, G. Bottesi, and V.A. Ramos. 2010. Neogene to quaternary extensional reactivation of a fold and thrust belt: The Agrio belt in the Southern Central Andes and its relation to the Loncopué trough (38°–39°S). *Tectonophysics* v. 492, p. 279–294.

Rojas Vera, E.A., J. Mescua, A. Folguera, T.P. Becker, L. Sagripanti, L. Fennell, D. Orts, and V.A. Ramos. 2015. Evolution of the Chos Malal and Agrio fold and thrust belts, Andes of Neuquén: Insights from structural analysis and apatite fission track dating. *Journal of South American Earth Sciences*, 47 p.

Rojas Vera, E.A., D.L. Orts, A. Folguera, G.Z. Valcarce, G. Bottesi, L. Fennell, F. Chiachiarelli, and V.A. Ramos. 2016. The transitional zone between the southern central and northern Patagonian Andes (36–39°S). *In* Folguera, A., et al. (eds.), *Growth of the Southern Andes*. Springer Earth System Sciences, Switzerland, p. 99–114.

Sagripanti, L., G. Bottesi, D. Kietzmann, A. Folguera, and V.A. Ramos. 2012. Mountain building processes at the orogenic front. A study of the unroofing in Neogene foreland sequence (37°S). *Andean Geology* v. 39 no. 2, p. 201–219.

Sánchez, N.P., M.M. Turienzo, L.V. Dimieri, V.S. Araujo, and F. Lebinson. 2014. Evolución de las estructuras Andinas en la faja corrida y plegada de Chos Malal: interacción entre el basamento y la cubierta sedimentaria de la Cuenca Neuquina. *Revista de las Asociación Geológica Argentina* v. 71 no. 2, p. 233–246.

Sánchez, N., M. Turienzo, F. Lebinson, V. Araujo, I. Coutand, and L. Dimieri. 2015. Structural style of the Chos Malal fold and thrust belt, Neuquén basin, Argentina: Relationship between thick- and thin-skinned tectonics. *Journal of South American Earth Sciences* v. 64, Pt 2, p. 399–417.

Sánchez, N.P., I. Coutand, M. Turienzo, F. Lebinson, V. Araujo, and L. Dimieri. 2018. Tectonic evolution of the Chos Malal fold-and-thrust belt (Neuquén Basin, Argentina) from (U-Th)/He and fission track thermochronometry. *Tectonics* v. 37, 23 p.

Saucier, H. 2019. Repeating a Miracle – The Permian Basin inspires Latin American Super Basin Ambitions. AAPG Explorer, August, p. 22.

Silvestro, J., P. Kraemer, F. Achilli, and W. Brinkworth. 2005. Evolución de las cuencas sinorogénicas de la Cordillera Principal entre 35°–36° S, Malargüe. *Revista de la Asociación Geológica Argentina* v. 60 no. 4, p. 627–643.

Smith Llinas, E. 2019. Argentina's Neuquén basin offers 'tremendous growth opportunities.' AAPG Explorer, April, p. 22–25.

Spacapan, J.B., O. Galland, H.A. Leanza, and S. Planke. 2016. Igneous sill and finger emplacement mechanism in shale-dominated formations: A field study at Cuesta del Chihuido, Neuquén Basin, Argentina. *Journal of the Geological Society*, London, 12 p.

Turienzo, M., L. Dimieri, C. Frisicale, V. Araujo, and N. Sánchez. 2012. Cenozoic structural evolution of the Argentinean Andes at 34°40'S: A close relationship between thick and thin-skinned deformation. *Andean Geology* v. 39 no. 2, p. 317–357.

Venari, V.V., P.P. Álvarez, and B. Aguirre-Urreta. 2012. A new species of *Andiceras* Krantz (Cephalopoda: Ammonoidea) from the Late Jurassic-Early Cretaceous of the Neuquén Basin, Mendoza, Argentina. Systematics and Biostratigraphy. *Andean Geology* v. 39 no. 1, p. 92–105.

Wikipedia. Neuquén Basin. Accessed 20 May 2023, https://en.wikipedia.org/wiki/Neuqu%C3%A9n_Basin.

Zamora Valcarce, G., T. Zapata, D. del Pino, and A. Ansa. 2006. Structural evolution and magmatic characteristics of the Agrio fold-and-thrust belt. *Geological Society of America Special Paper* 407, p. 125–145.

Zamora Valcarce, G., and T. Zapata. 2015. Building a valid structural model in a triangle zone: An example from the Neuquén fold and thrust belt, Argentina. *Interpretation* v. 3 no. 4, p. 117–131.

Zapata, T.R., G. Zamora, and A. Ansa. 2003. The Agrio fold and thrust belt: Structural analysis and its relationship with the petroleum system Vaca Muerta-Agrio-Troncoso Inferior (!), Argentina. VIII Simposio Bolivariano - Exploracion Petrolera en las Cuencas Subandinas, Cartagena de Indias, Colombia, p. 168–176.

Zapata, T., G. Zamora Valcarce, A. Folguera, and D. Yagupsky. 2008. Field trip guide: Andean Cordillera and backarc of the south-central Andes (~38.5°S to 37°S). *In* Kay, S.M., and V.A. Ramos (Eds.), Field trip guides to the Backbone of the Americas in the southern and central Andes: Ridge collision, shallow subduction, and plateau uplift: Geological Society of America Field Guide 013, p. 23–55.

Zeller, M. 2013. Facies, geometries and sequence stratigraphy of the mixed carbonate-siliciclastic Quintuco-Vaca Muerta system in the Neuquén Basin, Argentina: An integrated approach. PhD dissertation, University of Miami, 206 p.

3

From the Sea to the Sky: The Atacama Desert from Antofagasta to San Pedro de Atacama, Chile

Salar de Atacama looking east toward the Andes and Tumisa, Lejía, and Miñiques volcanoes. Courtesy of Francesco Mocellin, https://commons.wikimedia.org/wiki/File:Salar_de_Atacama.jpg.

Overview

This tour of the Atacama Desert consists of a multi-day loop starting and ending in Antofagasta, Chile. We visit the sea cliffs and the remarkable natural arch at La Portada, see some of the world's largest copper mines, salt-deformed sediments, classical Andean composite volcanoes, and salars (playa lakes) of the ultra-arid Atacama Desert. Stopping at the Atacama Fault, a regional transcurrent fault, we pass by and discuss the Mantos Blancos and Lomas Bayas open-pit copper mines, then transit the Cordillera Domeyko, a Jurassic volcanic arc. We see the Cordillera de la Sal Fold Belt and enter the Salar de Atacama Basin, a mostly dry lakebed that was a Jurassic back-arc and is a present-day forearc setting. Further east, we enter the Andean volcanic province at Volcán Lascar and visit the colorful Laguna en Salar de Aguas Calientes. At San Pedro de Atacama, we pass through Valle de la Luna (Valley of the Moon) with its dramatic landscape and unusual

DOI: 10.1201/9781351168281-3

salt-deformation features. We drive by a string of classic stratovolcanoes, from Volcán Licancabur to Volcán San Pedro, and visit the hot springs and fumaroles of El Tatio, at more than 4,300 m above sea level the highest geothermal field in the world. We continue through a volcanic landscape to the world's largest open-pit copper mine at Chuquimata. Returning to the coast, we pass through the Atacama nitrate district at María Elena and stop at the oasis at Balneario Coya Sur on the Río Loa, the longest river in Chile. From Tocopilla we travel south on a highway wedged between the stark Coastal Cordillera and the Pacific, past the giant stratiform copper mines of the Carolina de Michilla District, wide sandy beaches at Hornitos and Mejillones, rocky promontories, bounding faults, and uplifted beach terraces of the Mejillones Peninsula at Punta Angamos.

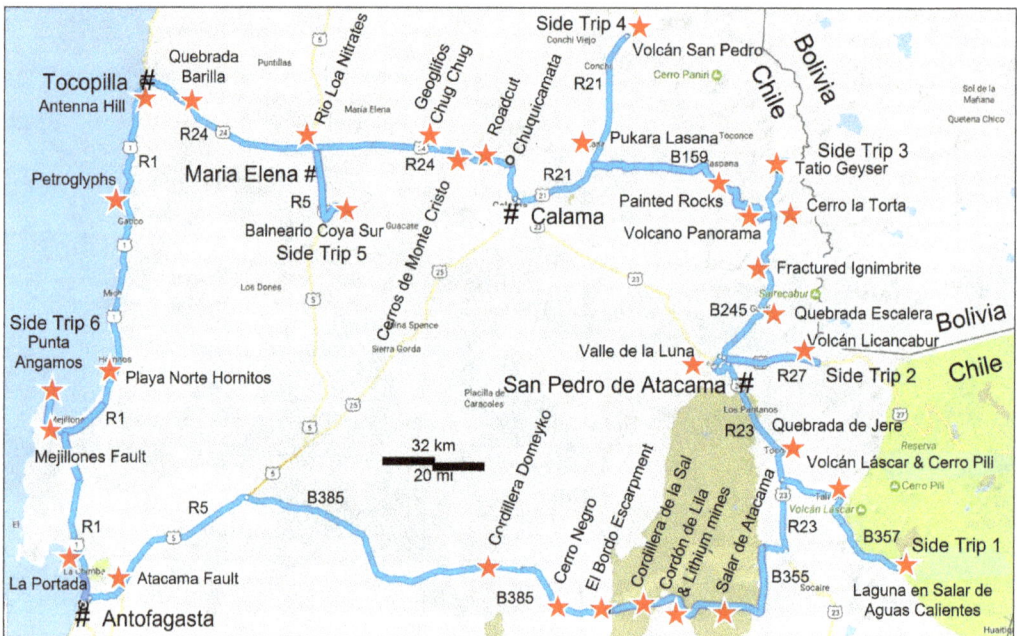

Antofagasta – San Pedro de Atacama – Calama – Tocopilla Loop.

Itinerary

Start Antofagasta Andrés Sabella Airport

  Stop 1 La Portada
  Stop 2 Antofagasta
  Stop 3 Atacama Fault
  Stop 4 Cordillera Domeyko (Precordillera)
  Stop 5 Cerro Negro, Eastern Cordillera Domeyko
  Stop 6 El Bordo Escarpment and Llanos de la Paciencia
  Stop 7 Cordillera de la Sal
  Stop 8 Cordón de Lila

The Atacama Desert and Copper Mining

This is one of the most arid spots in the world. It is a land of extremes. The terms driest, highest, and largest apply to so many aspects of the region. The highest geyser field, the largest ignimbrite fields, the largest copper mines, the clearest air, the darkest nights.

This is the driest non-polar desert on earth, averaging 15 mm (0.6 in) of rain per year, with some areas having less than 1 mm (less than one-half of one-tenth of an inch). Parts of this desert have not seen *any* rain in over four years. The rare cloudburst is an occasion for celebration, but it can also trigger horrendous flash floods and mudflows. There are huge swathes with not a blade of grass, not an insect, not a lizard to be seen. The oldest desert on earth, Triassic evaporites suggest the region has been semi-arid for the past 150 to 200 million years. Based on tectonics and ocean currents, the area has been hyper-arid for the last 3 to 15 million years (Lehman, 2019). The soil is so dry that NASA has compared it to Mars. Researchers tested the soil using the same instruments used by the Mars Viking landers. They were unable to detect any signs of life (Wikipedia, Atacama Desert). One of these sites is at María Elena, which we will pass by.

Perfect cone-shaped volcanoes reach up to 5,920 m (19,420 ft), and huge dry lakebeds, or "salars," lie 2,300 m (7,500 ft) above sea level. The Central Valley contains a number of endorheic basins, that is, basins that drain internally rather than to the sea. Snowmelt and rare rainfall flow into these basins and evaporate, forming playa (dry) lakebeds encrusted with salt and other minerals.

Massive nitrate and copper deposits are the economic drivers in the region. Chuquicamata is the largest (by volume) open-pit copper mine in the world. Not far away, a 20 km wide by 700 km long (12 by 430 mi) section of desert is known as the Nitrate Belt. It contains some of the world's largest deposits of sodium nitrate, used mostly for fertilizer. The source of these deposits was thought to be wind-swept sea spray, but recent work suggests it may have been deposited by evaporation of ancient groundwater.

The clearest, driest air and most cloud-free, darkest night-time skies on earth make the Atacama home to multiple observatories: the Cosmology Large Angular Scale Surveyor, Cerro Chajnantor Atacama Radio Telescope (ALMA), APEX Event Horizon Telescope, and the Atacama Large Millimeter Array are all within 50 km (30 mi) of San Pedro de Atacama. The European Space Agency's Paranal Observatory/Very Large Telescope is located 110 km (67 mi) south of Antofagasta. They are here because of the high altitudes, clear air, and lack of artificial light at night.

The Atacama Desert lies in a 240 km (150 mi) wide depression between the Coastal Cordillera and the Andes that extends 1,000 km (600 mi) south of the Peru border. It is in a double rain shadow, with the Coastal Cordillera blocking Pacific moisture and the even higher Andes blocking Atlantic rains. Add to this the Humboldt/Peru Current off the coast, characterized by upwelling of cold, deep Pacific waters, and you get a permanent inversion (cold air at the ocean surface capped by warm air above) that promotes coastal fog and a few clouds, but not rain. The main source of water is scarce snowmelt from the Andes.

And yet this is a mild desert, with average temperatures around 18°C (63°F).

The Chinchorro people occupied this region between 7,000 and 1,500 BC. They were mostly unremarkable fishermen along the coast but are famed for mummifying their dead. The interior was occupied by the Likanantaí/Atacameño tribe, known for their fortified towns called pukaras (Britannica, Atacameño). The mud-brick village of Tulor on the southern outskirts of San Pedro de Atacama, the oldest archaeological site in Chile, is a pukara dating from the eighth century BC (Leadbeater, 2017). The Atacameño were overrun by the Incas in the early 1400s.

Spaniards arrived in the 1540s. For 300 years nothing much happened in the area, until the development of "white gold," the sodium nitrate (saltpeter) deposits, in the 1830s to 1860s. At the time this area was part of Bolivia, but Chilean and British interests were keen on developing the area and, after a failure of diplomatic communications, Chile occupied the area in 1879 during the so-called "War of the Pacific." The reality on the ground was formally recognized by the *Treaty of Ancón* (1883), the *Treaty of Peace and Friendship between Chile and Bolivia* (1904), and the *Treaty of Lima* (1929). With the nitrate mining booms of the nineteenth century, the population in the region grew. Today there are over 170 nitrate mining ghost towns. Currently, the region's chief source of revenue is copper mining.

World-Class Copper Deposits

Chile continues to be the largest copper producer in the world, followed by Peru. Chile produced an estimated 5.6 million metric tons (6.2 tons) of copper in 2019. Peru produced an estimated 2.4 million metric tons (2.6 million tons) that year. Three of the world's ten largest copper mines (based on capacity) are located in Chile (Garside, 2020; Barrera, 2020).

The copper comes from a variety of different deposit types, with the most important being volcanic-hosted stratiform, or manto-type (tabular, or blanket) deposits, and disseminated porphyry/hydrothermal breccia deposits. Porphyry copper deposits are those formed by the interaction of hydrothermal (hot) fluids with a porphyry intrusive, that is, with an igneous rock formed at depth containing large crystals of feldspar. Hydrothermal breccia deposits consist of broken rock that has interacted with and been mineralized by hot fluids.

Volcanic-Hosted Copper

Copper deposits in the Coastal Cordillera are hosted in basaltic-andesitic volcanics (with minor amounts of limestone, marl, and conglomerate) of the Jurassic La Negra Formation (e.g. Carolina de Michilla District and Mantos Blancos Mine) and its northern equivalent, the Ofícina Viz Formation. These volcanic units host copper deposits with significant amounts of silver. Copper ores consist mainly of oxide minerals and sulfides such as chalcocite and digenite with minor bornite and chalcopyrite. Ore grades range from 1 to 3% Cu. The main copper mineralization event in the Michilla District is Late Jurassic (159 $\pm$ 16 Ma; Tristà-Aguilera et al., 2005). The Mantos Blancos deposit is hosted in Jurassic felsic volcanics and subvolcanic intrusions. At Mantos Blancos an older rhyolitic intrusion and hydrothermal brecciation occurred during Jurassic (~155 Ma); during Early Cretaceous (141–142 Ma), diorite and granodiorite stocks and sills intruded the breccias and sealed the hydrothermal system. When the fluids reached overpressure, they hydrofractured the surrounding units, causing decompression and mineral precipitation (Ramirez et al., 2006).

Some isotopic compositions of copper minerals imply magmatic sources or leaching of igneous minerals; other isotopes suggest an influx of seawater or marine carbonates provided the copper. In general, it can be stated that the interaction of near-surface groundwater and basinal brine with the volcano-sedimentary host rocks, driven by heat from deep-seated magma was the probable mechanism for mineralization. In North America these are called "volcanic redbed" deposits; they are found in northwest Canada (British Columbia, Yukon, Northwest Territories) and the Keweenaw Peninsula of northern Michigan (Kojima et al., 1998).

Porphyry Copper

About 460 million metric tons (507 tons, pre-mining) of copper is found in an Eocene-Oligocene trend, followed by a late Miocene-early Pliocene trend with about 220 million metric tons (243 million tons). The third largest endowment is in the Paleocene-Eocene tract of Chile-southern Peru, with about 98 million metric tons (108 million tons) (Cunningham et al., 2008).

"Porphyry copper deposits" refers to the texture of the igneous rock that hosts the deposits. A porphyritic texture is one where large crystals (e.g. feldspar) are found in a background of finer-grained minerals (quartz, mica, hornblende). This texture indicates early slow crystallization from a melt (forming large crystals), followed by magma injection into shallower zones where cooling proceeded rapidly (small crystals).

In this part of the Atacama Desert, magmatism created a Paleocene-early Eocene belt of small porphyry copper deposits (e.g. Lomas Bayas) associated with calc-alkaline (basalt, andesite, dacite, rhyolite) volcanics. This was followed by a later Eocene-early Oligocene trend of huge deposits (e.g. Chuquicamata) of mainly granodiorite affinity. It is thought that the granodiorites formed as a result of rapid convergence between the South American and Nazca plates, leading to flat subduction and partial melting of the down-going oceanic plate. Slab flattening has also been linked to subduction of buoyant oceanic features like mid-ocean ridges and volcanic seamounts. A period of no volcanism under conditions of intense shortening might have prevented the venting of SO_2 from the sulfur-rich, oxidized magma. This "closed porphyry system" favored the concentration of copper in large sulfide deposits. On the other hand, decreased compressional stress favored the rise of buoyant magma and allowed calc-alkaline-associated deposits to form during periods of volcanism. This "open system," with the outgassing of volatile constituents and metals, resulted in smaller copper accumulations. The younger, granodiorite-hosted deposits are up to ten times larger than the early, calc-alkaline ones (Oyarzun et al., 2001; Mpodozis and Cornejo, 2012).

Clusters or alignments of three or more porphyry deposits are common, as in the middle Eocene to early Oligocene trend extending from southern Peru into northern Chile. The greatest number of large porphyry copper deposits is in the Chuquicamata District and includes Radomiro Tomic, Chuquicamata, Ministro Hales/Mansa Mina, Quetena, Toki, Genoveva, and Opache. Porphyries in the Chuqui District all appear associated in some way with the West Fault System. These deposits all formed between ~44 and 30 Ma, a time characterized by a lack of volcanic activity and eastward shifting of the magmatic arc in northern and central Chile (Perello and Sillitoe, 2004).

We will pass by a number of these mineral deposits, including Mantos Blancos, Lomas Bayas, Conchi Viejo, El Abra, Ministro Hales, and Chuquicamata.

Age		Coastal Cordillera	Cordillera Domeyko	Altiplano		
Cenozoic	Pleistocene	Mejillones Fm	Vilama Fm		Paciencia Gp	Fore-arc Basin — Thrusting followed by normal faulting
	Pliocene	La Portada Fm	Campamento Fm			
	Miocene	Caleta Herradura Fm	El Yeso Fm			
			Tambores Fm	San Pedro Fm		Shortening & Inversion — Thin- and Thick-skinned Thrusting and basin inversion
	Oligocene	~35 Ma Pampa de Mula Fm		*Unconformity*		
	Eocene		Loma Amarilla Fm		Purilactis Gp	
	Paleocene		Naranja/Orange Fm			
Cretaceous	Late	~90 Ma	Totola Fm / Barros Arana Fm / Purilactis Fm	*Unconformity*		
	Early	Cerrillos Fm	Tonel Fm / Llanta Fm			Extension & Rifting — Steep and Listric Normal Faulting
			Santa Ana Fm			
		Chañarcillo Gp				
		Punta del Cobre Gp	Sierra del Fraga Fm			
Jurassic	Late	~153 Ma		*Unconformity*		
	Middle	La Negra Fm	Montandón Fm / Profeta Gp			Thermal Subsidence — Back-Arc Basin
		~190 Ma		*Unconformity*		
	Early	Mejillones Metamorphic Complex (basement)	Quebrada del Salitre Gp/Bardas Negras Fm			Extension & Rifting — Steep and Listric Normal Faulting
				Tinieblas Fm		
Triassic	Late	~215 Ma		*Unconformity*		
	Early	Tuina Fm	Cas Fm / El Bordo Fm / Barrancos Fm / Agua Dulce Fm			
		~248 Ma		*Unconformity*		
Paleozoic	Permian	Cerros de Cuevitas Fm				
	Carbonif	La Tabla Fm				
	Devonian	Sierra del Tigre Fm (basement)				

Legend:
- Volcanics (flows, tuffs, breccias)
- Clastics (conglomerate, sandstone, siltstone, shale)
- Limestone
- Metamorphic (basement)

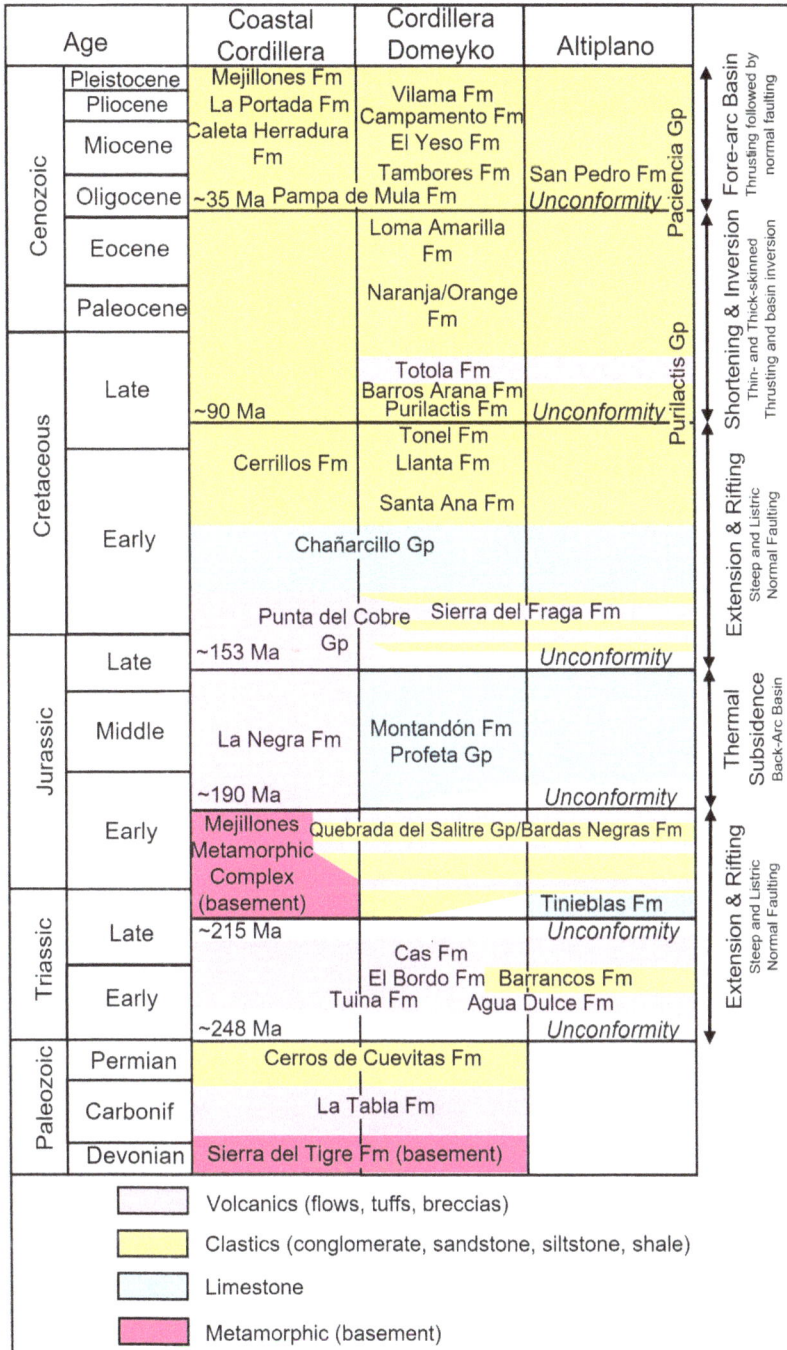

Stratigraphic and tectonic event chart for the Coastal Cordillera, Cordillera Domeyko, and Altiplano of northern Chile. Information derived from Mpodozis et al., 2004; Amilibia et al., 2008; Di Celma et al., 2013.

Begin: Andres Sabela Airport, Antofagasta, to La Portada*: Drive south on Ruta-1 (R-1) to B-446 exit and turn right (west) on B-446; drive to the parking lot overlooking **Stop 1, La Portada** (–23.499269, –70.428339), for a total of 8.6 km (5.4 mi; 10 min). Descend the sea cliffs and walk south about 475 m (1,550 ft).*

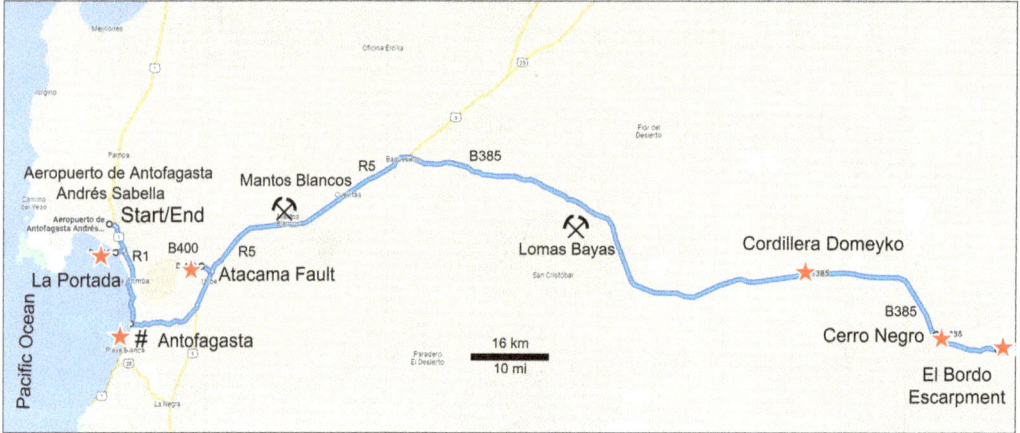

Stop 1 La Portada

La Portada (The Portal, or The Gateway) is a natural sandstone arch off the coast that was designated a Natural Monument in 1990 (Wikipedia, La Portada).

La Portada, sea cliffs, and Antofagasta in the background. View south.

The arch consists of yellow shelly sandstone of the Miocene-Pliocene La Portada Formation that lies unconformably over black Jurassic andesite of the La Negra Formation. The arch is 43 m (140 ft) high, 23 m (75 ft) wide, and 70 m (230 ft) long. The La Portada Formation also forms the sea cliffs that reach heights of 52 m (170 ft). The unit represents a marine transgression over eroded remnants of the Jurassic volcanic arc (Di Celma et al., 2013; González-Alfaro, 2013).

Walk along the beach. A close examination of this formation reveals normal faulting and the unconformity above the La Negra Formation.

> *La Portada to Antofagasta*: *Continue south on R-1 to* **Stop 2, Antofagasta** *(–23.661375, –70.403892) for a total of 21.9 km (13.6 mi; 33 min). Pull into the parking lane on the right.*

Stop 2 Antofagasta

Antofagasta is the capital of Antofagasta Province. The town, originally called La Chimba, was founded in 1868 as a port and capital of the Bolivian province of Mejillones. The town grew with the development of nitrate mining and later, copper mining. It had a population of 402,000 in 2015. The name was changed to Antofagasta around the time of the Chilean conquest in 1879. The origin of this name is hotly debated. It may be a Cacán (an extinct language spoken by northern Chilean tribes) word meaning "Town of the great salt lake." It could also be related to the Chango (another indigenous Atacama people) language, where Antofagasti means "Sun Gate," the name the Changos gave the natural arch now called "La Portada." Antofagasta is popularly known as "the Pearl of the North" (Wikipedia, Antofagasta).

Antofagasta's economy today is mainly based on mining copper, lithium, and nonmetallic minerals such as nitrate and iodine. It is an important port for exporting these resources.

When you arrive in Antofagasta, the first thing that strikes you is the complete and utter lack of vegetation in the surrounding countryside. This is because it is the world's driest town. This moonscape exists because Antofagasta receives an annual average rainfall of less than 0.1 mm (0.04 in) per year and has been known to go 40 years without any rain at all.

While in town be sure to rent a reliable vehicle, get lots of water, and tank up on gasoline. There is no water in the desert, and gas stations in the Atacama are few and far between: We will find them in San Pedro de Atacama, Calama, and Tocopilla.

Antofagasta is built on Jurassic La Negra Formation andesites. The La Negra represents the Jurassic volcanic arc that developed along the western margin of South America. These lavas are unconformably overlain by the Miocene-Pliocene La Portada sandstone that contains dense layers of shells and pebbles deposited on a transgressive marine terrace. About 165 million years of rock is missing between these two units.

Jurassic La Negra Formation volcanics on Antofagasta beach. This unit lies unconformably beneath the Miocene-Pliocene La Portada Formation. View northwest to the Mejillones Peninsula.

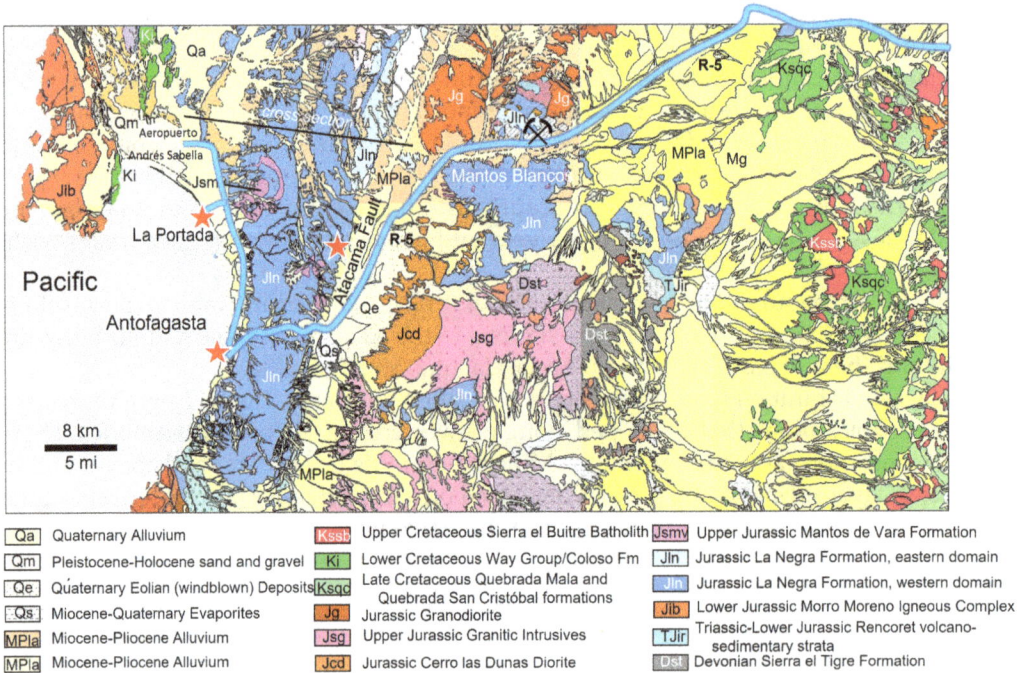

Qa	Quaternary Alluvium	Kssb	Upper Cretaceous Sierra el Buitre Batholith
Qm	Pleistocene-Holocene sand and gravel	Ki	Lower Cretaceous Way Group/Coloso Fm
Qe	Quaternary Eolian (windblown) Deposits	Ksqc	Late Cretaceous Quebrada Mala and Quebrada San Cristóbal formations
Qs	Miocene-Quaternary Evaporites	Jg	Jurassic Granodiorite
MPla	Miocene-Pliocene Alluvium	Jsg	Upper Jurassic Granitic Intrusives
MPla	Miocene-Pliocene Alluvium	Jcd	Jurassic Cerro las Dunas Diorite

Jsmv	Upper Jurassic Mantos de Vara Formation
Jln	Jurassic La Negra Formation, eastern domain
Jln	Jurassic La Negra Formation, western domain
Jlb	Lower Jurassic Morro Moreno Igneous Complex
TJir	Triassic-Lower Jurassic Rencoret volcano-sedimentary strata
Dst	Devonian Sierra el Tigre Formation

Geologic map of the Antofagasta area and Coastal Cordillera. Cross section is shown in the next figure. Modified after Sernageomin.cl, online geologic map of Chile.

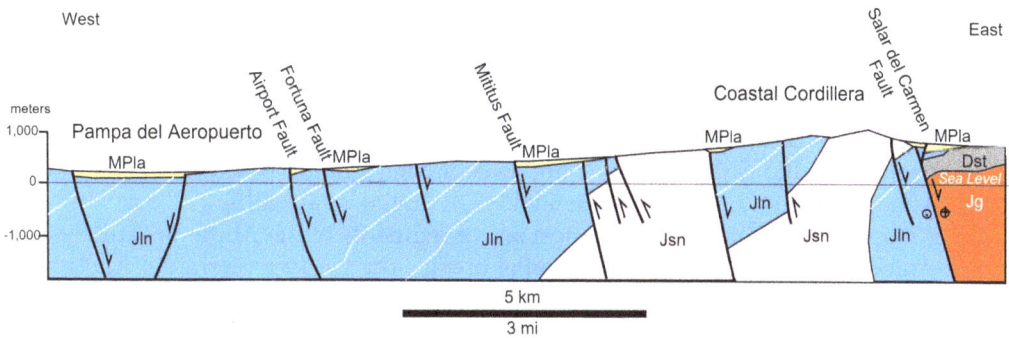

Cross-section through the Coastal Cordillera. MPla = Miocene-Pliocene alluvium; Jsn = Upper Jurassic intrusives; Jln = Jurassic La Negra volcanics; Jg = Jurassic granitic intrusive; Dst = Devonian Sierra el Tigre metamorphic basement. Modified after Mpodozis et al., 2014.

We traverse the Coastal Cordillera on our way to our next stop at the Atacama Fault. The Cordillera in this area consists of west-tilted fault blocks of Jurassic La Negra volcanics intruded by Upper Jurassic Mantos de Vara diorites and tonalites. The range is bounded on the east by the Salar del Carmen segment of the regional, left-lateral Atacama Fault Zone.

> ***Antofagasta to Atacama Fault****: Ruta-1 (R-1) north to Av. Salvador Allende; turn right (east) on Av. Salvador Allende/R-26; turn left (east) on R-26 and drive to R-5; turn left (north) on R-5 and drive to exit for B-400 to Mejillones; drive northwest on B-400 to **Stop 3, Atacama Fault** (–23.524462, –70.251682) and pull over on the right or left for a total of 30.8 km (19.1 mi; 43 min).*

Be mindful of speeding trucks on this road.

Stop 3 Atacama Fault

This 1,000 km (600 mi) long nearly north-south fault system separates the Coastal Cordillera from the Central Valley east of Antofagasta. The fault zone extends from near Iquique in the north to La Serena in the south. It is a system of overlapping left-lateral strike-slip faults that accommodate oblique convergence between the Nazca Plate and the South American Plate, much as the San Andreas System does along the margin of the Farallon and North American plates.

The Atacama Fault System is composed of three major segments, including a northern Salar del Carmen, a central Paposo, and a southern El Salado segment. The fault system is highly segmented, with each of the major segments comprising several parallel or subsidiary branches. Determination of offset is difficult because of a lack of piercing points. A total along-strike offset of 54 ± 6 km has been estimated across the El Salado segment (Seymour et al, 2021).

Strain partitioning is a term that refers to oblique convergence being broken into a shortening component and a translational (sideways-slip) component. Here the oceanic plate has a component of southward movement that is taken up by the Atacama Fault Zone. Any one branch can have tens to hundreds of meters of normal offset (here it is down-to-the-east) and as much as 5 km (3 mi) of lateral displacement. Normal offset was common along this zone during the Tertiary. Rarely, there is reverse offset along the faults. Subduction-related underplating, where partial melting is induced in the overriding South American Plate, is the process considered responsible for uplift and extension along the Coastal Cordillera and Atacama Fault Zone (Geen, 2015).

The Atacama Fault System was established during Early Jurassic to Early Cretaceous, simultaneous with the early magmatic arc. Recent convergence, based on GPS measurements, is about 6.5 cm/yr (2.6 in/yr) oriented N75E (Ewiak et al., 2015).

At this stop, we examine one of the most active segments of the Atacama Fault System. East of Antofagasta it consists of two parallel breaks, the western Sierra del Ancla Fault (extends north-south about 23 km, or 14 mi) that cuts bedrock, and the eastern Salar del Carmen Fault (about 60 km long, or 37 mi north-south) that cuts alluvium as a discontinuous east-facing, west-side-up escarpment along the east flank of the Coastal Cordillera. Recent fault activity is suggested by fault scarps in Quaternary fans, fresh fault surfaces, and abundant open tension cracks. Fault planes, cemented by salt and gypcrete (gypsum-based cement), are frequently preserved along the scarps. Various dating techniques put the age of displacement at between 12,000 and 424,000 years (Ewiak et al., 2015).

Ewiak et al. (2015) analyzed the Salar del Carmen Fault, and based on rupture lengths, knick points, and displacements suggests at least three or four seismic events. They determined the maximum cumulative lateral displacement on a single scarp is ~8.5 m (27.9 ft), with an average displacement of 5.5 m (16.4 ft). Along strike, the last vertical displacement has a maximum offset of ~4.7 m (15.4 ft) and an average displacement of ~1.8 m (5.9 ft). Surface rupture length near our stop is ~19 km (11.8 mi); estimates of associated earthquake magnitudes range from Moment Magnitude (Mw) 2.9 to 6.1, with the largest events possibly Mw 6.5 to 7.1 for this segment.

Alluvial fans in this area were eroded from the Coastal Cordillera. They contain debris from the Jurassic La Negra Formation (volcanics), and the Lower Cretaceous Way Group. The Way Group consists, from bottom to top, of the Caleta Coloso Formation (andesitic conglomerate and breccia), the conglomerate-gravel-sandstone Lombriz Formation, and the Tablado Formation (mainly marine limestone).

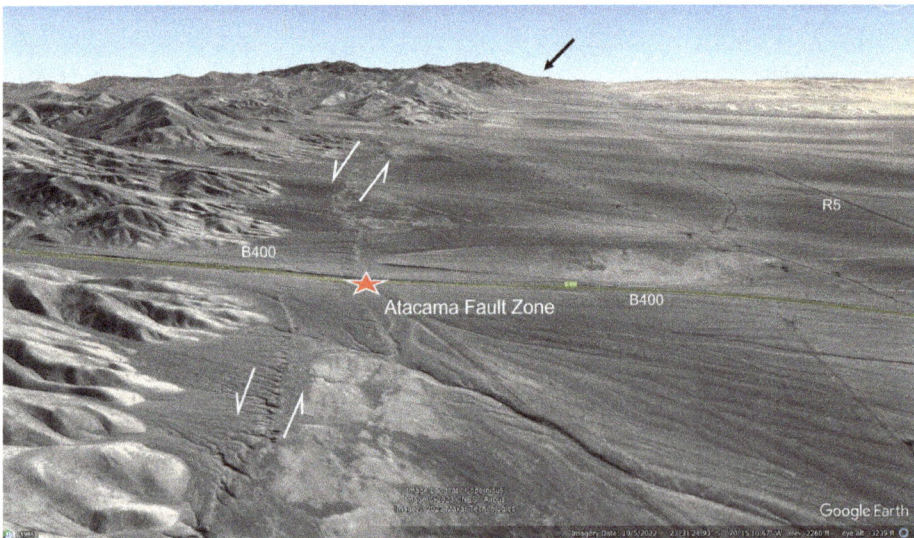

Google Earth oblique view north along the Atacama left-lateral strike-slip fault system. The fault strands lie at both the break-in-slope and in recently offset alluvium. Image copyright 2023 Maxar Technologies; CNES/Airbus; Landsat/Copernicus.

Looking north along the trace of the Atacama Fault (arrows). The offset alluvium indicates that this is very much an active fault.

The next two mines are described because we pass by them. They are not stops. *Do not bother trying to visit these mines unless you have an invitation or permit. If you just show up, they will not let you in.*

Atacama Fault to Mantos Blancos Mine: Continue east on R-5 for 22.5 km (14.0 mi; 18 min) to the Mantos Blancos Mine (−23.450937, −70.094221).

Mantos Blancos Mine

The Coastal Cordillera is host to Upper Jurassic to Lower Cretaceous strata-bound copper deposits. The volcanic hosted strata-bound ore bodies are mainly associated with hydrothermal breccias, in which the breccias contain most of the economic mineralization and the high-grade ore. The hydrothermal breccias are intruded by late mineralization diorite dikes. Sulfide mineralization consists of chalcocite, digenite, bornite, chalcopyrite, and pyrite. The Mantos Blancos ore body is one such strata-bound volcanic-hosted copper deposit (Ramirez et al., 2006).

Mantos Blancos tailings and roadcut. Photo courtesy of Chris Hunkeler, https://commons.wikimedia.org/wiki /File:Mantos_Copper_Mine_(32440311968).jpg.

A subduction-induced volcanic arc existed along the present Coastal Cordillera during Jurassic and Early Cretaceous time. The arc is represented by the 7,000 m (23,000 ft) thick basaltic to andesitic volcanics of the La Negra Formation, and granitic to dioritic intrusives of Lower Jurassic to Early Cretaceous age (200–130 Ma). The tectonic evolution of the Coastal Range during the Jurassic is interpreted in terms of sticking and slipping between the down-going oceanic tectonic plate and the overriding continental plate (Ramirez et al., 2006).

Geology

The Mantos Blancos deposit results from two superimposed hydrothermal events. An older rhyolitic intrusion and hydrothermal brecciation occurred during the Jurassic (~155 Ma); a younger, Early Cretaceous (141–142 Ma) event involved the intrusion of diorite, granodiorite stocks and sills, and diorite dikes. The main period of ore formation occurred during the second event, associated with the formation of hydrothermal breccias, stockworks (complex vein systems), and disseminated mineralization. A barren pyrite root zone is overlain by a pyrite-chalcopyrite zone, followed upward and outward by chalcopyrite-digenite or chalcopyrite-bornite zones. The tectonic setting during mineralization corresponded to extension related to transtensional faulting along the early Atacama Fault Zone. Diorite and granodiorite sills intruded the hydrothermal breccias and sealed the hydrothermal system. When the fluids reached overpressure, they hydrofractured the surrounding units, leading to decompression and mineral precipitation (Ramirez et al., 2006).

History

The Mantos Blancos mining company was formed in 1955, with open-pit operations starting in 1959. Underground mining began in 1974 and ended in 1996. Anglo American acquired almost 100% interest in Mantos Blancos in 2000 (Mining-Technology.com, 2020). Mantos Copper acquired the Mantos Blancos mine from Anglo American in 2015; it is a consortium led by Audley Capital Advisors and Orion Mine Finance Group (Jamasmie, 2019).

Google Earth oblique view northeast over the Mantos Blancos open-pit mine. Image copyright 2023 Maxar Technologies; CNES/Airbus; Landsat/Copernicus.

Production

Mine reserves were 454 million tonnes (500 million tons) of ore at 1.0% copper in 2006 (Ramirez et al., 2006). The mine contains facilities for processing oxide and sulfide ore and secondary leaching of ore. The main product is copper with minor silver. The operation is forecast to produce 45,400 tonnes (50,045 tons) of copper in 2019 and an average of 52,400 tonnes (57,800 tons) annually following the completion of its expansion in 2021. Silver concentrate production is increasing to 7.3 million tonnes (8.04 million tons) per year until the natural depletion of the oxide ore in 2023. Plant expansions are expected to extend the life of the open-pit mine until 2035 (Jamasmie, 2019).

> **Mantos Blancos Mine to Lomas Bayas Mine:** *Continue east on R-5 to B-385; turn right (southeast) and take B-385 to the* **Lomas Bayas Mine** *(−23.435114, −69.496625) on the right, a total of 72 km (44.7 mi; 55 min).*

Again, do not bother trying to visit this mine unless you have an invitation or permit. Mine Security will not let you in if you just show up.

Lomas Bayas Mine

First discovered in 1880, Lomas Bayas is now owned and operated by Compañía Minera Lomas Bayas SA, a subsidiary of Glencore plc. It is one of the world's lowest-grade copper operations, currently mining an average grade of 0.31% copper (https://thediggings.com/mines/27154; Mining Data Solutions, 2020).

Lomas Bayas double open-pit mine. Google Earth oblique view northeast. Image copyright 2023 Maxar Technologies; CNES/Airbus; Landsat/Copernicus.

Geology

The Lomas Bayas porphyry copper deposit is hosted by Upper Cretaceous volcanic-arc rocks (mainly dacites) and associated sediments. These were intruded by an Upper Cretaceous-Paleocene cluster of small granodiorite, granite, and diorite porphyries (the Lomas Porphyry). The original mineralized rocks were later exposed to leaching and supergene enrichment as well as *in situ* oxidation.

"Supergene" alteration and enrichment of metals occurs when sulfide-bearing deposits are exposed to near-surface, oxidizing groundwater. The sulfide minerals (pyrite, chalcopyrite, bornite) are oxidized, transforming the iron contained in these minerals to red, reddish brown, orange, and yellow-colored iron oxides. The sulfur mixes with groundwater to produce weak sulfuric acid. Copper in the rock is dissolved by this acidic solution, which then percolates down to the water table where it encounters reducing conditions (a lack of oxygen). Reduction reactions cause the copper to precipitate as chalcocite (a copper-bearing sulfide). Over time this process forms a near-surface leached zone and a thick, copper-rich blanket above the primary orebody.

Low-grade copper mineralization occurs in a concentric zone around a low-grade altered core. Structures, mainly faulting, influenced the emplacement of breccia pipes, hydrothermal alteration, and copper mineralization.

Both primary (hypogene) and secondary (supergene) mineralization occur at Lomas Bayas. The upper parts of the orebody (Lomas Bayas I) are mostly oxidized, with a few zones of mixed oxide-sulfide ore. High-grade oxide mineralization at Lomas Bayas occurs as water-soluble copper-sulfate minerals. Primary ore minerals include antlerite and brochantite, with minor chalcanthite, atacamite, chrysocolla, and malachite. Minor secondary enrichment occurs as chalcocite and covellite. Hypogene sulfides (Lomas Bayas III) occur below the oxide orebody and consist of chalcopyrite, bornite, and molybdenite. The Lomas Bayas II (Fortuna de Cobre) orebody is a porphyry copper deposit located 2 km (1.2 mi) south of the Lomas I pit. It is characterized by a dominance of water-soluble copper oxides (Glencore, 2015).

Production

There are two open pits in this mine reaching a maximum depth of 120 m (400 ft). Mine life is expected to last till 2029.

Copper is recovered using a solvent extraction-electrowinning process. Crushed ore is placed in heaps on impermeable pads for heap leaching and the metal is dissolved by repeated applications of sulfuric acid. The pregnant solution is collected for copper recovery by electrowinning. Electrowinning consists of passing direct current between a conductive anode and a stainless-steel cathode. This causes copper in the solution to precipitate on the cathode. Uncrushed run-of-mine ore is leached on separate pads with copper also being recovered by electrowinning.

In 2019, the Lomas Bayas operation produced ~75,000 tonnes (83,000 tons) of copper cathode from 70 million tonnes (77 million tons) of ore mined.

Proven and Probable Reserves in 2019 were 456 million tonnes (503 million tons) of 0.26% Cu (https://miningdataonline.com/property/140/Lomas-Bayas-Mine.aspx).

Lomas Bayas Mine to Cordillera Domeyko*: Continue east on B-385 to 63.1 km (39.2 mi; 48 min) to **Stop 4, Cordillera Domeyko** (–23.537228, –69.036622) and pull over on the right (south) side of the road.*

This is also 146 km (90.9 mi; 1 hr 42 min) from the Atacama Fault stop.

Stop 4 Cordillera Domeyko (Precordillera)

The Cordillera Domeyko is at least partly the result of movement along the Cordillera de Domeyko Fault System, including the Sierra de Varas Fault (SVF) in this area and the

West Fault near Calama. The ranges were uplifted and eroded mostly in middle Eocene time (Rubilar Contreras, 2015). The SVF crosses Route B-385 near this stop. The nearly north-south fault zone has up to 15.6 km (9.7 mi) of left-lateral displacement in the Sierra de Varas about 125 km (77.7 mi) to the south, and vertical offset as much as 4.9 km (3.0 mi) in a reverse sense, up on the west side (Niemeyer and Urrutia, 2009). Two episodes of faulting are recognized, a middle Eocene north-northeast to north-northwest left-lateral and basement-involved reverse event, and a smaller, superimposed post-Miocene right-lateral event on mainly north-south to north-northeast faults and northeast-verging thrusts. Left-lateral transpression along the west and northwest-verging thrusts and folds was dominant along the fault zone in the western Cordillera Domeyko during Eocene (Niemeyer and Urrutia, 2009).

The ranges of the Sierra Domeyko in this area consist of Late Carboniferous to Early Triassic uplifted metamorphic and igneous basement and Triassic-Jurassic sedimentary units within an overall background of Cretaceous-Tertiary volcanics (Amibilia et al., 2008). At this stop, you see an outcrop of highly fractured red granite. It is not clear if the granite is Carboniferous-Permian or Late Cretaceous-Eocene.

The easternmost (frontal) thrust of the Cordillera Domeyko is the east-verging Cerro Negro-Domeyko Thrust that puts the Triassic-Jurassic Agua Dulce Formation and Late Cretaceous Purilactis Group over Neogene strata near the El Bordo Escarpment. Uplift of the Cordillera led to sedimentation of mostly sands and gravels of the Loma Amarilla Formation in the Salar de Atacama Basin just east of here. Later back-arc compression led to structural inversion within the Salar de Atacama Basin (Mpodozis et al., 2004). Inversion means that previously downdropped blocks were uplifted, and normal faults were reactivated as reverse faults.

Google Maps image of roadcuts and outcrops in the Cordillera Domeyko near the trace of the Sierra de Varas left-lateral strike-slip fault. Image copyright 2023 Maxar Technologies; CNES/Airbus.

Highly fractured red granite, Cordillera Domeyko. View north.

Red granite of the Cordillera Domeyko.

Sierra de Varas Fault (SVF), Domeyko Fault Zone (DFZ), and West Fault (WF). The Cordillera Domeyko stop (star) is shown. Several significant mineral deposits are located along this left-lateral fault system. Modified after Niemeyer and Urrutia, 2009.

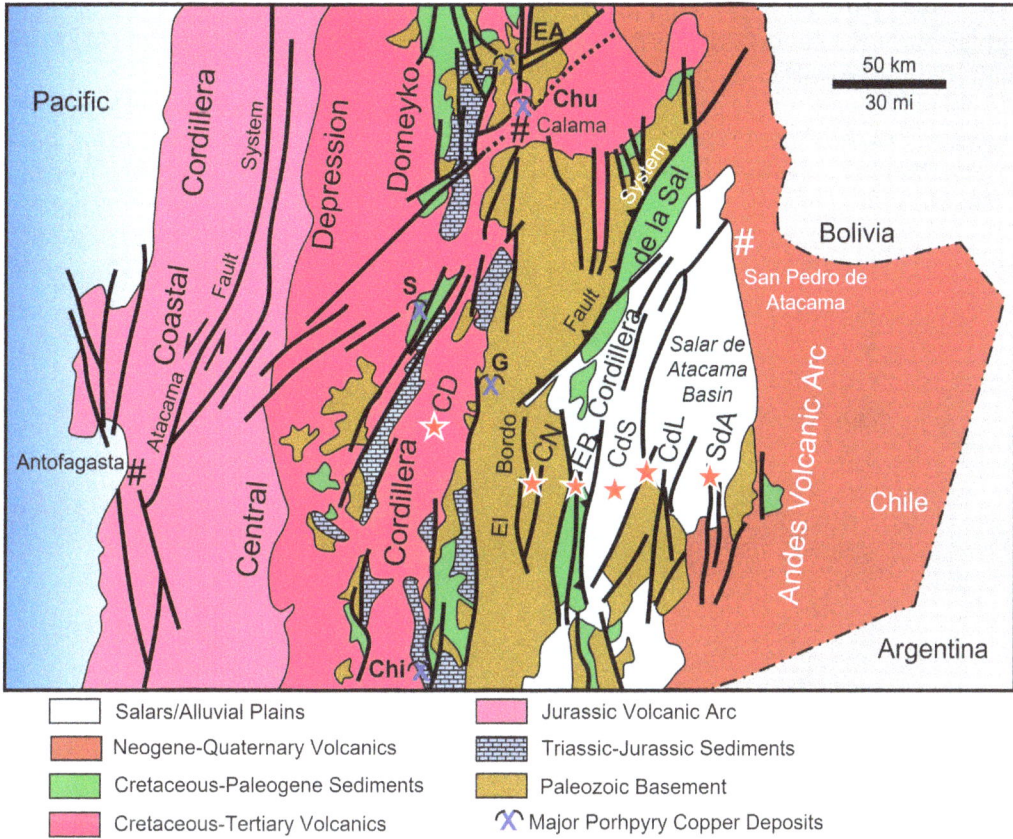

Geologic map of the Cordillera Domeyko (CD), Cerro Negro (CN), El Bordo escarpment (EB), Cordillera de la Sal (CdS), Cordón de Lila (CdL), and Salar de Atacama (SdA) stops. Major copper deposits include EA (El Abra), Chu (Chuquicamata), S (Spenz), G (Gaby), and Chi (Chimborazo). Modified after Amilibia et al., 2008.

*Cordillera Domeyko to Cerro Negro, Eastern Cordillera Domeyko: Continue east on B-385 for 33.6 km (20.9 mi; 25 min) to **Stop 5, Cerro Negro, Eastern Cordillera Domeyko** (–23.645276, –68.770556) and pull over on the right.*

Stop 5 Cerro Negro, Eastern Cordillera Domeyko

The El Bordo Escarpment is a topographic break between the basement outcrops in the Domeyko Uplift that overlook the Salar de Atacama salt flats. This stop is just south of Cerro Negro, a basement-cored anticline that has been uplifted and thrust eastward on the Cerro Negro Thrust. The uplift contains Late Triassic-Early Jurassic Agua Dulce Formation rift-related lavas and conglomerates, and Late Cretaceous to Cenozoic porphyries that contain copper-molybdenum (Cu-Mo) deposits at the Chuquicamata and Gaby mines. The Late Triassic synrift rocks on the upthrown side suggest that this steep, basement-involved thrust originated as a down-to-the-west normal fault that has been reactivated and inverted by Andean compression (Amilibia et al., 2008).

The rocks exposed at this stop are a dark, layered andesite porphyry or diabase cut by lighter dikes. A fault here trends 340° and dips 77° to the east, possibly a backthrust to the main Cerro Negro Thrust.

A few kilometers to the east, in the Llanos de la Paciencia and Cordillera de la Sal, several east-directed thin-skinned thrusts, detached in gypsum of the Purilactis Group, place folded Paleozoic through Triassic andesites, sandstones, and intrusives over deformed Cretaceous and Tertiary sedimentary rocks. These thrusts are likely a result of shallow detachment due to an eastward push (the "bulldozer effect") from the Cerro Negro and Llanos de la Paciencia basement uplifts to the west (Amilibia et al., 2008).

Google Maps image of Cerro Negro. Image copyright 2023 Maxar Technologies; CNES/Airbus; Landsat/Copernicus data.

Cerro Negro fractured volcanics of the Agua Dulce Formation. View south.

Agua Dulce Formation diabase at Cerro Negro.

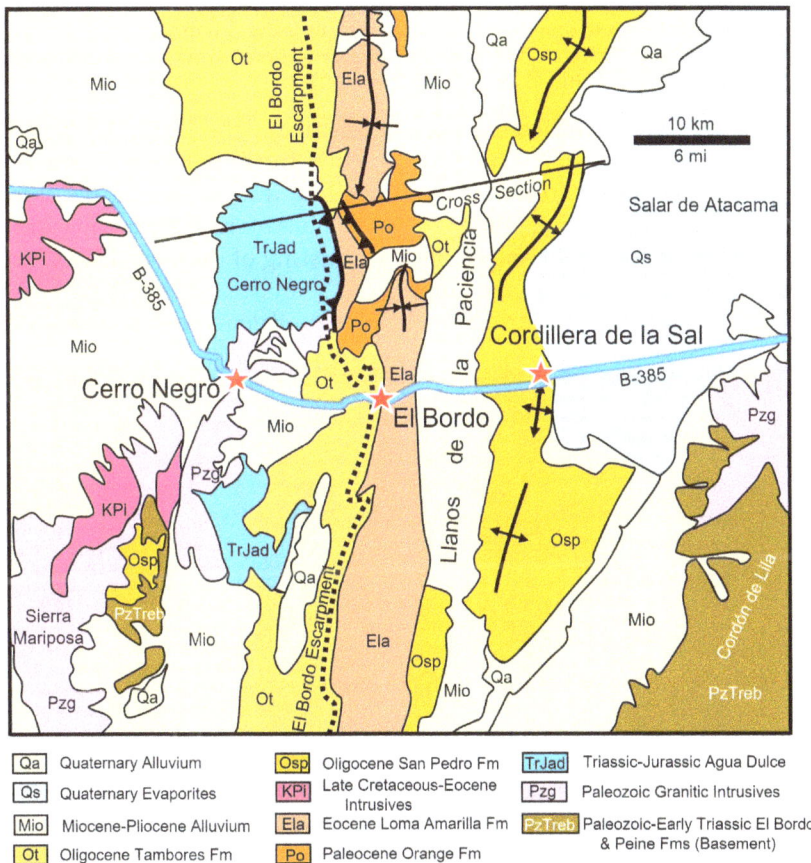

Geologic map of the Cerro Negro, El Borde Escarpment, and Cordillera de la Sal stops. A cross-section is shown in the next figure. Modified after Amilibia et al., 2008.

The boundary between the Domeyko Range and the Salar de Atacama Basin is the 800 m (2,600 ft) high El Bordo Escarpment, which exposes the Triassic-Jurassic Agua Dulce, the Cretaceous Purilactis Group (Totola, Purilactis, Barros Arana, and Tonel formations), the Paleocene Naranja (Orange) Formation, and Eocene Loma Amarilla Formation on the west (high) side. These units were uplifted and folded by a series of thrusts related to the Cerro Negro Thrust. This entire sequence is found in the subsurface beneath the Llano de la Paciencia and the Salar de Atacama (Ariagada et al., 2006).

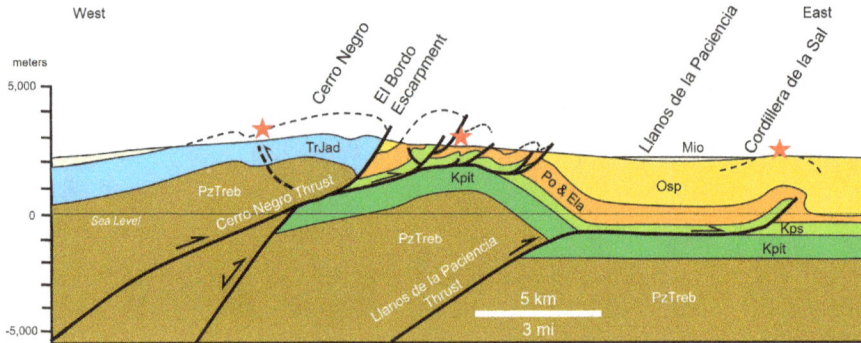

Cross-section from Cerro Negro to the Cordillera de la Sal. Mio = Miocene Alluvial deposits; Osp = Oligocene San Pedro Formation; Po and Ela = Eocene Loma Amarilla and Paleocene Orange/Naranja Formation; Kps = Cretaceous upper Purilactis Group; Kpit = Cretaceous lower Purilactis Group and Tonel Formation; TrJad = Triassic Agua Dulce Formation; PzTreb = Paleozoic-Early Triassic El Bordo Formation. Shortening is estimated at ~19%. Red stars indicate stops. Modified after Amilibia et al., 2008.

East of Cerro Negro a west-directed backthrust carries the Purilactis Formation over evaporites of the Tonel Formation. This thrust follows the El Bordo Escarpment along the western margin of the basin. Growth strata (units that thicken as a result of fault movements that create accommodation space) in the Loma Amarilla Formation indicate sedimentation occurred during uplift and eastward thrusting of the Cordillera Domeyko.

Evaporites in the Naranja/Orange and Loma Amarilla formations are intensely deformed below an east-verging thrust that brings the Paleozoic-Triassic basement to the

Google Earth oblique view north of the Cordillera Domeyko, El Bordo Escarpment, Llanos de la Paciencia, and Cordillera de la Sal. Image copyright 2020 Maxar Technologies; CNES/Airbus; Landsat/Copernicus.

surface. South and east of Cerro Negro the Loma Amarilla Formation is steeply west-dipping below the Cerro Negro Thrust.

> ***Cerro Negro, Eastern Cordillera Domeyko to El Bordo Escarpment and Llanos de la Paciencia***: *Continue east on B-385 for 14.2 km (8.8 mi; 12 min) to* **Stop 6, El Bordo Escarpment and Llanos de la Paciencia** *(–23.672489, –68.645551) and pull over on the right (emergency pullout).*

Stop 6 El Bordo Escarpment and Llanos de la Paciencia

This stop is just east of the El Bordo Escarpment and near the contact between the west-dipping Oligocene-Miocene Tambores and Eocene Loma Amarilla formations. These rocks are coarse, poorly sorted syntectonic sandstones and conglomerates of the Loma Amarilla Formation.

The El Bordo Escarpment extends 120 km (70 mi) and separates the Cordillera from the Salar de Atacama Basin. Up to 5 km (16,400 ft) of uplifted Cretaceous to Miocene basin fill (sedimentary and volcanic rocks of the Purilactis Group and overlying Naranja/Orange and Loma Amarila formations) is exposed along the escarpment (González et al., 2009; Rubilar Contreras, 2015).

Llanos de la Paciencia is a narrow sub-basin within the Salar de Atacama Basin that extends 80 km (50 mi) north-south and is 9 km (6 mi) wide with Quaternary alluvial fans at the surface (Rubilar Contreras, 2015). It is bounded on the west by the El Bordo Escarpment, and on the east by the north-northeast-trending Cordillera de la Sal.

Google Earth image of the El Bordo Escarpment and Loma Amarilla Formation stop in the footwall of the Cerro Negro Thrust. Image copyright 2023 Maxar Technologies; CNES/Airbus; Landsat/Copernicus.

Coarse, poorly sorted sandstone and conglomerate of the Loma Amarilla Formation is inclined about 60° west. View south.

El Bordo and Llanos de la Paciencia to Cordillera de la Sal: *Continue east on B-385 for 11.0 km (6.9mi; 10 min) to* **Stop 7, Cordillera de la Sal** *(–23.661777, –68.540881) and pull over on the right.*

Google Maps satellite image and geology of the Salar de Atacama area. Image copyright 2023 TerraMetrics.

Age	Lithology	Group/Formation
Pliocene		Vilama Fm (Ignimbrites, evaporites, clastics)
Late Miocene		Gravels and Ignimbrites
Oligocene- Lower Miocene		**Paciencia Group** — Tambores Fm conglomerate / San Pedro Fm shale, halite, gypsum
Time Gap		
Eocene		Loma Amarilla Fm conglomerate, sandstone, ash
Paleocene		Orange Fm conglomerate, sandstone
Cretaceous- Paleocene		**Purilactis Group** — Totola Fm sandstone, lava / Barros Arana Fm conglomerate / Purilactis Fm sandstone, mudstone, conglomerate / Tonel Fm sandstone, mudstone, evaporite
Paleozoic		Tuina Fm & Others volcanics

Stratigraphy of the Salar de Atacama region. Information is drawn from Mpodozis et al., 2004; Jordan et al., 2007; Rubilar Contreras, 2015.

Stop 7 Cordillera de la Sal

As you approach this stop you can see that you are climbing a gentle rise. In fact, you are climbing the Cordillera de la Sal, represented here by an anticline. The core of the anticline is 36 m (120 ft) above the surrounding plain. There is no other evidence of a fold belt, or an anticline at the surface: No bedding, no change in units. Satellite imagery shows that this stop is in the core of a low-relief anticline that exposes the Oligo-Miocene San Pedro Formation in the southern Cordillera de la Sal. The surface consists of salt-encrusted rubble of red sandstone and mudstone.

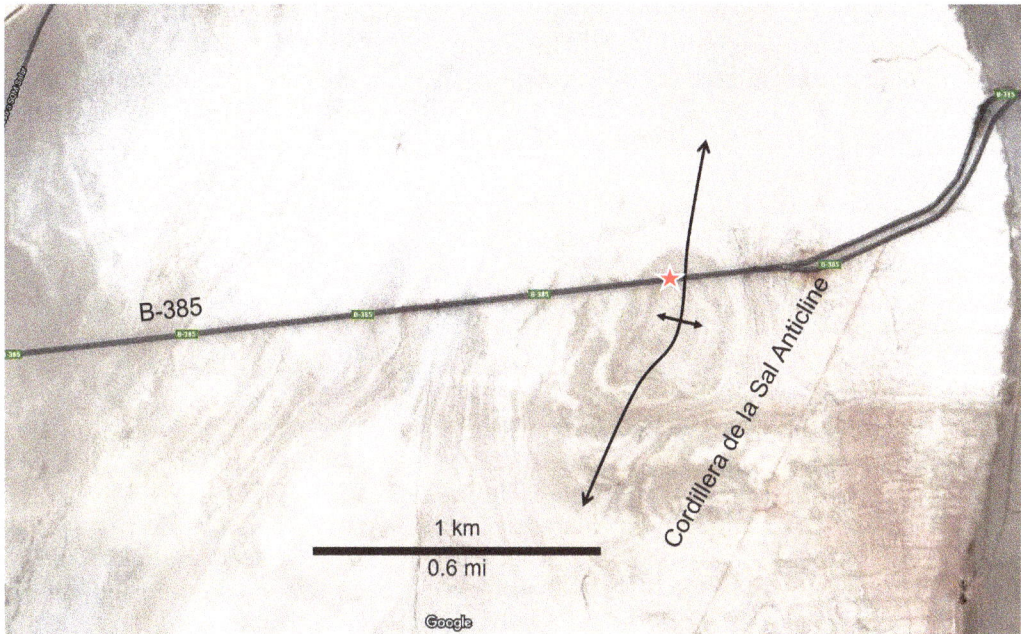

Google Maps image of the Cordillera de la Sal. Image copyright 2023 Maxar Technologies; CNES/Airbus.

Cordillera de Sal looking east to the Andes from the anticlinal core.

The Cordillera de la Sal is a ridge 5–10 km (3–6 mi) wide and 100 km (60 mi) long that extends south-southwest from San Pedro de Atacama and, at its highest, rises some 200 m (660 ft) above the valley floor. The Cordillera separates the Salar de Atacama proper on the east from the Llano de la Paciencia on the west.

The Cordillera de la Sal consists of over 3,000 m (10,000 ft) of continental sediments of the San Pedro Formation unconformably overlain by the upper Miocene-Pleistocene

Campamento Formation in the south, and the Pliocene-Pleistocene Vilama Formation in the north (Rubilar Contreras, 2015). The southern section of the ridge exposes a thin-skinned fold-thrust belt that uplifts Oligocene to Pleistocene continental deposits, including evaporite-rich sediments and ignimbrites (tuff, or volcanic ash deposited from a pyroclastic flow) of the San Pedro Formation, Paciencia Group, and overlying Vilama Formation (González et al., 2009).

The Cordillera de la Sal comprises a series of folds resulting from sinistral transpression (left-lateral strike-slip plus compression), thrusting, and diapirism (Rubilar Contreras, 2015). The southern Cordillera de la Sal has a number of anticlinal and synclinal folds. The La Paciencia Anticline on the west side is a fold overturned to the west that exposes evaporites of the lower San Pedro Formation. The Los Vientos Anticline on the east is a symmetric open fold that exposes upper members of the San Pedro Formation unconformably overlain by the Campamento Formation. The southern Cordillera de la Sal is bounded on the west by the La Paciencia reverse fault and on the east by the Los Vientos reverse fault. Thin-skinned thrusting in the Cordillera de la Sal is thought to be linked in the subsurface to thick-skinned structures of the Cordillera de Domeyko (Rubilar Contreras, 2015).

Diapirism becomes the dominant deformation style in the northern Cordillera de la Sal. Salt moving upward has deformed the strata into anticlinal and synclinal folds. The La Paciencia Anticline changes progressively northward from a west-overturned anticline, to an open and symmetrical syncline, and finally to a northeast-elongated salt dome 10 km (6 mi) long and 4 km (2.5 mi) wide. Further north, a series of anticlinal and synclinal folds occur in the Valle de la Luna area just west of San Pedro de Atacama. Diapiric salt flow initiated during early to middle Miocene above a deep reverse fault (Pananont et al., 2004). The detachment surfaces and root of diapirism are in evaporites of the Paleocene Naranja/Orange Formation and lower San Pedro Formation (Rubilar Contreras, 2015).

Because the climate is hyper-arid, the Orange/Naranja Formation evaporites outcrop along the western edge of the basin.

*Cordillera de la Sal to Cordón de Lila: Continue east on B-385 for 19.2km (11.9 mi; 15 min) to **Stop 8, Cordón de Lila** (−23.631389, −68.358611) and pull over on the right.*

Stop 8 Cordón de Lila

This roadcut at the northeast end of the Cordón de Lila is mapped as upper Miocene lacustrine deposits, but I believe it to be Oligocene-Miocene San Pedro Formation because the outcrops contain bedded gypsum that has been gently deformed, probably by soft-sediment deformation.

The Cordón de Lila proper consists of Ordovician to Pennsylvanian sedimentary and igneous rocks considered to be the basement rocks in this region. Apatite fission ages indicate that these rocks had been buried to depths of 3–4 km (9,800–13,000 ft) during the Eocene (38.6 ± 5.6 Ma; Ariagada et al., 2006).

Seismic reflections can be traced beneath the range. The Cerro Totola Formation and the Peine Group appear to be overthrust by the Cordón de Lila basement (Ariagada et al., 2006).

Bedded gypsum in roadcut, northern Cordón de Lila stop.

Google Maps image over the Cordón de Lila and lithium operations. Image copyright 2023 Maxar Technologies; CNES/Airbus; Landsat/Copernicus.

Lithium evaporation ponds and mineral piles. View southeast from the Cordón de Lila stop.

> ***Cordón de Lila to Salar de Atacama and Lithium Mines***: *Continue driving east on B385; continue straight onto B-367 for a total of 24.0 km (14.9 mi; 19 min) to **Stop 10**, **Salar de Atacama and Lithium Mines** (−23.653329, −68.136320).*

Stop 9 Salar de Atacama and Lithium Mines

A huge, high-altitude intermontane dry lake, the Salar de Atacama is the world's largest source of lithium, the main component used in lithium-ion batteries. Beneath the salty surface of this playa is a brine rich in the highly sought-after lithium. The lithium is used in rechargeable batteries, smartphones, computers, and increasingly, in electric vehicles.

Lithium minerals in the brines are thought to be derived from: (1) geothermal waters associated with active volcanism that produced huge amounts of rhyolitic ash-flow tuffs; (2) leaching of water-soluble salts from the volcanic rocks; (3) leaching of lithium-rich clays; and (4) saline waters of salars in the high Andes to the east that migrated into the Salar de Atacama Basin through permeable faults (Ide and Kunasz, 1989).

According to the US Geological Survey, this single ancient lakebed contains 27% of the world's reserves of the metal. Chile has been called "the Saudi Arabia of lithium." The brine lies 40 m (130 ft) below the salar's surface. When pumped from the ground, it comes up looking like yellow slush (lithiummine.com, 2018). The brines are pumped into evaporation ponds, which proceed to concentrate the lithium salts as the water evaporates in the dry air (USGS, Lithium Mining in Salar de Atacama).

Once the brine has reached an ideal concentration, it is pumped to a lithium-recovery facility for extraction. Extraction usually involves:

- Filtering or ion exchange to remove contaminants from the brine.
- Applying chemical solvents and reagents to isolate desirable products through precipitation.
- Filtering to separate out precipitated solids.
- Treating the brine with a reagent, such as sodium carbonate, to form lithium carbonate. The resulting product is then filtered and dried for sale. Depending upon the desired product, different reagents are applied to produce lithium hydroxide, lithium chloride, lithium bromide, and butyl lithium.

Once the lithium is extracted, the remaining brine is pumped back underground (Samco Technologies, 2018).

This part of the continental margin has undergone up to 500 km (310 mi) of shortening in the past 67 million years. Current plate convergence is about 70 mm/yr (2.75 in/yr; González et al., 2009). Back-arc compression led to inversion of the Jurassic-Early Cretaceous precursor basin and the accumulation of syntectonic sediments in the Salar de Atacama Basin. Inversion coincided with uplift of the Cordillera de Domeyko to the west (Mpodozis et al., 2004). The Salar de Atacama Basin has absorbed much of the shortening due to plate convergence. As such, it developed as a foreland basin in a back-arc setting behind the Cordillera Domeyko volcanic arc (Rubilar Contreras, 2015). The basin underwent thin-skinned and thick-skinned thrusting and sedimentation from mid-Cretaceous through Neogene times. With the eastward migration of the volcanic arc from the Cordillera Domeyko to the Andes in the past 40 Ma, the basin is today in a forearc setting (Reutter et al., 2005). It is a major depression 200 km (120 mi) long and 100 km (60 mi) wide on the western slope of the Andes. Seismic work indicates there are up to 8 km (26,000 ft) of sediments filling the basin. There is some evidence for Oligocene extension contributing to basin development (Pananont et al., 2004; Rubilar Contreras, 2015), but the basin we see today is largely an inverted, Neogene compressional basin (Ariagada et al., 2006; González et al., 2009; Rubilar Contreras, 2015).

The non-marine Salar de Atacama Basin is the largest and deepest sedimentary basin in northern Chile and has been for the past 90 million years. The basin contains up to 7,500 m (24,600 ft) of mainly siliciclastic (sand-silt-gravel) rocks, overlain by up to 1,600 m (5,250 ft) of evaporites (Pananont et al., 2004).

The surface of the Salar de Atacama Basin includes halite (salt) deposits surrounded by alluvial fans and fan-delta sediments. The Upper Cretaceous to lower Paleocene sequences (Purilactis Group) infilling the Salar de Atacama Basin reflect rapid local subsidence. The Tonel Formation consists of more than 1,000 m (3,300 ft) of continental red sandstones and evaporites that contain tectonic growth strata adjacent to the Cordillera Domeyko uplift. The Tonel Formation is capped by almost 3,000 m (10,000 ft) of Purilactis Formation alluvial, fluvial, and eolian sandstones and conglomerates, and lacustrine mudstones derived from multiple pulses of tectonic shortening. These are in turn covered by 500 m (1,600 ft) of alluvial fan conglomerates in the Barros Arana Formation. The top of the Purilactis Group consists of lava, welded tuffs, and redbeds of the Cerro Totola Formation (70–64 Ma). Early Paleocene deformation caused an angular unconformity that separates Paleocene sediments from the underlying Purilactis Group.

Above the Purilactis Group the Paleocene to Neogene fill includes the Naranja/Orange and the Loma Amarilla formations, the Paciencia Group (including the San Pedro and Tambores formations) and halite deposits partly correlated to the Vilama Formation. Neogene evaporite deposits contain 980 to 1,200 m (3,200 to 3,900 ft) of halite-gypsum interbedded with clastics and ignimbrites (González et al., 2009).

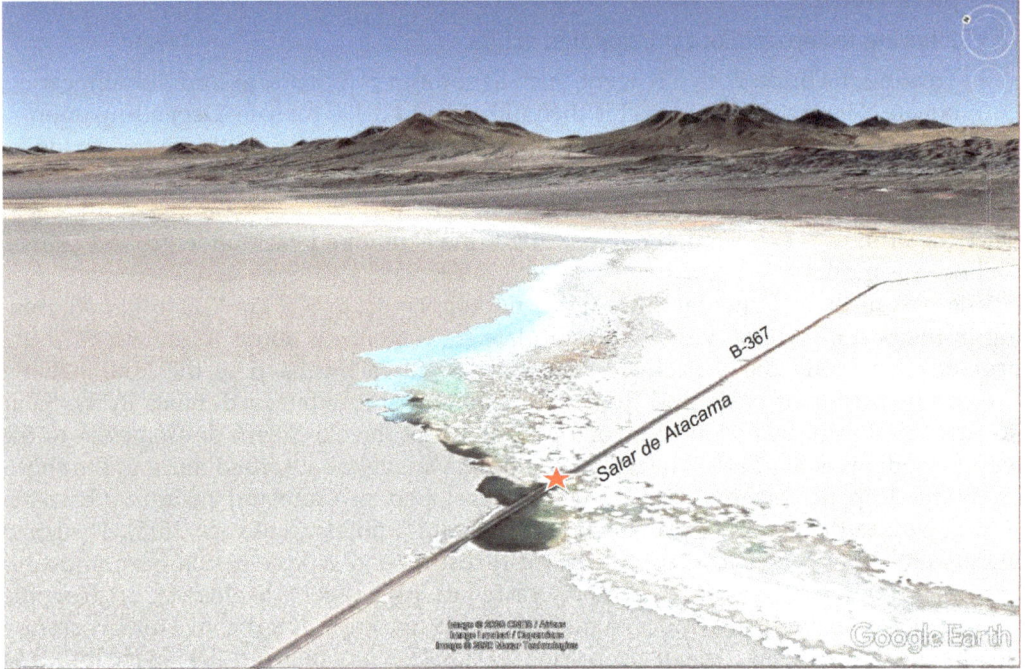

Salar de Atacama stop. Google Earth oblique view northeast. Image copyright 2020 Maxar Technologies; CNES/ Airbus; Landsat/Copernicus.

Salar de Atacama salt rubble looking southeast.

Steam eruption on Saturday, March 11, 2023, on Volcán Lascar as seen from the Salar de Atacama. View east.

As mentioned previously, the El Bordo Escarpment forms the western boundary of the greater Salar de Atacama Basin. The escarpment separates the basin from east-directed basement-cored uplifts of the Cordillera Domeyko, also called the Precordillera. The Cordillera de la Sal separates the Llanos de la Paciencia western sub-basin from the Salar de Atacama Basin proper.

Seismic data reveals that subsurface structures east of the Cordillera de la Sal are east-verging reverse faults and anticlines that formed mainly during Paleocene-Eocene compression and tectonic thickening.

Three phases of deformation are recognized in the Oligocene-Neogene development of the Salar de Atacama Basin. 1) The Paciencia Group was deposited under arid conditions during Oligocene to early Miocene time. Extension was concentrated along the northwest flank of the basin. 2) Between 17–10 Ma the Salar de Atacama Basin was part of the Andean forearc due to an eastward shift of the volcanic arc previously in the Cordillera de Domeyko. This shift coincides with compression and inversion of the Salar de Atacama Basin. Shortening was concentrated mostly in the southwest, and diapirism in the lower San Pedro Formation occurred in the northwestern Salar de Atacama Basin. The Cordillera de la Sal began to develop. East-directed compression had a left-lateral strike-slip component. 3) Compression-transpression tectonics shifted to the eastern basin margin during upper Miocene-Pleistocene time as the Vilama and Campamento formations were deposited above the folded San Pedro Formation (Rubilar Contreras, 2015).

Salar de Atacama and Lithium Mines to Quebrada de Jere*: Continue east on B-367 to B-355; turn left (north) on B-355 and drive to R-23; turn left (north) on R-23 and drive to Vilaco in Toconao (sign to Valle de Jere); turn right (east) on Vilaco and follow the signs to* **Stop 10, Quebrada de Jere** *(–23.191124, –67.993407), for a total of 94.4 km (58.6 mi; 1 hr 15 min) and pull into the parking area.*

Stop 10 Quebrada de Jere

The Quebrada de Jere recreation area (2,000 p/adult; 1,000 p/child 6–12 years) is a scenic canyon cut into the ignimbrites of the western slope of the Andes. Lush vegetation thrives where water is available. The stream is derived from snowmelt. Trails lead from the parking area both up and down the canyon.

A quarry near the end of the road has been used since 1940 by local artisans to extract building stones and to manufacture various crafts. The stone, known as Liparita, is white where freshly broken, weathers to a rusty reddish brown, and is very lightweight and porous. It is relatively soft and easy to work.

Since we are in ignimbrite country, this would be a good place to talk about ignimbrites. Ignimbrites are the result of widespread deposition and consolidation of explosive volcanic ash eruptions, or *Nuee Ardentes*. *Nuee Ardentes* are an awesome spectacle to behold: A flaming hot mass of gas-charged ash that is expelled with explosive force and moves at hurricane speed down a mountainside, they can cover hundreds of square kilometers and be several tens of meters thick.

Units on the Western Slope of the Andes comprise five sets of ignimbrite flows including, from base to top: (1) the Pujsa Ignimbrite (5.8 Ma), (2) the Atana-Toconao-La Pacana Ignimbrite (4.0–4.5 Ma), (3) the Tucucaro-Patao Ignimbrite (3.1–3.2 Ma), (4) the Talabre Ignimbrite (2–2.3 Ma), and (5) the Purico-Cajon Ignimbrite (1–1.4 Ma). The average thickness of the flows is about 50 m (160 ft). A strong northwest-oriented regional joint set, visible on satellite imagery, appears to have been imposed on the brittle ignimbrites.

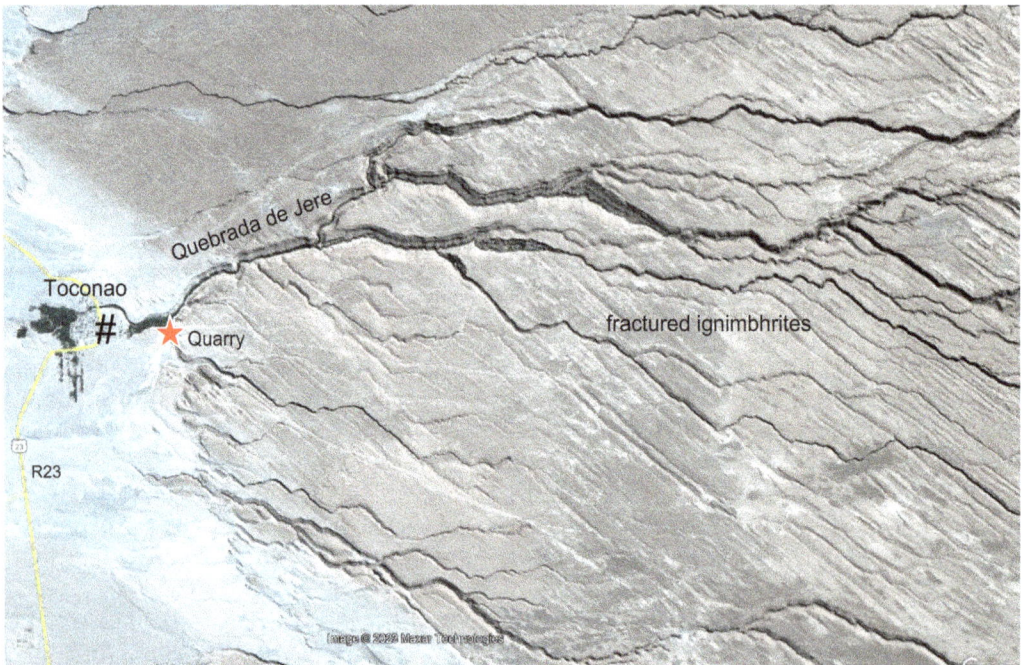

Google Maps satellite image of northwest regional jointing in ignimbrites east of Toconao. Image copyright 2023 Maxar Technologies.

Block of Liparita stone, part of the Pelón Ignimbrite, in the Toconao quarry.

Quebrada de Jere, view northeast.

To avoid Side Trip 1:

Quebrada de Jere to Las Tres Marias: *Return to R-23 and turn right (north); drive 35.4 km (22.1 mi) on R-23 to the turnoff to Calama; turn left (west) and drive 7.3 km (4.5 mi) to Ruta Valle de la Luna; turn left (west) on Ruta Valle de la Luna and drive 0.5 km to the park entrance. Pay 5,000 p/adult to enter, then drive to* **Stop 11.1, Las Tres Marias** *(−22.920593, −68.318818) for a total of 55.5 km (34.5 mi; 55 min).*

To go on Side Trip 1:

Side Trip 1, Quebrada de Jere to Volcán Lascar: *Return to R-23 and turn left (south); drive south on R-23 to B-357 and follow the sign to Talabre; follow B-357 for 13.6 km, through Talabre, to a pullout with a view on the far side of town. This is* **Stop ST 1.1, Volcán Lascar** *(−23.314088, −67.884172) for a total of 24.9 km (15.5 mi; 25 min).*

Side Trip 1 Altiplano – Puna Volcanic Complex

East of the Salar de Atacama Basin lies the "Western Slope" of the Andes, a 32-kilometer-wide (20 mi) zone of ignimbrites and alluvial fans shed off the Andes volcanic arc. We drive across these slopes as we head south to Laguna Lejía.

There are many dry lakes and shallow lagoons on the Chilean Altiplano, also known as the Puna Plateau or Puna de Atacama. These are fed by snowmelt, springs, and intermittent streams. Evaporation is greater than precipitation, so even as they have no outlet, they dry up. Minerals and salts are concentrated and precipitate on the dry lake bottoms. Sodium chlorite (salt) and calcium sulfate (gypsum) are the main minerals.

Most of these lakes were larger and deeper during the last ice age, but they dried out as a result of decreased precipitation and increased evaporation. Few organisms can survive in these lakes, but those that do include diatoms, algae, bacterial mats, ostracods, and crustaceans. Flamingos and shorebirds (specifically phalaropes) thrive here by feeding on these organisms.

We will visit two of these lakes, Laguna Lejía and Laguna en Salar de Aguas Calientes.

Stop ST1.1 Volcán Lascar

On the way to Laguna Lejía, we cross the western foothills of Volcán Lascar, an andesitic stratovolcano. This mountain is the most active volcano in the central Andes. It has six overlapping craters, with an actively steaming crater near the center. The last, and largest historical eruption occurred in 1993 and produced pyroclastic flows that extended up to 8.5 km (5.1 mi) from the summit. Pumice deposits from the 1993 eruption are still clearly visible (Calder et al., 2000; de Silva et al., 2009). There were ongoing active steam eruptions when we visited here in 2023.

A small steam eruption can be seen on the north flank of Volcán Lascar. View southeast from Talabre.

Side Trip 1, Volcán Lascar to Laguna Lejía: Continue driving southeast on B-357 for 37.5 km (23.3 mi; 36 min) and pull over where a short track leads off to the right. This is Stop ST 1.2, Laguna Lejía (−23.507991, −67.693367).

Stop ST1.2 Laguna Lejía

Laguna Lejía is surrounded by the volcanoes Aguas Calientes, Lascar, Tumisa, Lejía, Chiliques, and Cordón de Puntas Negras. This small saline lake lies at an elevation of 4,325 m (14,190 ft), has a surface area of 1.9 km^2 (0.73 mi^2), and has an average depth of 1.2 m (just under 4 ft). During the last ice age, the lake was up to 26 m (86 ft) deep: Higher lake levels are marked by shoreline terraces. Salt pillars line the shore, and the water is often a brilliant green, probably a result of algal mats. Air temperatures range from −6 to 7°C (21 to 45°F) with an average temperature of 2°C (36°F); water temperatures range from 3 to 11°C (37 to 51°F).

Water in the lake derives from two small streams and precipitation that averages 200 mm/yr (7.9 in/yr), mostly as snow. Evaporation is greater than 200 mm/yr, which is why it continues to shrink (Muñoz-Pedreros et al., 2013).

Laguna Lejía lies in a geologic depression thought related to movement along the Miscanti- Callejón de Varela Fault system (Wikipedia, Lejía Lake).

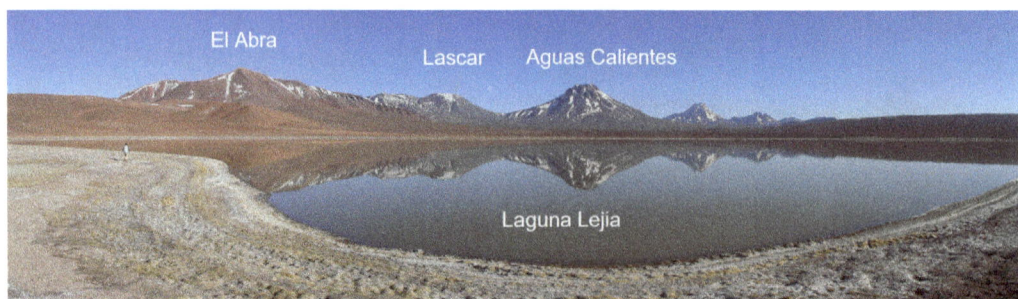

Panorama looking north over Laguna Lejía. Photo courtesy of RudiR, https://commons.wikimedia.org/wiki/File:Laguna_lejia.jpg.

Laguna Lejía, a small recent cinder cone, and Laguna En Salar de Aguas Calientes. Google Maps satellite image. Image copyright 2023 TerraMetrics.

> ***Side Trip 1, Laguna Lejía to Laguna en Salar de Aguas Calientes:*** *Continue driving east on B-357 for 10.1 km (6.3 mi; 27 min) to **Stop ST 1.3, Laguna En Salar de Aguas Calientes** (–23.537383, –67.606523). Walk about 1.8 km (1.1 mi) north to the lake.*

As you leave Laguna Lejía, notice the small and recent, perfectly conical black cinder cone on the south side of the road. A 600 m (2,000 ft) path leads to the crater rim.

Stop ST1.3 Laguna en Salar de Aguas Calientes

This is an unusually scenic playa, or salt flat, with an intermittent shallow pond nestled in the volcanic arc of the High Andes. Volcán Lascar and Volcán Aguas Calientes dominate the barren landscape to the north, and Chiliques rises to the south. The source of water is mainly snowmelt, with some derived from local springs.

There are at least three salars in Chile that go by the name Aguas Calientes. This one, at latitude 23.5 south, lies at an altitude of 4,200 m (13,780 ft) and has a surface area of 134 km² (52 mi²). Whereas precipitation is around 150 mm/yr (5.9 in/yr), potential evaporation is in the range of 1,500 mm/yr (59 in/yr; Stoertz and Ericksen, 1974; Garcés Millas and López Julián, 2012; Wikipedia, Laguna en Salar de Aguas Calientes). Which explains why this lake is mostly dry.

Laguna en Salar de Aguas Calientes, view northwest. Volcán Aguas Calientes (right) and Lascar (left). Photo courtesy of Silvio Rossi, https://commons.wikimedia.org/wiki/File:Volcanos_Lascar_left_and_Aguas_Calientes_right.jpg

> **Side Trip 1, Laguna en Salar to Las Tres Marias, Valle de la Luna:** *Return northwest on B-357 to R-23; turn right (north) on R-23 and drive toward San Pedro de Atacama; at the turnoff to Calama turn left (west) and drive 7.3 km (4.5 mi) to Ruta Valle de la Luna; turn left (west) on Ruta Valle de la Luna and drive 0.5 km to the park entrance. Pay 5,000 p/adult to enter, then drive to* **Stop 11.1, Las Tres Marias** *(–22.920593, –68.318818) for a total of 126 km (78.0 mi; 2 hrs 42 min).*

> *Caution*: The area south of the road leading to the park is dangerous because of the presence of land mines (de Waele et al., 2009). Do not wander off the beaten track!

Stop 11 San Pedro de Atacama and Valle de la Luna

This tourist town of 3,899 people (in 2012) lies at an altitude of 2,400 m (8,000 ft) above sea level. As an oasis in a hyper-arid desert, it has always been a place for travelers to stop. In pre-Columbian times this area was occupied primarily by the Atacameño people who farmed and hunted in the area. Rock carvings of guanacos and foxes dating to 500 BC occur in the area. Occasionally they built fortified towns, or *pukaras*, as this area was on the borderlands of the Inca Empire. Pukara de Quitor is located 3 km (1.8 mi) northwest of town.

After the wars of independence with Spain, and while it was still part of Bolivia, San Pedro became the capital of the province of Atacama. Chile took the region and the town during the War of the Pacific (1879), after which Antofagasta became the regional capital.

Today the town is one of the top three tourist destinations in Chile based mainly on the moonscape terrain at Valle de la Luna. Unusually clear and dark night skies have made it a mecca for astronomers as well. The Cosmology Large Angular Scale Surveyor, Cerro Chajnantor Atacama Radio Telescope (ALMA), APEX Event Horizon Telescope, and Atacama Large Millimeter Array are all within 50 km (30 mi) east of town (Wikipedia, San Pedro de Atacama).

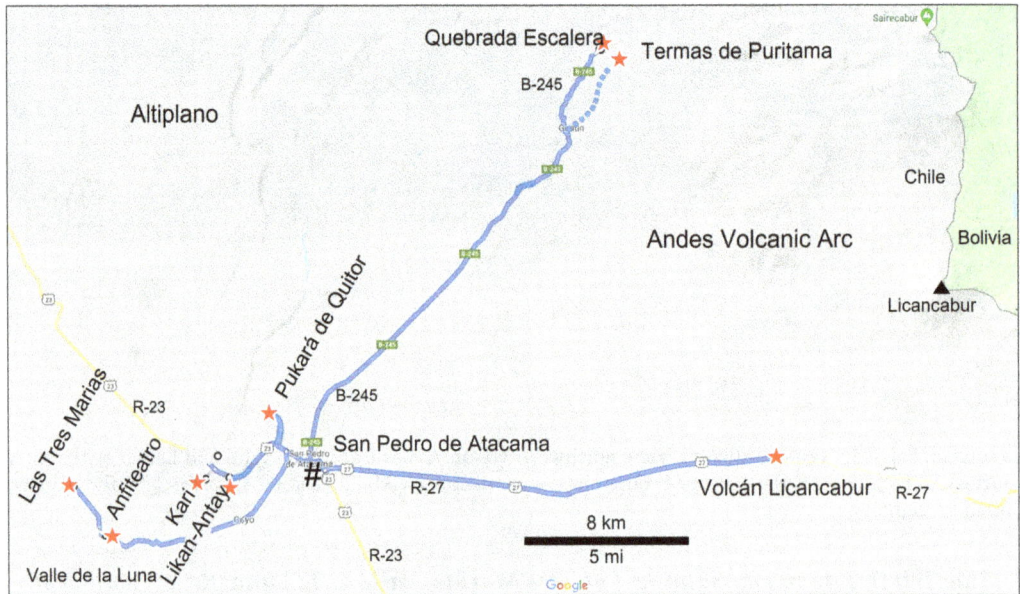

Valle de la Luna (Valley of the Moon) is located in the northern part of the Cordillera de la Sal. It has spectacular outcrops of folded Oligo-Miocene Paciencia Group (San Pedro Formation) sandstone, shale, and evaporites, dry salt-encrusted lakebeds, sparkling shards of anhydrite scattered on the ground, caves and salt karst collapse features, and large sand dunes. The park is part of the Reserva Nacional los Flamencos and was declared a National Nature Sanctuary in 1982 for its unique environment and otherworldly landscape.

The Cordillera de la Sal in this area consists of northeast-trending tight folds formed during Oligocene-early Miocene time. Timing of structural development is revealed by growth strata and thinning of units over the folds. Some workers (e.g. Panont et al., 2004) attributed these folds to diapirism, but others feel they are the result of thrusting on a shallow detachment (Arriagada et al., 2006; de Waele et al., 2009). Both mechanisms are probably active. Late Miocene ignimbrites and pyroclastic units thin over the earlier folds and are in turn gently folded into a broad anticline (de Waele et al., 2009).

Salt at the surface and the evaporites at depth are ultimately derived from Cretaceous-Paleogene marginal-marine deposits (Mpodozis et al., 2004; de Waele et al., 2009). In addition, there are abundant evaporites in the Oligocene lower San Pedro Formation due to the closing and initial filling of the interior seaway (de Waele et al., 2009; Rubilar Contreras,

2015). An east-verging thrust has been mapped at the eastern front of the Cordillera de la Sal (de Waele et al., 2009; Rubilar Contreras, 2015). The main features of the park are the large fold structures and the salt pinnacles that endure in this arid environment.

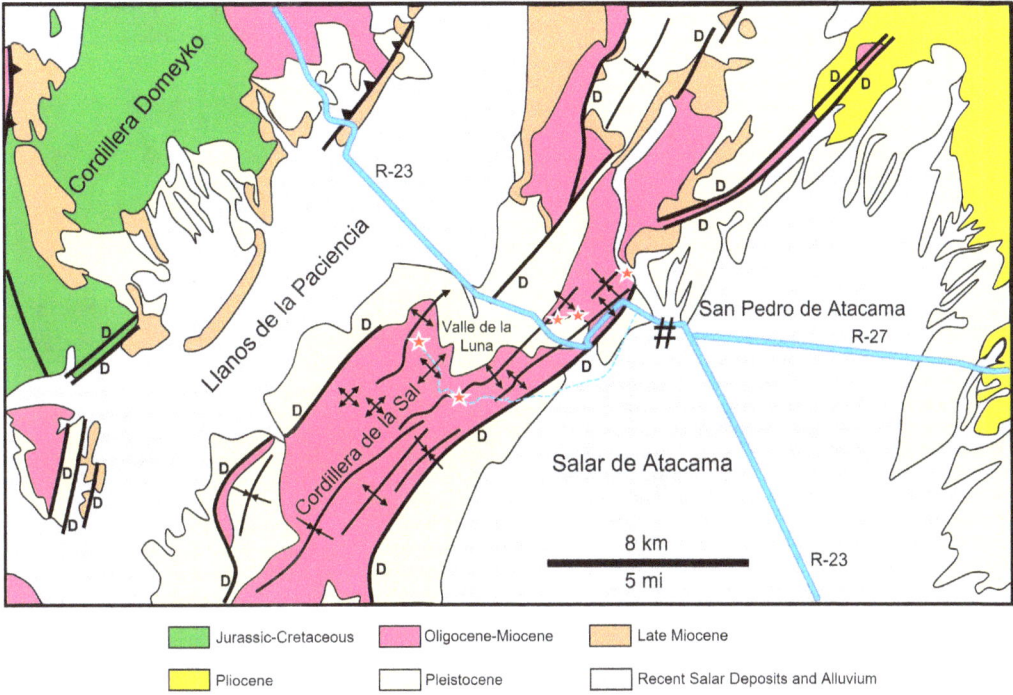

Geologic map of the Valle de la Luna area, San Pedro de Atacama. Modified after de Waele et al., 2009.

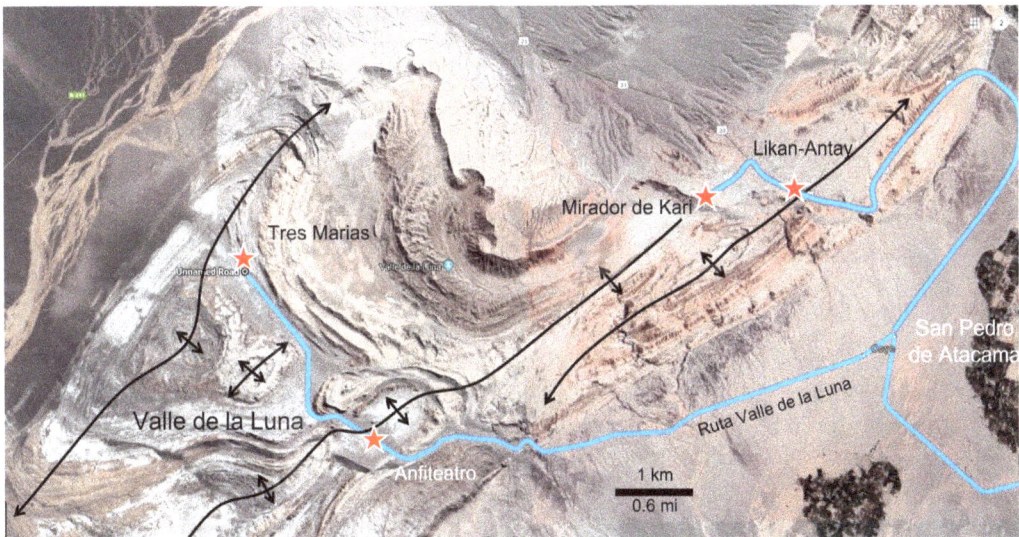

Google Maps satellite image of folding in the Cordillera de la Sal, Valle de la Luna. Image copyright 2023 TerraMetrics.

Stop 11.1 Las Tres Marias

This stop has halite (salt) pinnacles in the Oligo-Miocene San Pedro Formation that appear, to some, like sculpted, petrified ladies. The pinnacles *were* sculpted by the erosive power of wind-blown sand.

Tres Marias, salt pillars in Valle de la Luna. Almost Biblical: During the destruction of Sodom an angel warned Lot to run and not look back. "But his wife, from behind him, looked back, and she became a pillar of salt" (Genesis 19:26). View looking west.

> *Las Tres Marias to Anfiteatro*: Backtrack southeast for 3.2 km (2 mi; 7 min) to *Stop 11.2, Anfiteatro* (−22.942377, −68.303050).

Stop 11.2 Anfiteatro

As you approach this stop you see a broad, gentle syncline to the west developed in San Pedro Formation sandstones. The actual amphitheater is in the eroded core of an anticline. Both features may have formed by upward salt movement in the subsurface, probably related to a deep thrust fault. The reddish-brown salt actually outcrops in some roadcuts here.

Syncline just west of the Anfiteatro, Valle de la Luna. View to the west.

Salt outcrop in roadcut, Anfiteatro.

*Anfiteatro to Mirador Likan-Antay: Continue driving east through Coyo to R-23; turn left and drive to **Stop 11.3, Mirador Likan-Antay** (–22.917011, –68.240061) for a total of 16 km (23 min).*

Stop 11.3 Mirador Likan-Antay

This stop offers a panoramic view into the core of an eroded, probably salt-cored anticline.

Folding related to probable salt movement, as seen from Mirador Likan-Antay. View south. The trace of bedding is shown as dashed lines.

*Mirador Likan-Antay to Mirador de Kari: Continue NW on R-23 for 900 m (1 min) to track (–22.913219, –68.247680); turn left and follow the track south to **Stop 11.4, Mirador de Kari** (–22.917389, –68.251950).*

Stop 11.4 Mirador de Kari

This stop provides another great view into the core of a breached salt anticline. Although harder to discern, the stop is itself in the core of an anticline. There are also stunning views of towering Volcán Licancabur to the east.

Folding from Mirador de Kari, view south.

Volcán Licancabur rises like a ghost above the Puna Plateau. View east from Mirador de Kari.

*Mirador de Kari to Pukara de Quitor: Return to R-23 and turn right (east); take R-23 to the turnoff to Cordillera de la Sal on the left; take the graded dirt road to the Río San Pedro and turn left (north) onto Camino a Quitor; drive to **Stop 11.5, Pukara de Quitor** (−22.888492, −68.219746), parking area on the left for a total of 6.7 km (4.2 mi; 12 min).*

Stop 11.5 Pukara de Quitor

Pukara de Quitor is a scenic tourist stop only indirectly related to the local geology. The twelfth-century fortress of Pukara de Quitor was built by the Atacameño people, taken and held under Incan rule until 1540, when it was captured by Pedro de Valdivia, a Spanish conquistador and the first royal governor of Chile. The fort was well protected due to its location on a hillside overlooking the fertile valley of the Río San Pedro. It took the Spanish 20 years to take the city and kill the defenders. Valdivia established an outpost that became the town of San Pedro de Atacama. The pukara was made a national monument in 1982 (Wikipedia, Pukara de Quitor).

The building stones are of the local Liparita stone, from the Miocene Pelón Ignimbrite, so were fairly easy to work. They blend well into the hillside of the same material.

To avoid Side Trip 2:

*Pukara de Quitor to Quebrada Escalera: Return east on Camino a Quitor; at the bridge, (0.9 km) turn left (north) and cross the Río San Pedro; across the river immediately turn right (east) and drive to the junction with B-245; bear left (northeast) onto B-245 and drive to **Stop 12.1, Quebrada Escalera** (−22.726077, −68.054051) parking area on the left for a total of 27.7 km (17.2 mi; 38 min).*

The ruins of Pukara de Quitor blend in well with the landscape.

To go on Side Trip 2:

> **Side Trip 2, – Pukara de Quitor to Volcán Licancabur:** *Return to R-23 and drive east to San Pedro; turn left onto R-27 and drive to* **Side Trip 2, Volcán Licancabur** *(–22.909258, –67.978592) for a total of 27.4 km (17 mi; 30 min).*

Side Trip 2 Volcán Licancabur

Located on the border between Chile and Bolivia and 40 km (24 mi) east of San Pedro de Atacama, Volcán Licancabur is a classic cone-shaped stratovolcano with 30° slopes that rises 1,500 m (4,900 ft) above the surrounding terrain. The summit is at 5,916 m (19,409 ft). Its neighbor to the east, Juriques, has a 5,710 m (18,730 ft) high symmetrical cone. The Chile-Bolivia border runs along the crest.

Part of the Central Andean Volcanic Zone, Licancabur is built on Pleistocene ignimbrites. The stratovolcano consists of three episodes of interbedded andesitic lava and ignimbrite flows. The youngest flows are dated at 13,240 ± 100 years before the present. The last eruption occurred in October 2015 (as of March 2016; Wikipedia, Licancabur).

Juriques is dated to the Pleistocene. The magmas at Licancabur and Juriques were derived from partial melting of oceanic crust subducted at the Peru-Chile Trench and mixing with overlying continental crust (University of Oregon; Volcano Discovery).

The lack of glacial features indicates that the mountain as we see it today is more recent than the last glaciation (less than 10,000 to 12,000 years old).

The summit crater contains Licancabur Lake, one of the highest lakes in the world. The lake is less than 4 m (13 ft) deep with a temperature of 6°C (43°F) at the bottom. The lake is kept above freezing due to geothermal heating from vents in the crater (Oregon State University, 2020; Volcano Discovery, 2020).

The name "Licancabur" derives from the Kunza language of the Atacameño people and is translated as "mountain of the people." Inca ruins exist on the crater rim, and it is thought that the peak may have been used as a watchtower.

Licancabur is considered a holy mountain by the Atacameño and attempts to climb it by Europeans were discouraged: It is said that Licancabur would punish people who climbed it. Despite that, the first recorded ascent of the volcano was in 1884 by Severo Titicocha, a local guide.

The volcano is considered to be the mate of Quimal, a peak in the Cordillera Domeyko, since their shadows cover one another at the solstices. According to native myth, this "copulation" fertilizes the earth (Legends of the Atacama; Wikipedia, Licancabur).

The high-altitude, arid volcanic terrain around Licancabur has been described by NASA as the best Earth-based analog for conditions on Mars billions of years ago, during a time when it's possible that the surface of Mars held icy lakes and rivers (NASA Earth Observatory, Licancabur).

> **Side Trip 2, Volcán Licancabur to Quebrada Escalera:** *Return west on R-27 to San Pedro; turn right (north) on B-245 (graded dirt road) and drive to* **Stop 12.1, Quebrada Escalera** *(–22.726077, –68.054051) parking area on the left for a total of 50.7 km (31.5 mi; 50 min).*

Licancabur and a quebrada cut into the ignimbrites, seen looking north from R27.

Stop 12 The Puna Plateau

Driving north from San Pedro de Atacama we enter the valley of the volcanoes. This valley lies between the Puna Plateau/Altiplano of Chile, the second-largest high plateau on Earth (after the Tibetan Plateau) on the west and the Andean volcanic arc on the east. The boundaries are a bit fuzzy. The high plateau extends 1800 km (1,100 mi) along the Central Andes with a width of 350–400 km (220–250 mi) between northern Chile, Argentina, and Bolivia. The central Altiplano region began uplifting about 25 Ma, although some areas could have begun rising as early as Eocene (53–34 Ma). Uplift continues today. The average altitude is 4,200 m (13,800 ft). Geologists think the high elevation is a result of unusually thick continental crust (about 80 km, or 50 mi). The thick crust, in turn, may be the result of shortening by thrusting in the continental crust and "flat plate subduction" and underplating of the continent (Allmendinger et al., 1997: Galland and Sassier, 2018). This is where the Nazca oceanic plate extends sub-horizontally beneath the South American Plate as opposed to diving into the mantle at a steep angle.

Volcanism has been active in this area since the magmatic arc migrated here over the past 40 million years (since the Eocene). Besides the majestic cone volcanoes, the area is known for gigantic, extensive ignimbrites. As mentioned previously, ignimbrites are volcanic ash and rock fragments deposited in a pyroclastic flow, that is, as part of fast-moving flow of hot gas and volcanic material. The large Miocene to Quaternary ignimbrite fields in the Puna-Altiplano plateau area form one of the largest ignimbrite provinces on Earth (Kay et al., 2010). High-volume ignimbrite fields linked to giant calderas are distinctive to the Altiplano-Puna plateau. The injection of mantle-derived melts into the deep crust led to partial melting of the crust and more silica-rich melts, magma that rose to mid-crustal levels (20–25 km; 12–15 mi), and magmatic differentiation. The Puna-Altiplano plateau is bracketed on both the west and east by folded thrust (contractional deformation) belts. The magmas thus erupted through a shortened and thickened crust over a shallow subducting Nazca plate. Gas and silica-rich melts are prone to explosive volcanism. Features of the large ignimbrites suggest they formed by partial emptying and collapse of magma chambers, followed by rapid replenishment leading to post-collapse magmatic resurgence. Large volume eruptions of high viscosity, crystal-rich, low-volatile generally silicic to intermediate composition magmas require violent gas release and emptying of the magma-chamber to initiate the collapse of the overlying roof rocks. Pressures before eruption indicate magma chamber depths of 4 to 8 km (2.5 to 5 mi; Kay et al., 2010).

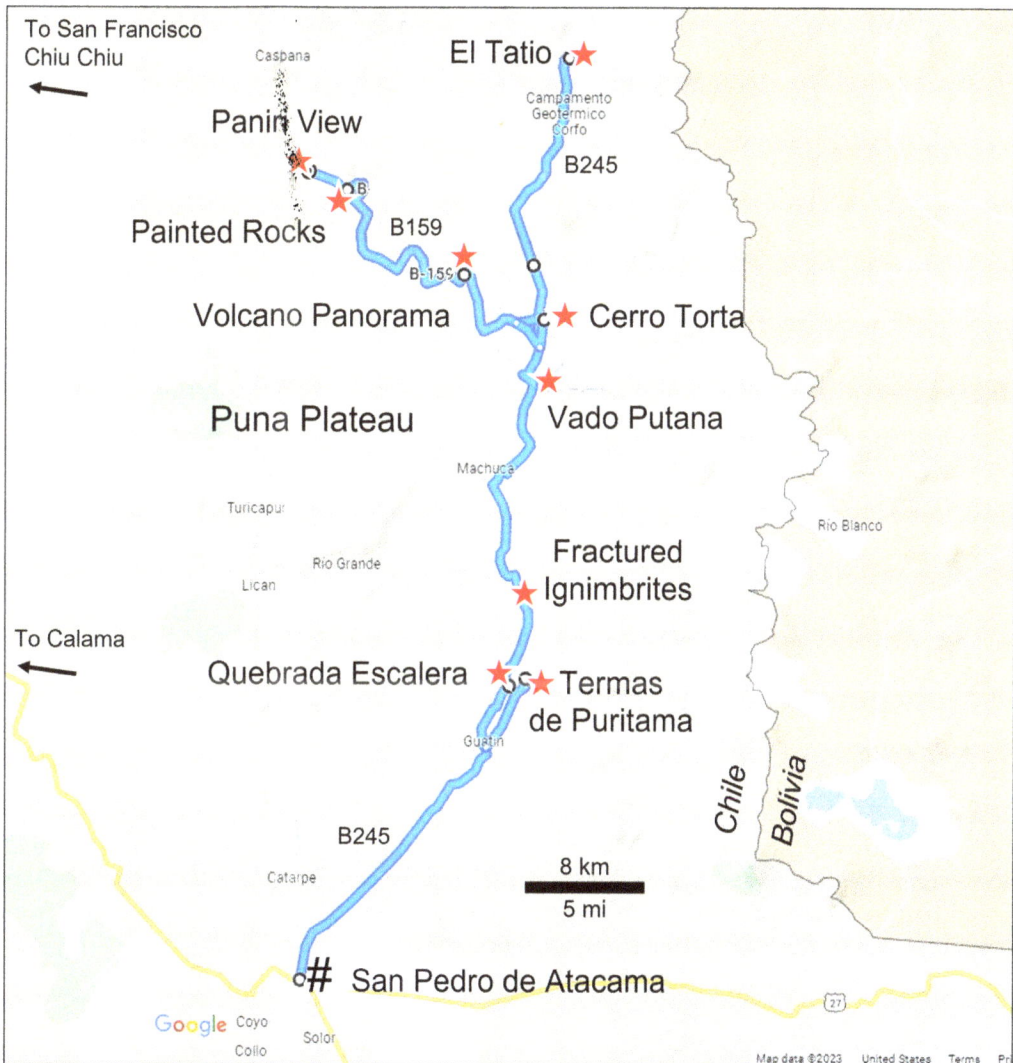

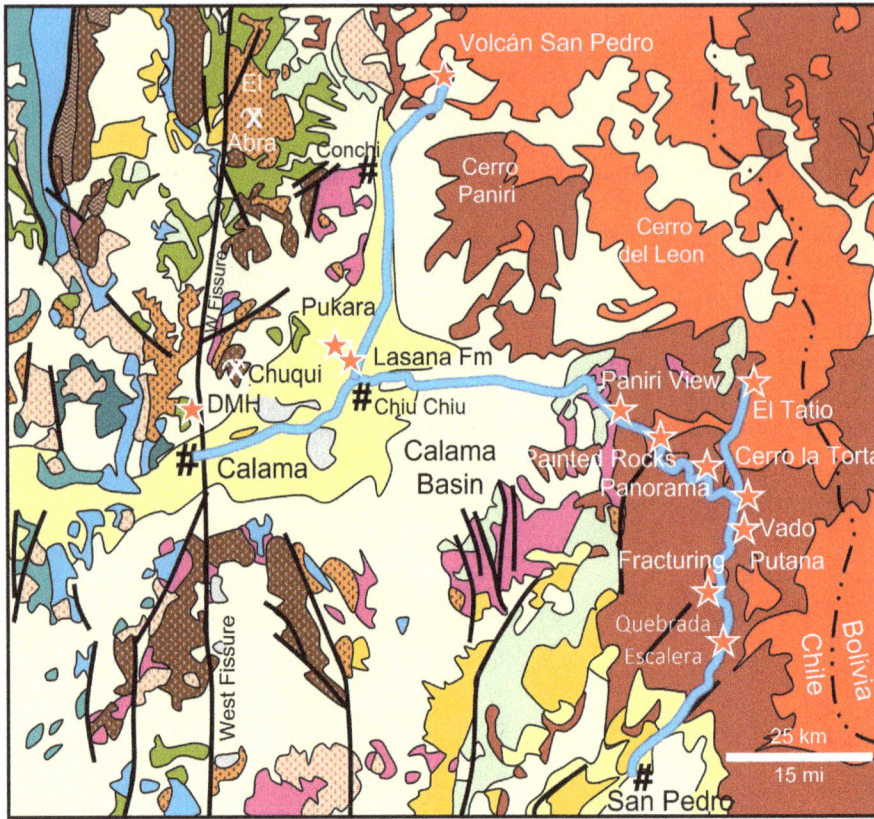

Geologic map of the Puna Plateau (mostly red) between San Pedro de Atacama, Volcán San Pedro, and Calama. Modified after Hartley and May, 1998; Reutter et al., 2005; Tomlinson et al., 2018.

Stop 12.1 Quebrada Escalera

This scenic arroyo is cut into the Miocene to Pleistocene ignimbrites of the Puna Plateau. Be aware of the dizzying altitude when approaching the canyon rim, as there are no established paths and no guardrails. The stream that flows in the bottom gets its water from snowmelt and springs, including some local hot springs.

The author standing at the rim of Quebrada Escalera. View west.

> *Quebrada Escalera to Termas de Puritama*: *Return south on B-245 5.6 km (3.5 mi) to the turnoff to Termas on the left (east); take the dirt road 6.2 km (3.9 mi) to* **Stop 12.2, Termas de Puritama** *(−22.719987, −68.041394) parking area for a total of 11.8 km (7.3 mi; 18 min). Take the trail 1 km (0.6 mi) down the rim into the canyon to the hot springs. You should get reservations in advance.*

Stop 12.2 Termas de Puritama

Puritama is the indigenous word for "hot water." The hot springs, at an elevation of 3,475 m (11,398 ft), are administered by the Explora Hotel. Facilities include changing rooms and a series of eight large outdoor pools and waterfalls extending about a kilometer (0.6 mi) along the Puritama River and connected by wooden footpaths.

The mineral-rich (calcium, magnesium, sodium sulfate, boron) water emerges at 31 to 33°C (88 to 91.4°F). The indigenous Atacameño have used the springs for centuries for their purported healing properties.

Visit

Contact information for the Explora Hotel:

Address: Domingo Atienza Sn, 1410000 San Pedro de Atacama, Chile

Phone: +56 2 2395 2800

Reservations and payments can be made through the website:
 https://termasdepuritama.cl/en/shop/ .

For reservations with indigenous certification discounts, write to puritama@explora.com.

Hours: Monday to Sunday from 9:30 am to 6:30 pm.

General admission is 30,000 p.

Children under 12 years and seniors pay 15,000 pesos.

Children under 3 years are free.

Puritama hot springs.

> **Quebrada Escalera to Fractured Ignimbrites:** *Continue driving north on B-245 for 4 km (2.5 mi; 5 min) to* **Stop 13, Fractured Ignimbrites** *(–22.697222, –68.044167) and pull over anywhere.*

Stop 13 Fractured Ignimbrites

An intense regional joint set oriented ~N30W is exposed as eroded furrows in gently south-sloping Miocene-Pleistocene ignimbrites at this stop. Ignimbrites, being brittle units, are easily fractured. The origin of these joints, however, is not well understood. They may be the result of stress relaxation following east-northeast-directed Andean folding and thrusting.

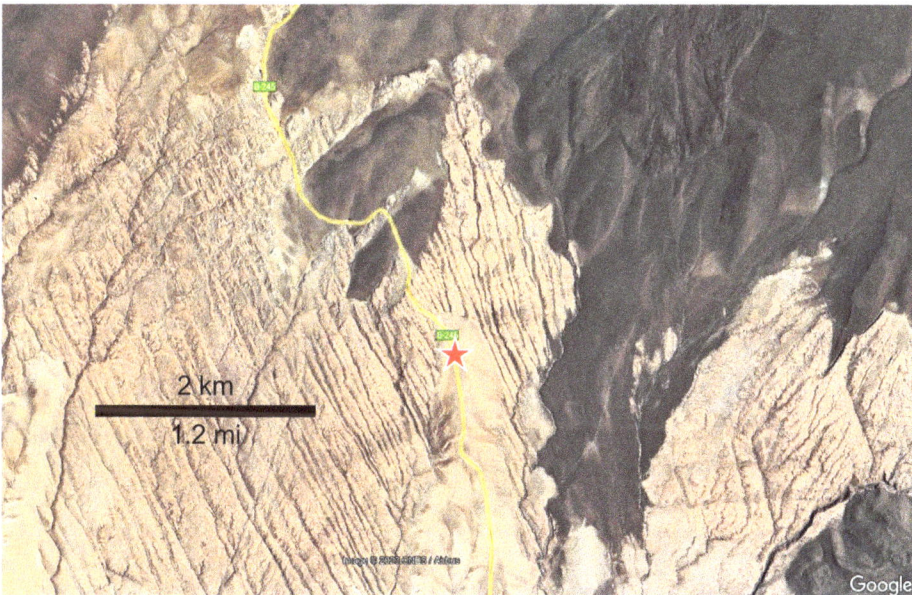

Google Earth satellite image of regional fractures in ignimbrites. Image copyright 2023 CNES/Airbus.

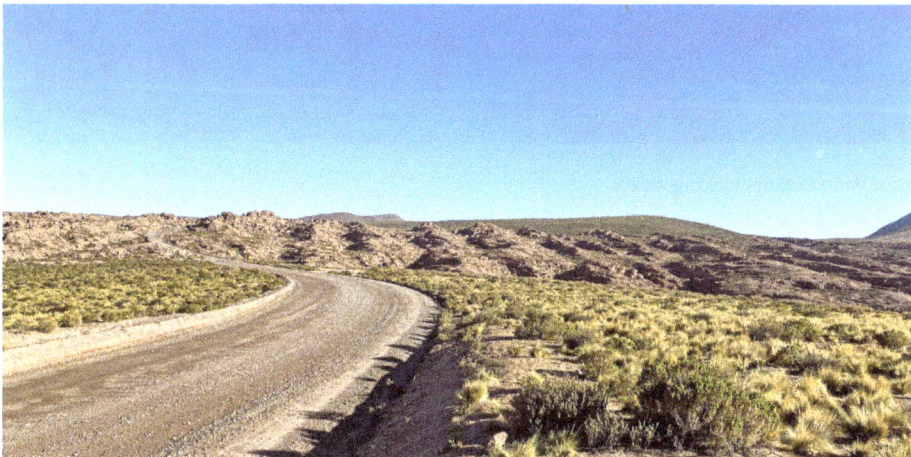

Fractured ignimbrites, view north.

Fractured Ignimbrites to Vado Putana: *Continue north on B-145 for 22.4 km (13.9 mi; 27 min) to* **Stop 14, Vado Putana** *(–22.533697, –68.042751).*

Stop 14 Vado Putana

Vado Putana is a desert oasis with meadows and springs along the Río Putana. The source of the springs is snowmelt in the High Andes several kilometers to the east. Watch for the flamingos and vicuñas that frequent the meadows.

Springs of Vado Putana, looking east.

Wildlife is drawn to the abundant fresh water and vegetation of the Vado.

Flamingos in the alpine ponds.

Vado Putana to Cerro la Torta*: Drive north on B-245 for 4.9 km (3 mi; 5 min) to turnoff on right to* **Stop 15, Cerro la Torta** *(−22.497592, −68.034614) and pull over anywhere.*

Stop 15 Cerro la Torta

Silica-rich (rhyolitic) magmas are inherently viscous, sticky lavas that do not flow far from their source. Cerro la Torta (Cake Mountain) is so named because it is a flat-topped, steep-sided rhyolite lava dome that looks like a large cake. The dome is 34,000 years old (Hernandez Prat, 2017). The lava is over 60% phenocrysts, that is, large crystals of quartz, plagioclase feldspar, hornblende, and biotite. The matrix is glassy (de Silva et al., 2009). Rhyolite domes are not common but do occur in this region. Rhyolites are thought to be a result of magmatic differentiation: The molten rock in a magma chamber separates into magma with heavy minerals near the bottom that crystallize first, and lighter minerals that remain in the magma near the top of the chamber. If those light minerals (mainly quartz and feldspar) erupt, they form a rhyolite dome. Because rhyolite lava is more viscous than other lavas (think cold honey as opposed to warm maple syrup), it forms flat-topped, steep-sided domes.

Google Earth oblique view northeast to Cerro la Torta. Image copyright CNES/Airbus; Landsat/Copernicus.

To avoid Side Trip 3:

> **Cerro la Torta to Volcano Panorama**: *Continue north on B245 for 145 m (500 ft); turn left (west) on a graded dirt track and drive 1.6 km (1 mi) to the intersection with B-159 (paved); continue west on B-159 for 8.2 km (4.9 mi) for a total of 10 km (6 mi; 12 min). This is **Stop 16, Volcano Panorama** (−22.470278, −68.085833).*

To go on Side Trip 3:

> **Cerro la Torta to El Tatio**: *Continue north on B-245 for 25.3 km (15.7 mi; 32 min). The turnoff is on the right. This is **Stop ST 3, El Tatio** (−22.331914, −68.013040).*

View northeast to Cerro la Torta. The large volcanoes to the right (south) are Los Cerros de Tocorpuri.

Side Trip 3 El Tatio

Steaming cauldrons of hot water and mud on a high-altitude plateau. At 4,320 m (14,173 ft) elevation, this is the highest geyser field in the world, the third largest in the world, and the largest in the southern hemisphere. The field contains geysers, boiling water fountains, fumaroles, hot springs, mud pools, and steaming mud volcanoes. Extremophile micro-organisms, bacteria that thrive in boiling water and the excessive ultraviolet radiation of high altitude, make this environment an analog for early life on Earth, and perhaps past life on Mars (Atlas Obscura, El Tatio Geysers).

Geology

The geyser field is named for the adjacent volcano of the same name. "El Tatio" comes from the Atacameño word, el tata, "the grandfather." Volcán El Tatio lies 10 km (6 mi) southeast of the geysers. Legend has it that The Grandfather protected the people of the high plateau and provided the steam for their benefit (Glennon and Pfaff, 2003).

The history of Volcán El Tatio includes two major eruptive periods during Miocene and Pleistocene. Local structure is characterized by roughly north-northeast-oriented, east-dipping, and west-verging thrust faults of middle-late Pleistocene. The easternmost thrust may have served as a magma pathway for the 34,000 year-old Cerro La Torta rhyolite dome, the youngest volcanic activity in the El Tatio area. The intense fluid discharge in the El Tatio geyser basins is apparently controlled by the westernmost thrusts (Lucchi et al., 2009).

Over 80 true geysers have been identified, as well as 30 "perpetual spouters." Despite the conventional wisdom that geyser activity only occurs in the morning, activity occurs throughout the day in all parts of the field. The contrast between the steam and ambient air is greatest, however, during the cold hours of the morning. Whereas some geysers can reach 7 m (23 ft) in height, eruptions are generally lower than 1 m (3 ft; Glennon and Pfaff, 2003).

The geysers are distributed in three separate basins, each with slightly different characteristics. The Upper Geyser Basin (or Main Terrace) lies on the floor of a gently sloping valley and is characterized by well-developed sinter terraces and relatively low water discharge. The Upper Basin is the largest, contains the greatest number of erupting springs, and has the highest erupting geyser. Most geyser eruptions are erratic, although some appear to have predictable intervals. The Middle Geyser Basin, a flat sinter plain, lies immediately south of the Upper Basin. It contains a number of 3-meter-deep (10 ft) pools with frothy fountain eruptions. The Lower Geyser Basin (River Group) lies along the Río Salado about 2 km (1.2 mi) downstream from the Middle Basin. At least ten springs erupt in and near the river to heights of 1 to 3 m (3 to 10 ft; Glennon and Pfaff, 2003).

The hot springs issue at 60–85°C (140–185°F; Galand and Sassier, 2018). The thermal springs are located in the Tatio graben, a 20 km (12 mi) north-south-oriented fault-bounded downdropped block. The graben is limited on the west by the Serrania de Tucle-Loma Lucero horst (uplifted fault block) and appears to be bounded on the east by modern volcanoes. The horst and graben are a result of Pliocene east-west extension (Glennon and Pfaff, 2003).

The hot springs flow from Pleistocene ignimbrites and lavas overlying a Mesozoic sedimentary basement. The Puripica and Salado ignimbrites, forming the lower aquifer, are overlain by impermeable Tucle Tuffs in the Upper Geyser Basin. A lower aquifer spanning the central, southern, and southeastern basins, occurs in the Tucle Dacite, which is overlain by the impermeable Tatio Ignimbrite (Glennon and Pfaff, 2003).

The ultimate source of water at Tatio is snowmelt on volcanoes 15 to 20 km (9 to 12 mi) east and southeast of the field. Discharge from the El Tatio area coalesces to form the Río Salado (Glennon and Pfaff, 2003).

El Tatio fumaroles, view east.

El Tatio fumaroles.

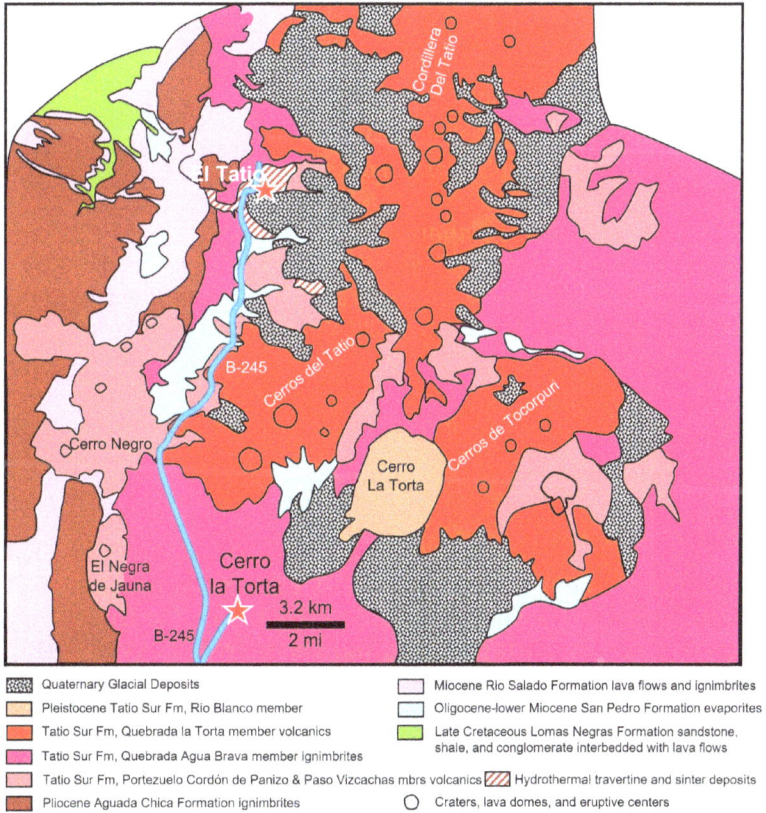

Geologic map of the Cerro la Torta-El Tatio area. Modified after Lucchi et al., 2009.

The area has been considered for geothermal energy projects since at least the 1920s, but drilling affected geyser activity, and technical and economic challenges prevented the development of economic geothermal power. The last attempt was abandoned after a spectacular well blow-out in 2009 that generated a 60-meter-high (197 ft) steam plume. The blowout aroused the public against geothermal energy in Chile, much to the relief of indigenous people and the gratification of tourism operators (Galand and Sassier, 2018).

There are no active glaciers on the peaks in this area today, but moraines, striations, and cirques indicate that there was glaciation in the recent past (Lucchi et al., 2009).

Visit

The geysers have been a protected area since 2010 (Atlas Obscura, El Tatio Geysers). The Geysers Complex, known as Tatio Mallku, is administered by the indigenous communities of Toconce and Caspana, and there is an entrance fee. Protection became necessary because several springs on the Main Terrace appeared altered for tourism and vandalized by tourists (Glennon and Pfaff, 2003).

When visiting the geyser field, it is recommended that you eat light, drink lots of water, get proper rest, and drink the local chachacoma tea to help you get accustomed to the cold and altitude. Most visitors arrive early in the morning to see the sunrise and the geysers shooting steam into the air, most clearly visible between 6 and 7 am. Although, because of the altitude and cold air, you can see steam any time of day. Walk carefully, as tourists have been scalded, and some fatalities have occurred at El Tatio (Atlas Obscura, El Tatio Geysers).

Entrance fee:	General admission	15,000 p.
	Indigenous admission	5,000 p.
	Children under 8 years	Free.

*Side Trip 3, El Tatio to Volcano Panorama: Return south on B-245 12.5 km (7.8 mi); at the intersection bear right onto a shortcut to B-159; drive an additional 6 km (3.7 mi) to the intersection with B-159 (paved). This is **Stop 16, Volcano Panorama** (−22.470278, −68.085833) for a total of 18.5 km (11.5 mi; 20 min) and pull off anywhere.*

Stop 16 Volcano Panorama

This may literally be the high point on the tour: At 4,505 m (14,780 ft) you are higher than the summit of Mt Whitney, the highest peak in the contiguous 48 United States. You may feel a little light-headed, as much from the altitude as from the view. And the view is spectacular. Snow-covered volcanic peaks stretching to the north and south as far as the eye can see, some of them smoking. From north to south are Volcán San Pedro, Volcán San Pablo, Cerro Paniri, Cerro del Leon, Cerro Toconce, Volcán Tatio, Cerros de Tocorpuri and Cerro la Torta, Volcán Putana, and Volcán Colorados.

This area has been mapped as Pliocene Aguada Chica Formation ignimbrites (Lucchi et al., 2009). Satellite imagery suggests that this stop is on the crest of a gentle north-south-trending anticline. The Miocene to Recent ignimbrites may be gently draped over more intense folding that is a northern extension of the Cordillera de Sal that outcrops 37 km (23 mi) south of here.

Google Earth oblique view north over the Volcano Panorama stop. Image copyright 2023 Maxar Technologies; CNES/Airbus; Landsat/Copernicus.

View north from Volcano Panorama stop.

Cerro la Torta sits at the base of Los Cerros de Tocorpuri. View east-northeast from the Volcano Panorama stop.

Cerro Putana with active fumaroles on the peak. View east-southeast from the Volcano Panorama stop.

Volcano Panorama to Painted Rocks: Continue west on B-159 for 18 km (11.2 mi; 20 min) to Stop 17, Painted Rocks (–22.417222, –68.167778) and pull over on the right.

Stop 17 Painted Rocks

The colorful rocks seen across the valley to the south are ignimbrites, probably the pink to gray Pliocene Puripica tuffs.. The colors are a result of ancient hot springs altering primary minerals in the ignimbrite to secondary clay and iron oxide minerals. The layers dip gently west.

View south to altered ignimbrites at the Painted Rocks stop.

Painted Rocks to Cerro Paniri View: Continue west on B-159 for 3.1 km (1.9 mi; 3 min) to Stop 18, Cerro Paniri View (–22.406153, –68.194350) and pull over anywhere.

Stop 18 Cerro Paniri View

The view north to Cerro Paniri reveals a magnificent stratovolcano with a recent lava dome sitting in the snow-covered crater on top.

Cerro Paniri (Aymara for "he who comes/visits") is 5,960 m (19,550 ft) high. The volcano sits on Oligo-Miocene San Pedro Formation sediments and Miocene ignimbrites. Volcanic units range from basaltic andesite to dacite. Early eruptions formed a shield volcano, whereas later eruptions formed the main cone. The oldest flows are around 325,000 years old; the youngest are around 150,000 years old.

Cerro Paniri is sacred to the village of Ayquina. The first recorded ascent was in 1972 by Claudio Lucero and Nelson Muñoz.

View north to Cerro Paniri from Paniri View stop. Note the recent snow-covered dome in the caldera.

Sharp-eyed observers can find vicuñas in the tall brush along B159.

To avoid Side Trip 4:

> ***Cerro Paniri View to Pukara de Lasana***: *Continue west on B-159; continue straight onto B-165; turn right (north, then west) onto B-169 at the sign for 'Chiu Chiu 11 km'; in Chiu Chiu turn right (north) on R-21 and drive to the sign to Turi and Cupo on the right; at the sign, turn left (west) onto B-151 and drive across the Río Loa to the village of Lasana; this is* **Stop 19, Pukara de Lasana** *(–22.269669, –68.631635) for a total of 63.2 km (39.3 mi; 57 min) and pull over on either side.*

To go on Side Trip 4:

> ***Side Trip 4, Cerro Paniri View to Volcán San Pedro and Volcán La Poruña***: *Continue west on B-159; continue straight onto B-165; turn right (north, then west) onto B-169 at the sign for 'Chiu Chiu 11 km'; in Chiu Chiu turn right (north) on R-21 and drive to* **Stop ST4, Volcán San Pedro and Volcán La Poruña** *(–21.935556, –68.526667) for a total of 102 km (63.3 mi; 1 hr 20 min).*

Side Trip 4 Volcán San Pedro and Volcán Poruña

Volcán San Pedro is visible from Calama 85 km (53 mi) away. Its neighbor and twin, San Pablo, is just 6 km (3.7 mi) to the east. San Pedro is a Holocene composite volcano, and at 6,145 m (20,161 ft), is one of the tallest active volcanoes in the world. The stratovolcano has lavas ranging from basaltic andesite to andesite to dacite. The volcano was built on a base of Miocene ignimbrites and other volcanics.

The large "Old Cone" has been dated at around 168,000 to 68,000 years old. The Old Cone was modified by a large landslide that effectively removed its northwest side. Dark material in the foreground is the debris avalanche caused by the landslide that left a horseshoe-shaped scarp and crater seen today. A similar flank collapse occurred during the 1980 eruption of Mount St. Helens in Oregon. Within the landslide scar, lava flows and pyroclastic flows built the Young Cone. The Young Cone is dated at around 100,000 years old (Wikipedia, San Pedro Chile Volcano; de Silva et al., 2009).

Scars at elevations of 5,500 m (18,000 ft) on the side of the mountain show the origin of mudflows that occurred between 110,000 and 36,000 years ago. The largest was the Estación debris flow: It covers much of the southern and western base of San Pedro to distances of 30 km (19 mi). The debris flow reached the Loa and San Pedro rivers and created temporary dams.

At least one Plinian (explosive) eruption occurred at San Pedro around 10,000 years ago. It was as large as the AD 79 eruption of Vesuvius in Italy. The extremely violent eruption generated an ignimbrite (hot gas and ash debris flow) that covers much of the southern, southwestern, and western slopes of the mountain and reaches thicknesses of 3 m (10 ft). Several small eruptions have been recorded over the past 300 years. Presently, only fumaroles are active and are often seen puffing away on the eastern summit. San Pedro is one of the most active volcanoes in the region.

Evidence of Pleistocene glaciation on the volcano includes cirques, moraines, rock glaciers, and striated boulders. There are currently no glaciers on the mountain.

Hydrothermally altered rocks and sulfur deposits are found in the summit crater and were mined into the 1930s. Old tracks left by sulfur miners lead to the top of the mountain (Wikipedia, San Pedro Chile Volcano).

Volcán La Poruña is a 140-meter-high (460 ft) scoria cone (elevation 3,577 m, or 11,733 ft) on the western flank of San Pedro Volcano. It is considered a "parasitic vent" whose location may have been influenced by a normal fault in the area. It sits on Miocene lava flows and has been dated as 103,000 years old. The west side of the cone slumped and lava flows extending up to 8 km (5 mi) issued from the breach (de Silva et al., 2009; Wikipedia, San Pedro Chile Volcano).

Google Earth oblique view east toward Volcán San Pedro and Volcán Poruña. Image copyright 2023 CNES/Airbus; Landsat/Copernicus.

Volcán San Pedro (right) and La Poruña (left). View northeast.

West of here we leave the present-day Andes Volcanic Arc and once again enter the Precordillera/Cordillera Domeyko. The Precordillera is characterized by a pre-Jurassic basement, usually metamorphic or igneous, Jurassic to Cretaceous sediments and volcanics, Triassic to Eocene intrusives, and extensive Cenozoic alluvial cover.

Conchi Viejo and Mina El Abra

The Conchi Viejo and El Abra deposits occur in the hills to the west as you drive south from Volcán San Pedro. They form the northernmost extension of the Chuquicamata mineralized trend. We do not stop at these sites, but they are described because they are such an important part of the region's geology.

Conchi Viejo

The now-closed Conchi Viejo mine, at the small village of Conchi Viejo, was a relatively small open-pit copper mine with some underground workings. Underground workings extended to depths around 85 m (279 ft). The orebody is 400 m by 400 m (1,312 by 1,312 ft) and is a high-grade supergene (shallow, concentrated at the paleo-water table) chalcocite -chalcopyrite-molybdenite-brochantite manto (blanket) deposit on the eastern margin of the El Abra intrusive complex. The host rock is Oligocene (34–23 Ma) diorite. Argillic and propyllitic alteration and mineralization are thought to be localized along N60W faults (USGS; The Diggings).

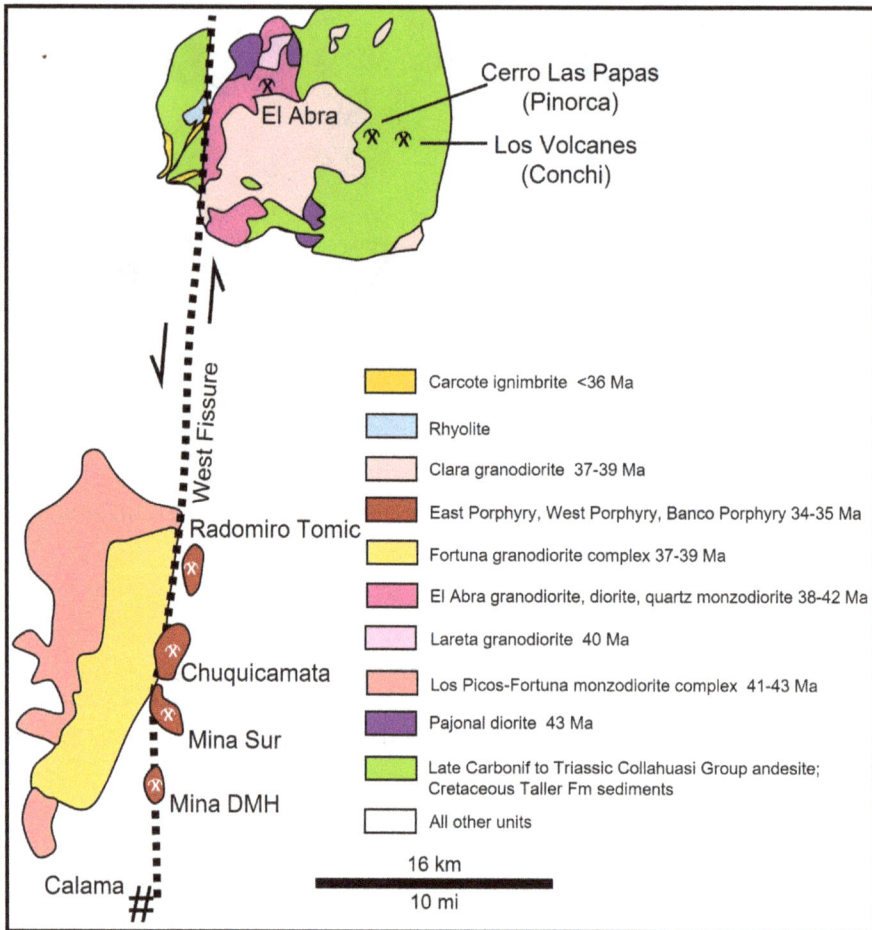

Map showing locations of El Abra, Conchi, and Chuquicamata deposits with respect to the West Fissure. The distribution of Eocene-Oligocene intrusive units is shown. Modified after Correa et al., 2016.

El Abra

El Abra mine is about 40 km (24 mi) north of the Chuquicamata mine. Incas mined turquoise and chrysocolla from surface manifestations of the deposit. The British Compañia Minera de Calama mined vein deposits in the area in the early 1900s. Anaconda Copper Mining Co. explored the area between 1945 and 1969 and, through a drilling program, found the first indications of a large, low-grade deposit. Codelco (the Chilean government company) took over and mapped and drilled out a significant copper-oxide deposit by 1975. Cyprus Amax (now part of Freeport McMoRan) bought 51% of the deposit in 1994 and the new partnership began mining in 1995. The first cathode copper was produced in August 1996. Historical production for El Abra was 300 tonnes (331 tons) of ore at 0.65% Cu. The remaining reserves as of January1, 2003, were 536 tonnes (591 tons) at 0.41% Cu (Gerwe et al., 2003).

Production in the period 2010–2012 was expected to be about 152,000 to 168,000 tonnes (168,000 to 185,000 tons) of copper. The underlying sulfides were mined starting in 2010:

This was expected to extend mine life past 2020 (Wikipedia, Codelco). In 2022, the last year with available data, El Abra produced 90,700 tonnes (100,000 tons) of copper.

The El Abra deposit in the northern Chuquicamata District is a large, low-grade deposit containing significant copper-oxide mineralization in a granodiorite porphyry (Ambrus, 1977; USGS, El Abra Porphyry Cu Deposit; The Diggings, El Abra Mine). El Abra is a porphyry copper-molybdenum-type deposit. The deposit lies about 3 km (1.8 mi) east of the West Fissure (in the north-south-oriented Domeyko Fault system) and is associated with a north-west trending fault zone that might be an off-shoot of the Fissure. Paleozoic-Triassic age sedimentary and volcanic rocks are intruded by 45–36 million-year-old dioritic to granodioritic bodies (the El Abra Granodiorite Complex) that host mineralization (Gerwe et al., 2003; Correa et al., 2016). Mineralization occurred at the time of porphyry intrusions around 36 Ma (Ambrus, 1977; Correa et al., 2016).

An extensive cap of leachable oxide and sulfide mineralization existed at the surface (Wikipedia, Codelco). Copper oxide mineralization at El Abra crops out in an area of about 1,700 by 1,000 m (5,577 by 3,280 ft). Mineralization consists of an upper oxide mineral zone, which extends from the surface down to depths of 90 to 300 m (295 to 984 ft). Chrysocolla, pseudomalachite, and tenorite are the principal oxide minerals (Gerwe et al., 2003).

Within the main orebody, alteration grades outward from a large potassic core zone (characterized by orthoclase-biotite-magnetite) closely associated with the copper mineralization, to a phyllic (quartz-sericite-clay) and potassic zone that contains hydrothermal breccias. The phyllic zone appears to be associated with northwest-trending faults. Propylitic alteration (calcite-epidote-chlorite) affects the margins of the deposit (Gerwe et al., 2003).

A "mixed zone" 10 to 20 m (33 to 65 ft) thick lies at the transition between the oxide and sulfide zones. In addition to the oxide minerals mentioned above, this zone also contains cuprite, native copper, secondary chalcocite, bornite, and chalcopyrite (Gerwe et al., 2003).

The sulfide zone extends more than 700 m (2,300 ft) below the oxide zone. Bornite, chalcopyrite, chalcocite, and pyrite are the principal sulfide-ore minerals. Molybdenite is common as veinlets in porphyry dikes. Secondary chalcocite occurs as veins and in quartz-sericite breccias (Gerwe et al., 2003).

Dates from the El Abra and Toki mineralized zones, along with tectonic reconstruction along the Domeyko Fault System, suggest that a single, ~30 km (18 mi) diameter deposit had developed by late Eocene. This deposit was subsequently split and separated ~35 km (22 mi) along the West Fault (Correa et al., 2016).

Visit

You will have to stop at a main gate and show that you have permission to enter the area. Contact Codelco or Freeport McMoRan for a permit before attempting to visit the mine.

> *Side Trip 4, Volcán San Pedro and Volcán Poruña to Pukara de Lasana*: *Return south on R-21 to the sign that says 'Chiu Chiu straight ahead'; immediately after the guardrail turn right (west) onto B-151 and drive across the Río Loa to the village of Lasana; this is* **Stop 19, Pukara de Lasana** *(−22.269669, −68.631635) for a total of 42.7 km (26.5 mi; 35 min) and pull over on either side.*

Stop 19 Pukara de Lasana

This stop is more historical and touristic than geological. Lasana is a small village located 40 km (25 mi) northeast of Calama along the banks of the Río Loa.

The Pukara de Lasana, (in Quechua *pukara* is a fortress) is a pre-Columbian fortress town built in the tenth to twelfth centuries. It is very much like the Pukara de Quitor we visited at San Pedro de Atacama. The cliffside ruins vary in size and design. In addition to being living quarters, some rooms were storage silos for corn, meat, and herbs (GoChile.com, 2020). The ruins are the main attraction of the village. The site was declared a National Monument in 1982.

Petroglyphs can also be found in the area (Wikipedia, Pukara de Lasana).

The ancient Inca fortress Pukara de Lasana.

*Pukara de Lasana to Valle de Lasana and Lasana Formation: Return north on B-151, cross the Río Loa, and drive south along the river on B-175 to **Stop 20, Valle de Lasana** (−22.315653, −68.648108) for a total of 6.5 km (4 mi; 13 min) and pull over on the right.*

Stop 20 Valle de Lasana and Lasana Formation

The Lasana Valley provides an opportunity to view the upper Miocene to upper Pliocene Chiquinaputo Formation which lies unconformably over the middle Miocene Lasana Formation. Looking south down the valley, the bedding bends downward along the Chiu Chiu Monocline.

The bulk of the cliffs here are of the Lasana Formation, a clastic unit of ephemeral fluvial and alluvial marsh origin, containing coarse sands and conglomerates that interfinger laterally with fine-grained facies of the Jaluinche Formation in the center of the basin. The Lloradero member, seen here, is mainly conglomerates and dark gray coarse sands with subordinate pink siltstone.

The Chiquinaputo Formation, at the top of the cliffs, is a clastic unit consisting of medium to coarse conglomerates, coarse and pebbly sands, siltstones, and diatomites partially cemented by carbonate. They were deposited over the Sifón Ignimbrite (mostly missing here).

View southwest at the Chiu Chiu monocline and the Chiquinaputo Formation over Lasana Formation, Valle de Lasana.

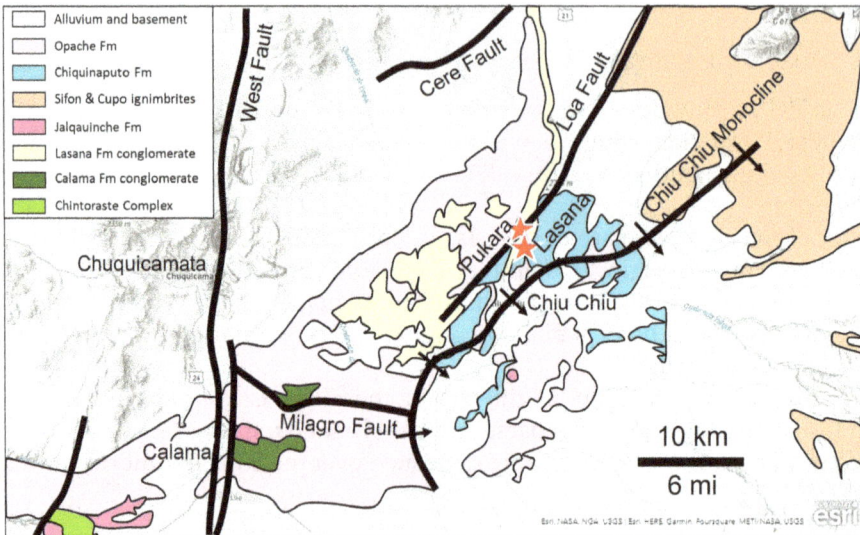

Map showing the distribution of units and major structural features along the Río Loa and Lasana Valley. Red stars indicate the pukara and Valle de Lasana outcrop stop. Modified after Jordan et al., 2015.

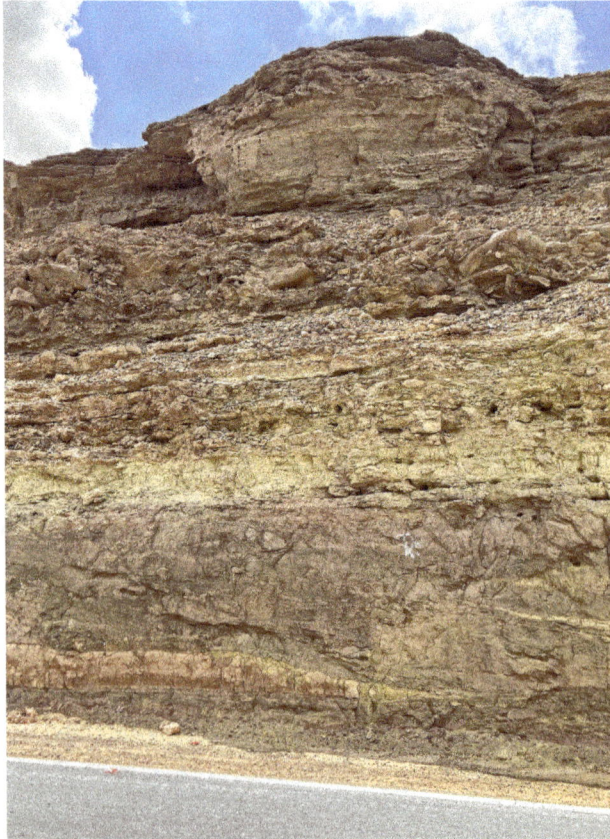

Lasana Formation, mostly coarse fluvial sandstone and conglomerate, as seen on the east side of Valle de Lasana.

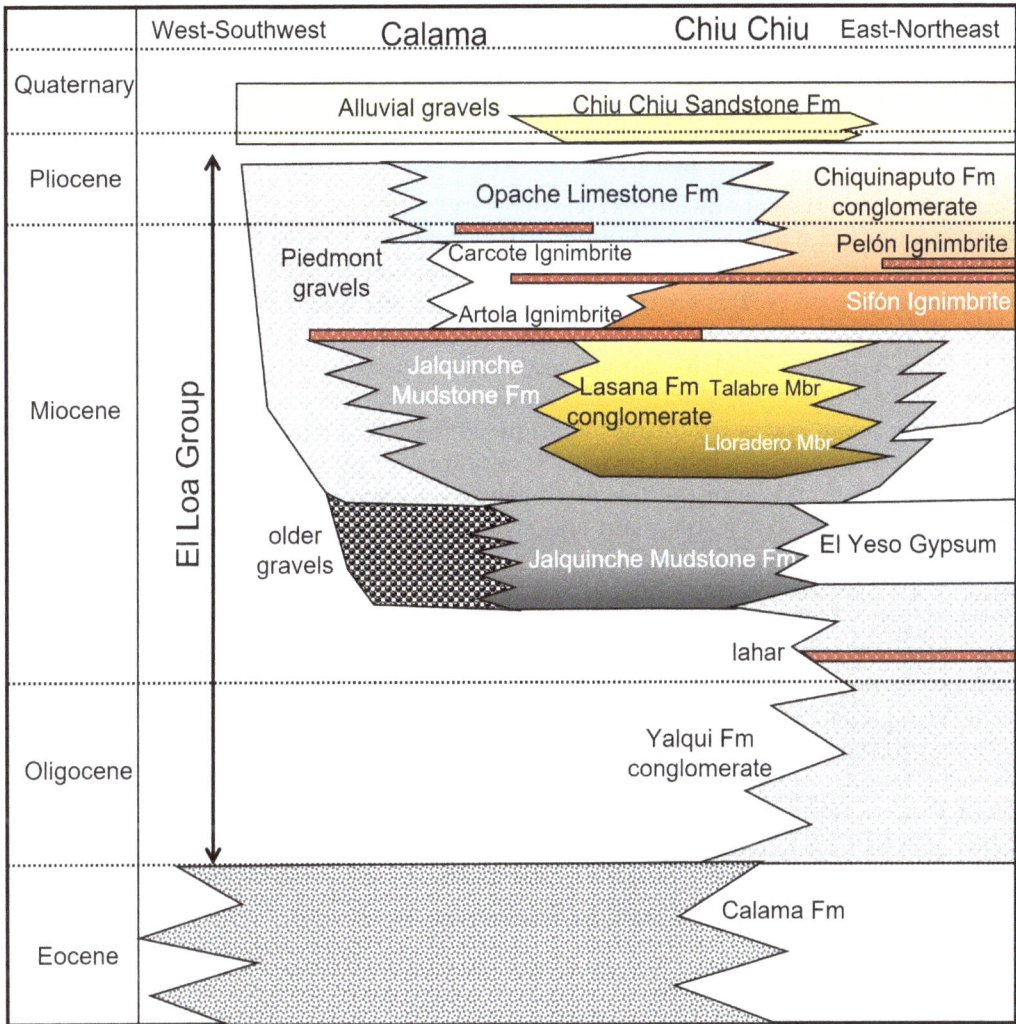

Tertiary stratigraphy of the Río Loa Valley near Calama. Stratigraphic information is derived from Rech et al., 2010 and Jordan et al., 2015.

*Valle de Lasana and Lasana Formation to Calama, Mirador DMH, and the Chuquicamata Mine: Continue south on B-175; merge with R-21 going south and drive to Calama; in the eastern outskirts of Calama use the right lane to get on R-25 north to Tocopilla and Chuquicamata; turn right (north) onto R-24/Av Balmaceda and drive to **Stop 21, Calama, Mirador DMH, and the Chuquicamata Mine** (–22.381297, –68.929539) parking area on the right.*

Stop 21 Calama, Mirador DMH, and the Chuquicamata Mine

Welcome to copper country. This is the heart of Chile's extraordinary copper-mining region.

This stop overlooks the open pit of the División Ministro Hales (DMH) deposit. The roadcuts on the west side of the highway are highly fractured Eocene Fortuna granodiorite. The parking area contains a large boulder of oxide ore that is worth examining.

Calama

Calama is a vibrant company town in one of the most inhospitable places on Earth. It is the third-sunniest place in the world. Sitting at an elevation of 2,260 m (7,410 ft), Calama is one of the driest cities on Earth, with an annual precipitation of just 5 mm (0.20 in). This city of 147,886 people (in 2012) is the capital of El Loa Province and serves as a company town and bedroom community to the gigantic Chuquicamata Copper Mine just north of town (Wikipedia, Calama Chile). Mainly on account of this mine, this is the premier copper-producing province in the world (de Silva et al., 2009).

There are two theories regarding the origin of the town's name, both from the Kunza language spoken by the original people. "Ckara-ama," means "town in the middle of the water." Until the middle of the twentieth century, Calama was surrounded by the Río Loa on two sides and a fertile plain and swamps on the other sides. Another guess is that Calama comes from the Kunza word "Ckolama," meaning "place where partridges abound" (Wikipedia, Calama Chile).

The oasis here was settled in pre-Columbian time, as evidenced by the petroglyphs of Yalquincha and the pre-Columbian cemeteries east of the city. There were extensive lands for growing corn to supply food for Inca messengers (Chasquis) and to pay tribute to the Incas (Wikipedia, Calama Chile).

The hostile climate impeded Spanish colonization. Calama was, and is, a main supply station along the routes crossing the desert. After Bolivia's Declaration of Independence (1825), Calama was one of the main towns in Atacama Province. In 1840, the provincial capital moved from Chiuchiu to Calama. In 1886, the Antofagasta-Bolivia Railway built a station in Calama (Wikipedia, Calama Chile). The town grew and became a major mining center with the development of nearby copper mines in the early twentieth century.

Mirador DMH (DMH Overlook)

This is one of the newer mines along the Chuquicamata trend. The buried DMH copper deposit was discovered by drilling in 1989 under a blanket of about 1,000 m (3,000 ft) of Cenozoic gravels. It is on the west side of the West Fault that extends to and truncates the Chuquicamata deposit. Formerly the Mansa Mina (MM) deposit, it was renamed in 2004 after a former Minister of Mines, Alejandro Hales. Owned by Codelco, the DMH Mine has been operating since 2013.

The deposit has mineral resources of around 237 million tonnes (261 million tons) at an average grade of 1.1% Cu (Wikipedia, Minister Hales). Estimated total resources at 0.96% Cu are 1.3 billion tonnes (1.4 billion tons) of sulfide ore and 26 million tonnes of 0.54% Cu as supergene oxides, providing a total of 12.6 million tonnes of metallic copper. This qualifies the accumulation as a supergiant porphyry-type copper deposit.

The richest, upper zone of the deposit, containing 227 tonnes (250 tons) of sulfide reserves at 1.06% Cu, is presently being open-pit mined. The remaining deeper Cu and Mo mineralization will be exploited by underground mining (Boric et al., 2009).

DMH is primarily a porphyry Cu-Mo deposit with superimposed hydrothermal brec-cias/stockwork bodies containing copper and subordinate silver-arsenic (Ag-As) mineral-ization in the upper parts. The deposit is located west of the West Fault System. The West Fault System is a set of strike-slip faults with early right-lateral and later left-lateral offset and active in Cenozoic time. Paleozoic metamorphic, intrusive, and volcanic basement rocks occur east of the fault. West of the fault, a volcanic basement is intruded by Eocene granodiorites and granite porphyries. The basement rocks are andesitic flows and breccias of the Triassic Collahuasi Formation.

Deep mineralization occurs in an array of near-vertical breccia bodies that are parallel to the West Fault. Most of the ore is hosted by the pervasively altered MM Granodiorite of Triassic age. The MM Granodiorite is considered to be the equivalent of the Este Granodiorite at Chuquicamata. The MM Porphyry is deep in the deposit and adjacent to the West Fault; it is late Eocene (39 Ma) and has undergone potassic alteration. Ore miner-als include bornite-chalcopyrite-digenite. Dacite dikes intrude all of the above units along faults. Hydrothermal breccias above the MM Porphyry contain the highest-grade copper, as well as high silver (Ag) values. The breccias comprise isolated angular dacite fragments within a matrix of sulfides (chalcocite, enargite, and pyrite), fault gouge, silica, and alunite. The breccias grade laterally into mineralized stockworks (complex vein systems). The brec-cias are characterized by advanced argillic alteration (silica, alunite, pyrophyllite, sericite, and dickite; Boric et al., 2009).

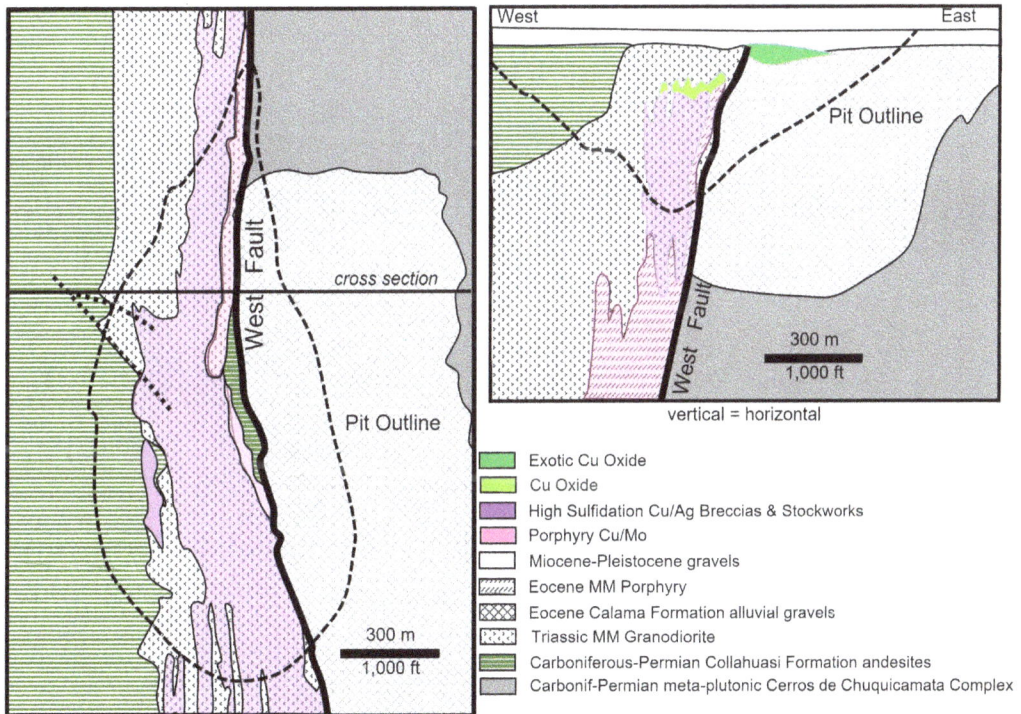

Geologic map and cross-section of the DMH deposit. Modified after Zúñiga Bilbao, 2012.

A 10 to 50 m (33 to 164 ft) thick blanket of exotic copper mineralization is found near the base of the Eocene MM Gravel on the east side of the West Fault. The gravels contain chrysocolla, copper-bearing cryptomelane, malachite, azurite, and conichalcite (Boric et al., 2009).

The DMH ore system began with intrusion of the MM Porphyry ~39 Ma, and the Quartz Porphyry ~36 Ma. Three phases of hypogene mineralization are recognized: 1) Cu-Mo porphyry with potassic alteration (dominated by potassium feldspar and biotite) forms a bornite-chalcopyrite core and a chalcopyrite-pyrite halo; 2) Phyllic alteration characterized by quartz, sericite, and abundant pyrite; and 3) Late stage high-sulfidation copper and subordinate Ag-As mineralization, with advanced argillic alteration (characterized by alunite, anhydrite, kaolinite, halloysite, dickite, pyrophyllite, andalusite, topaz, and diaspore) and the introduction of hydrothermal breccias with pyrite, enargite, chalcocite, bornite, and tennantite. High-sulfidation deposits form in geothermal systems where hot acidic fluid comes directly from the intrusion, undiluted by groundwater. This last stage is dated at 31.4 to 32.2 Ma. After at least 1 km (3,280 ft) of cover was eroded, *in situ* oxidation and weak supergene (shallow depth) enrichment occurred during early Miocene. 'Exotic' copper ores formed in the Cenozoic gravels (Boric et al., 2009).

The pit is located about halfway between Calama and Chuquicamata, within a mining complex that is part of the Chuquicamata Mine. MMH extends 7 km (4.2 mi) north-south, 200 to 320 m (656 to 1,050 ft) east-west, and is more than 1,200 m (3,940 ft) deep. The work facilities cover an approximate area of 4,709 ha (11,636 ac). Ore is taken to a primary crusher adjacent to the pit, and then by means of a conveyor belt to a concentrator plant located about 15 km (9 mi) north of Calama.

South end of the DMH open pit and tailings. View east from Mirador DMH.

Copper--oxide minerals in the Eocene MM Porphyry, Mina DMH.

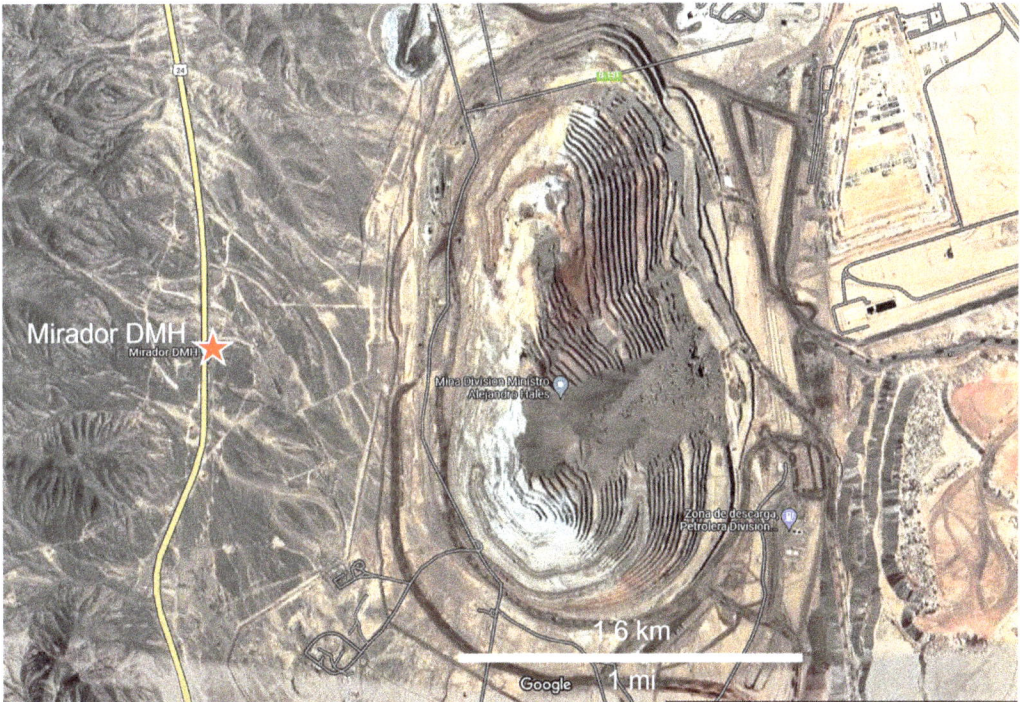

Google Maps satellite image of the DMH pit. Note the landslide on the east side. This pit is smaller than Chuquicamata. Image copyright 2023 Maxar Technologies; CNES/Airbus.

Chuquicamata

Chuquicamata is a place of pilgrimage for many geologists interested in mining and resource extraction. The mine is an engineering marvel and a thing of beauty to behold. It is the largest (by volume) open-pit copper mine in the world. The pit is 4.3 km (2.7 mi) long, 3 km (1.9 mi) wide, and 900 m (3,000 ft) deep (Galland and Sassier, 2018). The original surface was at 2,850 m (9,350 ft) elevation 16.8 km (10.5 mi) north of Calama in the hyper-arid Atacama Desert (Ossandon et al., 2001). The mine accounts for about one-third of Chile's foreign trade (Wikipedia, Chuquicamata).

History

High-grade copper oxide ore (green and blue minerals) at the surface here was first worked in a small way by the Incas and later the Spanish. From 1879 to 1912, Chilean and English companies worked rich veins of brochantite, enargite, and chalcocite. They then followed the veins underground along a N10E fault zone. In 1915, Chile Exploration Co., a branch of the Chile Copper Co., bought up most of the properties and started open-pit mining of disseminated oxide ore averaging 1.9% Cu. Production started in 1925 at 9,070 tonnes (10,000 tons) per day of 2.4% Cu. Anaconda Copper Mining Co. purchased the property from the Guggenheims in 1923 and expanded the operation over the next 48 years. Production increased and shifted from mainly oxide to sulfide ores in the 1950s. In 1957, the Exótica deposit (now South Mine) was discovered beneath a tailings dump. It is still the largest exotic copper deposit discovered to date. "Exotic copper deposits" are those where copper has been dissolved, transported by groundwater, and redeposited in the groundwater oxidation zone. Anaconda built a plant, concentrator, smelter, refinery, and town next to the mine. Chuquicamata soon became the world's largest copper producer. In 1971, the mine was nationalized and taken over by Codelco, the Chilean national mining company. Drilling programs extended the orebody to 1 km (3,280 ft) below the original surface. Today the Chuquicamata District encompasses ore deposits that extend nearly 40 km (25 mi) north-south, from the Toki Cluster in the south to Radomiro Tomic in the north (Ossandon et al., 2001; Barra et al., 2013).

Panoramic view of Chuquicamata open-pit mine. White rectangle is enlarged in the next figure. Courtesy D. Delso, https://commons.wikimedia.org/wiki/File:Mina_de_Chuquicamata,_Calama,_Chile,_2016-02-01,_DD_110-112_PAN.JPG

Detail of the Chuquicamata open pit enlarged from the previous figure to give a sense of scale. The Codelco trucks (circled) are some of the largest mine equipment in the world, approximately the size of a single-family house and can hold up to 327 tonnes (360 tons) of ore. When loaded, the Komatsu trucks weigh up to 762 tonnes (840 tons). Smaller vehicles fly a flag on a tall antenna so that they are not accidentally overrun by the giant trucks. Courtesy D. Delso, https://commons.wikimedia.org/wiki/File:Mina_de_Chuquicamata,_Calama, _Chile,_2016-02-01,_DD_110-112_PAN.JPG.

Geology

The Chuquicamata District is located in the Precordillera of northern Chile (Ossandon et al., 2001).

The deposits of the district occur in a porphyry copper belt that extends for about 1,400 km (870 mi), from 18°S to 31°S. The main structural feature in this belt is the Domeyko Fault System. Within the Calama area the West Fault (or West Fissure), part of the Domeyko Fault System, divides the district into eastern and western domains (Barra et al., 2013).

Chuquicamata is a typical porphyry copper deposit. It consists of a magmatic intrusive complex of dioritic to granodioritic compositions. Copper and molybdenum minerals are disseminated and in veins in a zone many kilometers wide (Galland and Sassier, 2018).

Mineralization is related to late Eocene-early Oligocene porphyry intrusions localized by the Middle to Late Cenozoic West Fault System (Ossandon et al., 2001). At Chuquicamata, Eocene magmatism occurred between 44 and 37 Ma. Magma intruded Jurassic and Lower Cretaceous sedimentary rocks, Upper Cretaceous volcanic and volcaniclastic rocks, and early Eocene andesitic lavas (Somoza et al., 2015).

The Domeyko Fault is a regional structure extending roughly north-south for several hundred kilometers. It is interpreted as a Cenozoic arc-parallel series of strike-slip and reverse faults. The West Fault segment was active prior to and following the intrusion of the Chuquicamata Porphyry Complex, changing its sense of movement at least twice. It played a role in localizing the host intrusions, in directing the mineralizing fluids, and in the post-mineral offset of the orebodies (Ossandon et al., 2001).

In the Chuquicamata pit, post-mineral West Fault movement separates the mostly barren Fortuna Complex to the west from the intensely mineralized Chuquicamata Porphyry Complex to the east. The Fortuna Granodiorite Complex is a multiphase pluton with published ages between 43.1 and 37.6 Ma. The Chuquicamata Porphyry Complex was emplaced between 34.5 and 33.1 Ma (Barra et al., 2013). Rocks with textures essentially identical to those of the Chuquicamata Porphyry Complex extend northward at least 9 km (5.6 mi) through the Radomiro Tomic mine (Ossandon et al., 2001).

The East Porphyry, West Porphyry, Banco Porphyry, and Fine Texture Porphyry make up the Chuquicamata Porphyry Complex. The East Porphyry is the main host rock. Potassic alteration affects all of the porphyries. A band of quartz-potassium feldspar alteration lies along the southward extension of Banco Porphyry dikes and coincides with bornite-digenite mineralization at the center of the sulfide zone. Sulfides in the potassic alteration zone are abundant only where there is intense brecciation. Quartz-molybdenite veins up to 5 m (16 ft) wide penetrate all of the porphyries. Veins were emplaced between the early and the main stages of mineralization. Main-stage veins were focused in a zone adjacent to the West Fault. This stage may have involved remobilization of earlier mineralization. Main-stage veins with quartz, pyrite, chalcopyrite, and bornite were formed during right-lateral shear of the West Fault System. The last phase of the main stage mineralization was enargite, digenite, covellite, pyrite, and minor coarse sphalerite emplacement. Some enargite veins formed after the displacement on the West Fault System changed to left-lateral. Vein and veinlet-filled faults and fault-related shatter zones contain most of the copper at Chuquicamata. Practically all of these fractures were opened and mineralized more than once. A supergene blanket enriched in chalcocite developed at the surface (Ossandon et al., 2001).

A partially preserved leached cap and copper oxide ore overlies a high-grade supergene chalcocite body that extends to depths around 800 m (2,625 ft) in the fault breccia zone. Net left-lateral displacement of about 35 km (22 mi) is suggested by regional mapping (Ossandon et al., 2001).

Dating indicates that supergene oxidation and enrichment occurred from 44 to 6 Ma (Hartley and Rice, 2005). Recent work (Barra et al., 2013) indicates that Porphyry Cu-Mo mineralization occurred episodically between 35 and 31 Ma in the Chuquicamata District.

Production, Resources, Reserves

The Chuquicamata District includes the Chuquicamata and Radomiro Tomic deposits, the exotic copper deposit Mina Sur, the newly developed Ministro Hales mine (formerly Mansa Mina) located 5.5 km (3.4 mi) south of Chuquicamata, and the deposits of the Toki cluster (Toki, Genoveva, Quetena, Miranda, and Opache) located about 15 km (9 mi) south-southwest of Chuquicamata. Total identified resources for the Chuquicamata District were estimated at 107.4 million tonnes (118.4 tons) of fine copper at a cutoff grade of 0.2 wt% copper (Barra et al., 2013).

The Chuquicamata deposit itself is one of the largest-known copper resources in the world. In 1997, the combined total historical production from the Chuquicamata and Exótica (South Mine) orebodies was 644,000 tonnes (710,000 tons) of fine copper (Ossandon et al., 2001). For many years, it was the mine with the largest annual production in the world. It was recently overtaken by Minera Escondida, another Chilean giant. Still, it remains the mine with the world's largest total production of approximately 29 million tonnes (32 million tons) of copper to the end of 2007 (Wikipedia, Chuquicamata).

Proven reserves are 638 million tonnes (703 million tons) of 0.88% copper, or 5.6 million tonnes (6.2 million tons) of fine copper. Probable reserves are an additional 695 million tonnes (766 milllion tons) of 0.59% Cu, or 4.1 million tonnes (4.5 million tons) of fine copper (Codelco, 2016). As of 2016, the mine had an expected lifetime of another 80 years.

Visit

Pre-pandemic, you could reserve a tour by writing or calling ahead. Tours were given on weekdays at 1 pm and ended between 4 and 4:30 pm. The tour was free, but they requested a donation to the fund that helps children of injured workers.

It is unfortunate and more than a little sad that Codelco has not resumed public tours since the end of the COVID crisis. It was a great personal disappointment that, despite months of trying to contact the company to arrange a tour and obtain information, no response was forthcoming. Perhaps in the future they will once again provide this service. It is wonderful publicity for the company and helps inform the public about the contributions made by the mining industry to society.

Should you want to contact the company, you can try the address and emails that worked pre-COVID.

Address: Avenida Granaderos 4025, Calama.

Phone: 232-2122.

Email: visitas@codelco.cl.

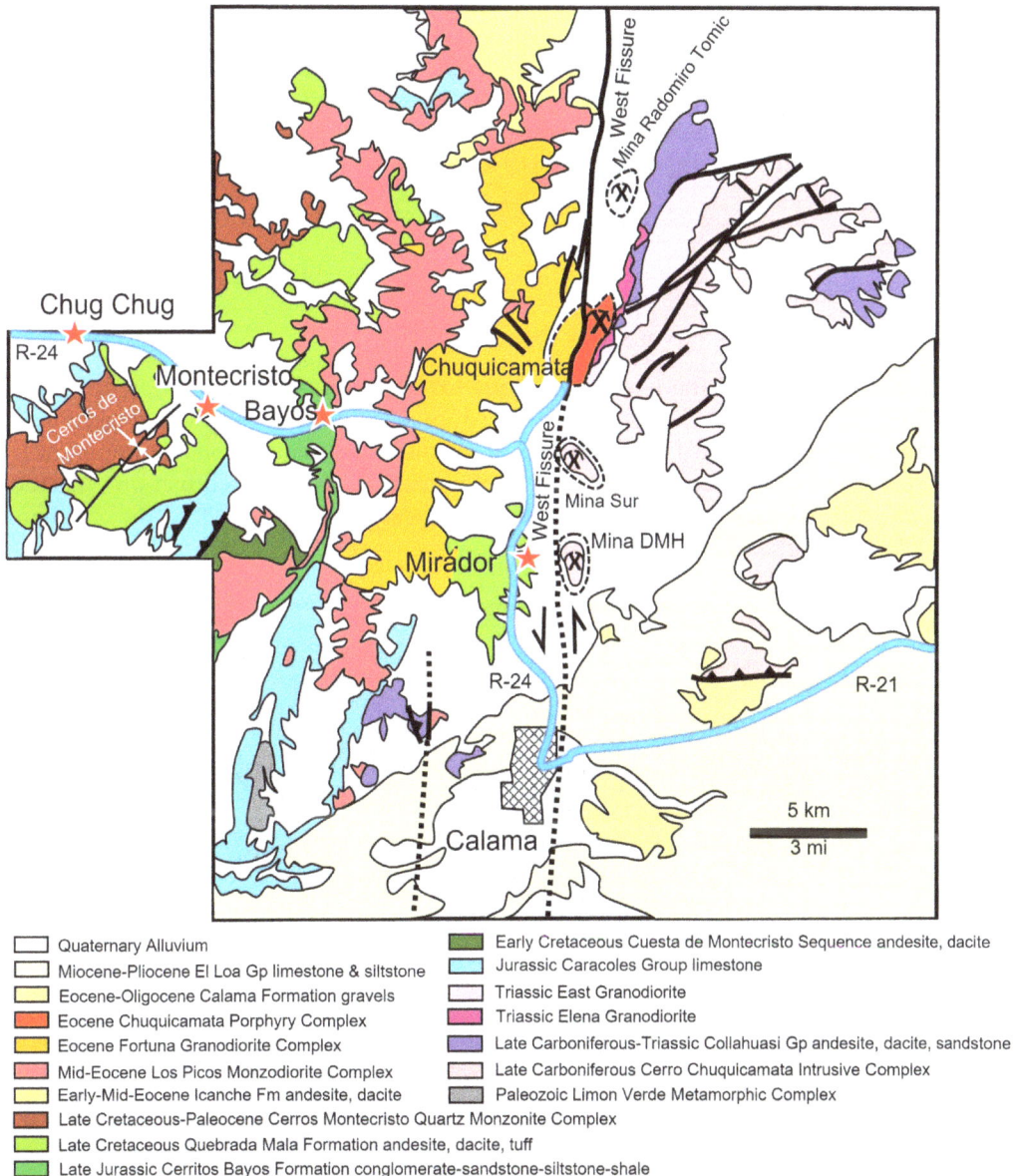

	Quaternary Alluvium		Early Cretaceous Cuesta de Montecristo Sequence andesite, dacite
	Miocene-Pliocene El Loa Gp limestone & siltstone		Jurassic Caracoles Group limestone
	Eocene-Oligocene Calama Formation gravels		Triassic East Granodiorite
	Eocene Chuquicamata Porphyry Complex		Triassic Elena Granodiorite
	Eocene Fortuna Granodiorite Complex		Late Carboniferous-Triassic Collahuasi Gp andesite, dacite, sandstone
	Mid-Eocene Los Picos Monzodiorite Complex		Late Carboniferous Cerro Chuquicamata Intrusive Complex
	Early-Mid-Eocene Icanche Fm andesite, dacite		Paleozoic Limon Verde Metamorphic Complex
	Late Cretaceous-Paleocene Cerros Montecristo Quartz Monzonite Complex		
	Late Cretaceous Quebrada Mala Formation andesite, dacite, tuff		
	Late Jurassic Cerritos Bayos Formation conglomerate-sandstone-siltstone-shale		

Geologic map of the Calama and Chuquicamata area. Modified after Somoza et al., 2012; Barra et al., 2013.

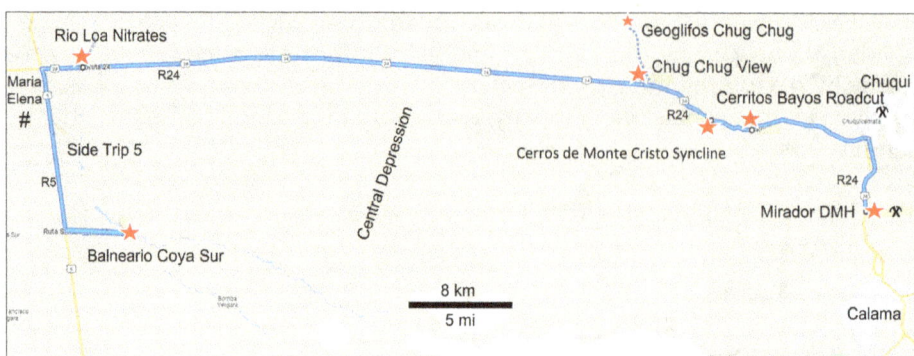

Calama, Mirador DMH, and the Chuquicamata Mine to Cerritos Bayos Formation Roadcut*: Drive north then west on R-24 for 16.7 km (10.4 mi; 15 min) and pull over on the right or the left. This is **Stop 22, Cerritos Bayos Formation Roadcut** (–22.322685, –69.022226). Walk west about 80 m (260 ft) to the roadcut. Be mindful of the large trucks that go roaring by.*

Stop 22 Cerritos Bayos Formation Roadcut

The hills here are part of the Cordillera Domeyko. The Upper Jurassic Cerritos Bayos Formation consists of a marine succession transitional upward to continental sediments. The marine unit is well stratified, with a basal pebble conglomerate followed upward by thin-bedded brown sandstones with ripples, siltstones, shales, and reddish-brown mudstones at the top. The beds here are inclined to the west about 40° to 60° on the east flank of the Cerros de Monte Cristo Syncline.

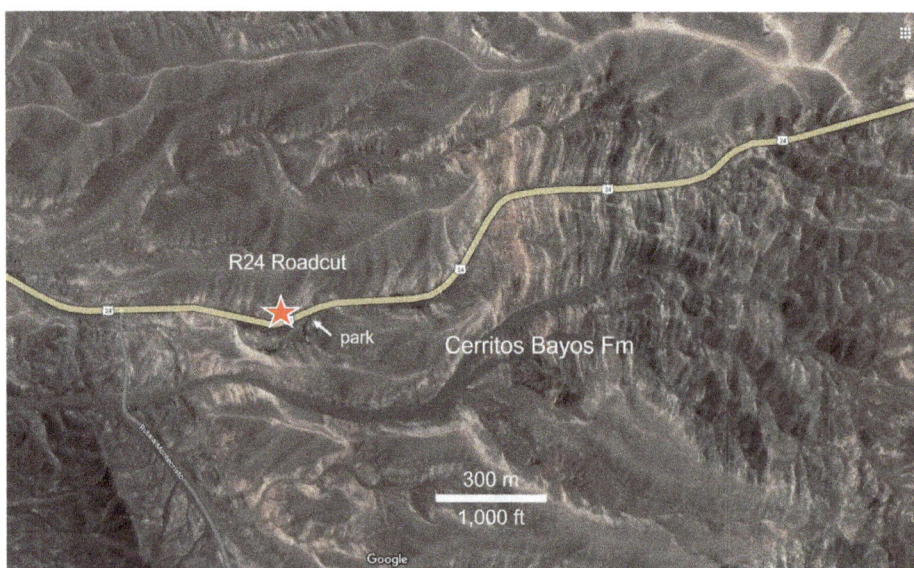

Google Maps image of the R-24 Roadcut stop and Cerritos Bayos Formation. Image copyright 2023 Maxar Technologies; CNES/Airbus.

Late Cretaceous Cerritos Bayos Formation dips steeply west. R-24 roadcut looking west.

Cerritos Bayos Formation Roadcut to Cerros de Montecristo Syncline and Quebrada Mala Andesite: Continue driving west on R-24 for 3.6 km (2.2 mi; 3 min) to Stop 23, Cerritos de Montecristo Syncline and Quebrada Mala Andesite (−22.315973, −69.054709).

Stop 23 Cerros de Montecristo Syncline and Quebrada Mala Andesite

Outcrops of Late Cretaceous Quebrada Mala Formation andesites/dacites/andesitic tuffs dip gently west near the axis of the Cerros de Montecristo Syncline.

The Quebrada Mala Formation, which unconformably overlies Late Jurassic–Early Cretaceous sediments and Late Cretaceous (98–94 Ma) volcanic rocks, consists of 3,700 m (12,100 ft) of conglomerates and sandstones interbedded with andesitic lavas and welded tuffs dated between 86 and 66 Ma. It was deposited in a small basin bounded to the east by the north-northeast-trending, west-dipping, high-angle Sierra El Buitre Fault. The Sierra El Buitre Fault formed in the Late Cretaceous as a normal fault along the southeastern edge of the basin. The Late Cretaceous Quebrada Mala Basin, part of the Central Depression of the Antofagasta region, was inverted prior to the accumulation of Paleocene lavas and tuffs of the Cinchado Formation (Mpodozis et al., 2004). The basin inversion at the Cretaceous-Paleogene boundary indicates a change from regional extension to compression.

Beds of the Quebrada Mala Formation on the east flank of the Cerros de Monte Cristo Syncline dip gently west.

Cerros de Montecristo to Geoglifos Chug Chug and San Salvador Formation:
Continue driving west on R-24 till you see the sign for Geoglifos Chug Chug; bear right
on a graded dirt track and drive to **Stop 24, Geoglifos Chug Chug and San Salvador**
Formation *(−22.285953, −69.110614) and pull off anywhere for a total of 6.8 km (4.3 mi;*
5 min). This stop is to see the San Salvador Formation.

If you wish to see the geoglyphs, continue on the dirt road another 11.8 km (7.3 mi) to Mirador
Geoglifos (−22.199410, −69.173639).

Stop 24 Geoglifos Chug Chug and San Salvador Formation

This stop provides a view looking south to the continental mudstones, siltstones, and
fine-grained quartz sandstones of the Upper Jurassic to Lower Cretaceous San Salvador
Formation and the overlying Quebrada Mala Formation.

The folding that formed the Cerros de Montecristo Syncline is thought to have occurred
during the Late Cretaceous Peruvian contractional deformation phase (Somoza et al., 2012).

A look south at the west flank of the Cerros de Monte Cristo syncline. The San Salvador and Quebrada Mala beds dip moderately to the east here.

If you continue on this track to the geoglyphs at Chug Chug, you will see a series of ancient patterns carved into the soil. The area is in an archeological preserve that includes 23 sites with nearly 500 geoglyphs. They are thought to be located along an old caravan route between the oasis village at Calama and the pre-Spanish town of Quillagua on the Río Loa 70 km (44 mi) northwest of here.

The oldest glyphs date to 1,000 BC, while most are thought to have been created between AD 900 and 1550 by the Atacama and Tarapacá tribes. Similar to the Nazca Lines, the geoglyphs of Chug Chug include human figures and animals such as birds and llamas, as well as circles, stars, and rhombuses. They may have been used as signs for llama herders and to guide caravans across the desert, although no one really knows. The geoglyphs suggest that lines of communication existed between the desert and the Pacific Coast, as they show men on rafts and men harpooning fish.

Continuing west, we leave the Cordillera Domeyko and enter the Gravas de Pampas, or gravel plains. This is part of the Central Depression, previously a back-arc basin, and home to extraordinary world-class nitrate deposits.

Geoglyphs at Chug Chug. Photo courtesy of Jan Woreczko, https://commons.wikimedia.org/wiki/File
:Geoglyphs_Chug-Chug.jpg.

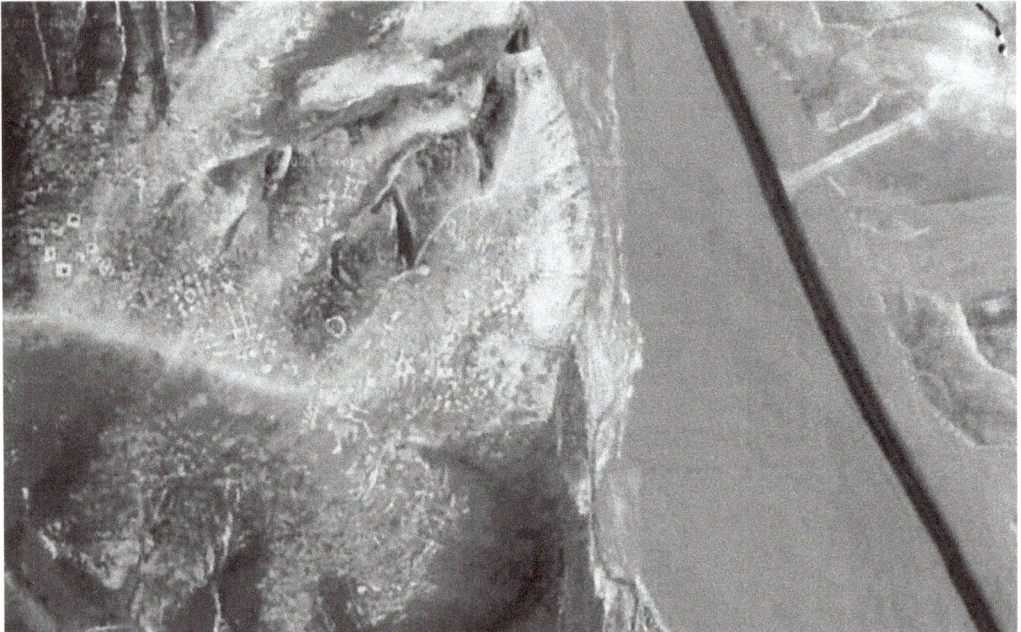

Google Maps image of geoglyphs at Chug Chug. Image copyright 2023 Maxar Technologies; CNES/Airbus.

Geoglifos Chug Chug and San Salvador Formation to Río Loa Nitrates: *Continue
west on R-24 for 46.4 km (28.8 mi; 32 min) to **Stop 25, Río Loa Nitrates** (−22.278283,
−69.562815), and pull over on a road on the right (north) side of the highway.*

Stop 25 Río Loa Nitrates

Rising in the High Andes, we joined the Río Loa at Volcán San Pedro. From there, it flowed south to Calama, turned west, and here it flows north to enter the Pacific approximately 75 km (47 mi) north of Tocopilla. The Río Loa wends its way across Chile in the shape of a giant U: here we are on the western, north-flowing segment. Río Loa is the longest river in Chile, a total of 440 km (273 mi), and the main river in the Atacama Desert. In the upper reaches the river is fed by snowmelt; in the lower reaches it is brackish and fed mainly by rainfall in January and February. The river is considered polluted after passing through the mining areas of El Abra and Calama (Jordan et al., 2015; Wikipedia, Río Loa).

Río Loa canyon looking north from Puente Posada. Outcrops are Miocene-Pleistocene Quillagua Formation marls, diatomites, and limestones with interbedded gravel and shale.

It may not look like much, but this is nitrate country! The Atacama nitrate deposits are unique in the world. They exist because of a rare and favorable environment, rather than a rich source rock. Essential conditions include: (1) extreme aridity; (2) slow accumulation; and (3) few nitrate-using plants and microorganisms.

The fact that they are widely distributed has been explained by a process of atmospheric transport and deposition followed by concentration. The consensus is that the salts originated as seawater spray that blew inland. The spray accumulated at or near the sites of the deposits chiefly as dry fallout and as solutions in condensate droplets from fogs prevalent here in winter. These saline materials were later leached, redistributed, and enriched in nitrate at deeper soil levels and low on hillsides. Ore-grade nitrate deposits, evidently formed by slow accumulation on old land surfaces, have had little or no modification since the Miocene. They accumulated on lower hillsides as the result of leaching and redeposition by rare rainwater, and by accumulation in playas. Chloride, sulfate, and nitrate, the dominant components of the nitrate deposits, are ultimately derived from the Pacific, although significant amounts may have been supplied during times of intense Andean volcanism (Ericksen, 1981).

Pod of nitrate ore (bright white) exposed in R24 roadcut just west of the Río Loa bridge. View southeast.

Either early, soft-sediment deformation, or recent deformation caused by variable dissolution of nitrates. R24 roadcut west of Río Loa bridge.

Recent work suggests that the nitrates may have been deposited by evaporation of ancient nitrate-charged groundwater.

To avoid Side Trip 5:

> **Río Loa Nitrates to Quebrada Barriles**: *Continue driving west on R-24 to* **Stop 26, Quebrada Barriles** *(–22.118333, –70.088333) for a total of 60 km (37.3 mi; 40 min) and pull over on the right.*

To go on Side Trip 5:

> **Side Trip 5, Río Loa Nitrates to Balneario Coya Sur**: *Continue west on R-24 to the intersection with R-5; turn left (south) on R-5 and drive to the sign for Coya Sur on the left; turn left (east) onto Ruta S/R Balneario Coya and drive to* **Side Trip 5, Balneario Coya Sur** *(–22.397252, –69.529002) for a total of 22.2 km (13.8 mi; 18 min) and pull into the parking area on the west side of the river.*

Side Trip 5 Balneario Coya Sur

On our way to Balneario Coya Sur we pass the dusty town of María Elena. West of the Pan-American Highway (R-5), Maria Elena was founded around 1920 in the heart of the Chilean Nitrate Belt. It was named after Mary Ellen Comdon, wife of the first manager of the saltpeter refinery (Oficina Salitre). The town's population was 7,530 in 2012 (Wikipedia, María Elena).

The town was built for the workers of a nitrate plant that opened in 1926. The plant used an extraction process patented by the Guggenheim brothers. Together with the former refinery at Pedro de Valdivia just south of town, by 1931 the town was the largest saltpeter (potassium nitrate) producer in the world with a combined output of over 907,000 tonnes (one million tons) per year. The product is used in gunpowder (black powder), explosives, and fertilizers.

In 1965 both works were nationalized through the Sociedad Química y Minera (SQM or "Chemicals and Mining Company"). SQM was subsequently privatized in 1980. This privatization explains why most land and buildings are still owned by SQM. In 1996, it became the only active nitrate mining town in the country after the closure of the saltpeter works at Pedro de Valdivia.

Continued mining in the area has turned the countryside into a series of massive furrows as if churned by a giant rototiller.

This region is occasionally rocked by large earthquakes. On November 14, 2007, the area was devastated by a magnitude 7.7 temblor. Almost all of the town's old buildings collapsed or were made uninhabitable (Wikipedia, María Elena).

Balneario Coya Sur

The Balneario is a spa on the Río Loa near the confluence with the Río San Salvador. Adjacent wetlands attract waterfowl and other wildlife. Bluffs of the Miocene-Pleistocene Quillagua Formation rim the river valley. Deposited in river and lake environments, the Quillagua consists of mostly horizontal, cream-colored marl, white diatomite, and limestone, with brown shale and sandy, gray gravel interbeds.

This stop is special if only because it reveals a slender thread of water in one of the driest places on earth.

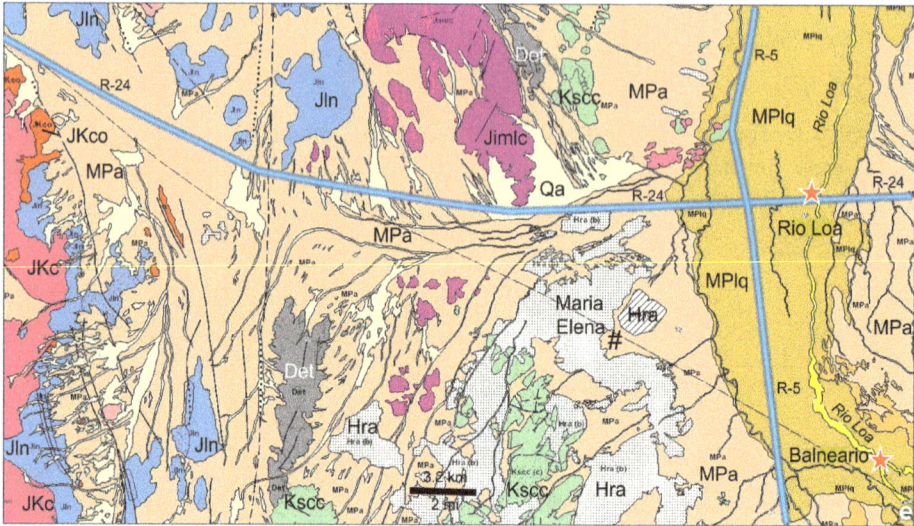

Geologic map of the María Elena-Balneario Coya Sur area. Hra = Holocene artificial fill and tailings; Qa = Quaternary alluvium; MPlq = Miocene-Pleistocene Quillagua Formation marls; MPa = Miocene-Pliocene alluvium and gravels; Kscc = Upper Cretaceous Cerro Cortina Formation lavas and tuffs; JKco = Jurassic-Cretaceous Colupo diorite and gabbro; JKc = Jurassic-Cretaceous Colupito monzonite and monzodiorite; Jln = Jurassic La Negra volcanics; Jimlc = Lower-Mid-Jurassic La Cruz Monzodiorite; Det = Devonian El Toco Formation meta-graywacke turbidites. Modified after Sernageomin.cl, online geologic map of Chile.

Río Loa at Balneario Coya Sur.

Continuing west we leave the Central Depression and enter the Coastal Cordillera.

*Side Trip 5, Balneario Coya Sur to Quebrada Barriles: Return west to R-5; turn right (north) and drive to R-24; turn left (west) on R-24 and drive to **Stop 26, Quebrada Barriles** (–22.118333, –70.088333) for a total of 75.4 km (46.9 mi; 50 min) and pull over on the right.*

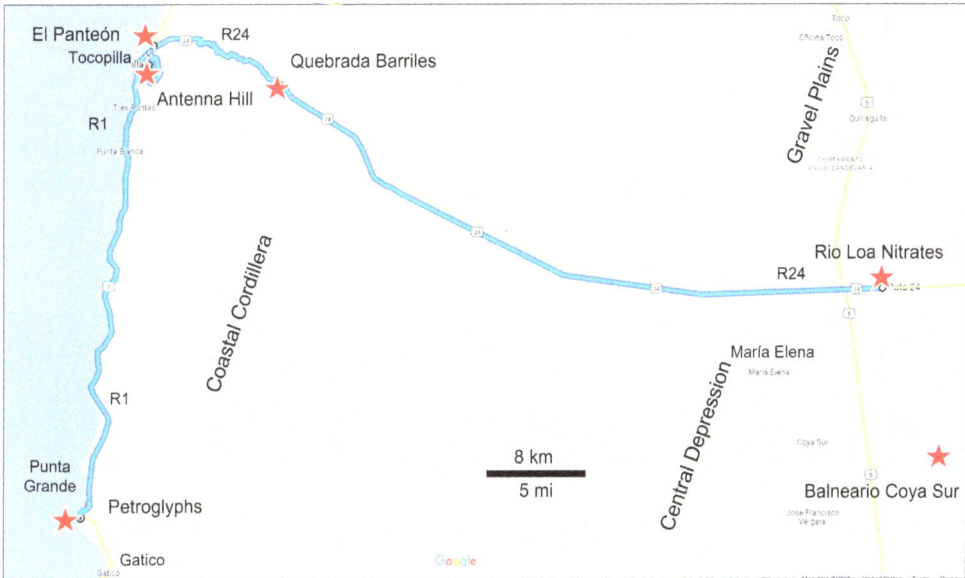

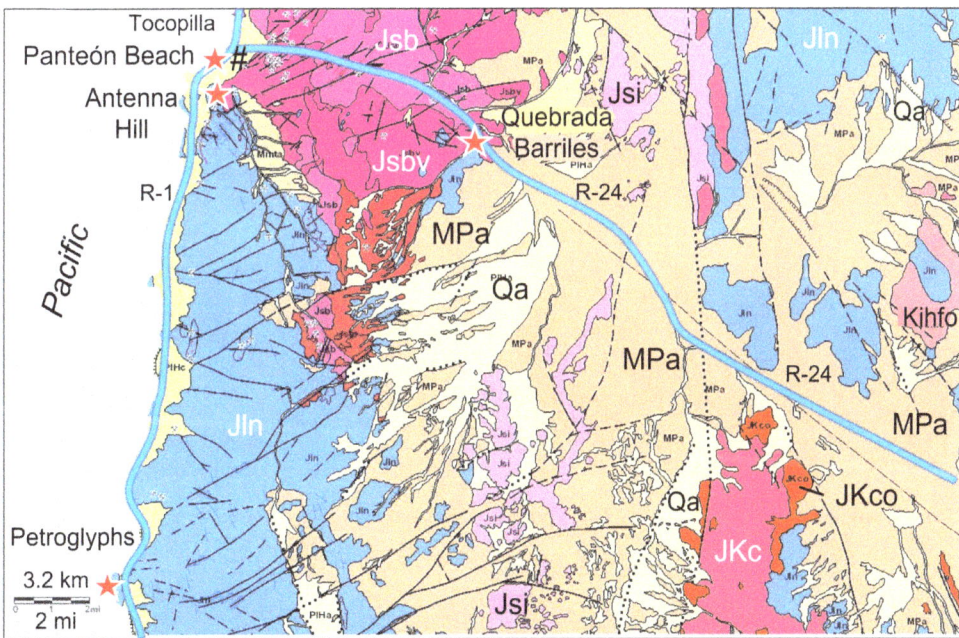

Geologic map of the Tocopilla area. Qa = Quaternary alluvium; MPa = Miocene-Pliocene alluvium and gravels; Kihfo = Lower Cretaceous Farellón-Ojeda diorites; JKc = Jurassic-Cretaceous Colupito Monzonite and Monzodiorite; JKco = Jurassic-Cretaceous Colupo Diorite and Gabbro; Jsb = Upper Jurassic Barriles Diorite and Monzodiorite; Jsbv = Upper Jurassic Buena Vista Monzodiorite; Jsi = Upper Jurassic Irene Diorite and Monzodiorite; Jln = Jurassic La Negra volcanics. Modified after Sernageomin.cl, online geologic map of Chile.

Stop 26 Quebrada Barriles

Ruta-24 follows a steep canyon through the Coastal Cordillera as it drops from the elevated gravel plains of the Central Depression, roughly 1,200 m (4,000 ft) to the sea at Tocopilla.

The Coastal Cordillera is dominated by Middle to Upper Jurassic basaltic to andesitic volcanics of the La Negra Formation. The La Negra represents the Jurassic volcanic arc. The easterly dip of the La Negra Formation is thought to be the result of block rotation on normal faults during extension of the Coastal Cordillera (Moreno and Gibbons, 2007). In addition, the coast ranges contain the Upper Jurassic Barriles Diorite and Buena Vista Monzodiorite of the Coastal Pluton. The highway crosses both the Barriles and Buena Vista intrusives.

Fractured Upper Jurassic Barriles Diorite, Quebrada Barriles. The fractures appear to impart a bedding, but this is an illusion. View southwest.

Quebrada Barriles to El Panteón, Tocoplla: *Continue west on R-24 to Av 18 de Septiembre; at the sign for Iquique on the right, bear left (southwest); continue on Av 18 de Septiembre; at the roundabout continue straight onto R-1; turn right (west) onto Cienfuegos; turn left (south) onto Barros Arana and park on the right. This is* **Stop 27, El Panteón** *(−22.088459, −70.197824) for a total of 15.3 km (9.5 mi; 18 min).*

Stop 27 El Panteón, Tocopilla

This is a scenic stop with views of the local geology thrown in. Tocopilla is a city of 24,247 people (2012) and is the capital of Tocopilla Province. Tocopilla means "the devil's corner" in the aboriginal language (Wikipedia, Tocopilla). That is probably an appropriate description. Stark is a word that comes to mind. Not usually a tourist destination.

This is a gritty industrial port on an alluvial fan wedged between the Coastal Cordillera and the sea. It is divided into the main city in the north and La Villa Sur south of the power plant, factories, and port. A number of beaches, among them the large beach called "Covadonga," a small beach called "Caleta Boy," and a sandy beach called "El Panteón" allow an escape from the dryness of the desert.

Most people living here work outside of town. The city generates electricity for the entire region and is therefore known as "the city of energy." When nitrate production was booming, Tocopilla was a significant export center. Tocopilla is still the home office of the nitrate companies.

Fishing, along with mining, is a main industry. There are fish-meal and canned-fish plants. But the town is primarily a port city. Tocopilla exports copper from Chuquicamata and nitrates from the María Elena District. The town also has metallurgical, chemical, and nitrate processing plants (Wikipedia, Tocopilla).

The beach at El Panteón is a local favorite. The name derives from the shape of the breakwater that encloses the beach, open in the center of the arc similar to the Pantheon in Rome.

East-dipping outcrops of the Jurassic La Negra volcanics occur prominently on the towering cliffs both north and south of the city. More modest outcrops occur along the beach north of El Panteón.

El Panteón, a protected beach to enjoy the Pacific.

Jurassic La Negra Formation volcanics outcrop on the coast just north of El Panteón.

Chile has been battered by subduction-related earthquakes over the centuries. The Nazca Plate is plunging beneath the South American Plate at a rate of 6–7 cm/yr (2.4–2.8 in/yr), setting off truly monstrous quakes. The 1960 Valdivia earthquake was the most powerful earthquake ever recorded, at 9.4–9.6 on the moment magnitude scale. The resulting tsunamis pounded the Chilean coast with waves up to 25 m (82 ft), devastated Hilo, Hawaii, and sent waves as high as 10.7 m (35 ft) as far as Japan and the Philippines. Death toll estimates from the earthquake and tsunamis ranged from 1,000 and 7,000 killed (Wikipedia, 1960 Valdivia Earthquake). The slightly smaller 1995 Antofagasta earthquake had a moment magnitude of 8.0. The Antofagasta Region in Chile was affected by a moderate tsunami, with three people killed, 58 or 59 injured, and around 600 made homeless (Wikipedia, 1995 Antofagasta Earthquake).

The most recent "big one" struck on November 14, 2007, between Tocopilla and Antofagasta. The Mw 7.8 quake caused extensive damage to Tocopilla. Maximum vertical displacement was about 40 cm (16 in) on the Mejillones Peninsula. Maximum lateral slip was about 2.5 m at a depth of between 30–55 km (Motagh et al., 2010; Schurr et al., 2012).

Watch for signs throughout town that indicate whether you are in or above the tsunami zone. Have a plan, just in case.

El Panteón, Tocopilla to Antenna Hill, Tocopilla: Return to R-1 and turn right (south); drive to Bolivar and turn left (southeast); drive to Matta and turn right (southwest); drive to B-160 and turn left (south); drive to **Stop 28, Antenna Hill, Tocopilla** *(–22.097450, –70.195844) and pull over where it is safe, for a total of 2.7 km (1.7 mi; 7 min). You may have to walk a few meters for the view and outcrops.*

Stop 28 Antenna Hill, Tocopilla

For expansive views of Tocopilla and the Pacific, and roadcuts in the La Negra Formation, drive up Antenna Hill (informal name).

East-dipping La Negra Formation exposed on Antenna Hill above Tocopilla. View south.

Tocopilla harbor at sunset. Waves are breaking on outcrops of the La Negra Formation.

*Antenna Hill, Tocopilla to Petroglyphs at Punta Grande: Return to R-1 and turn left (south); drive to **Stop 29, Petroglyphs at Punta Grande** (−22.457016, −70.261993) and pull out on the right just south of the curve. You can drive most of the way to the site on a sandy track. Four-wheel drive is not required.*

Stop 29 Petroglyphs at Punta Grande

To drive along the coast is to drive through a twice-baked landscape. First, it is almost entirely La Negra volcanics. Second, it is brutal, bleak, harsh black rock cliffs that loom above the deep blue sea. Most of the way the highway is on a thin ribbon of alluvium that snakes its way along the coast.

This changes at Punta Grande, a low-lying promontory that juts into the ocean. The rocks at Punta Grande are the dark, coarsely crystalline Barriles Diorite of the Coastal Batholith. The diorite is overlain by Pleistocene to Holocene marine gravels, mainly of andesitic composition, derived from the nearby uplands.

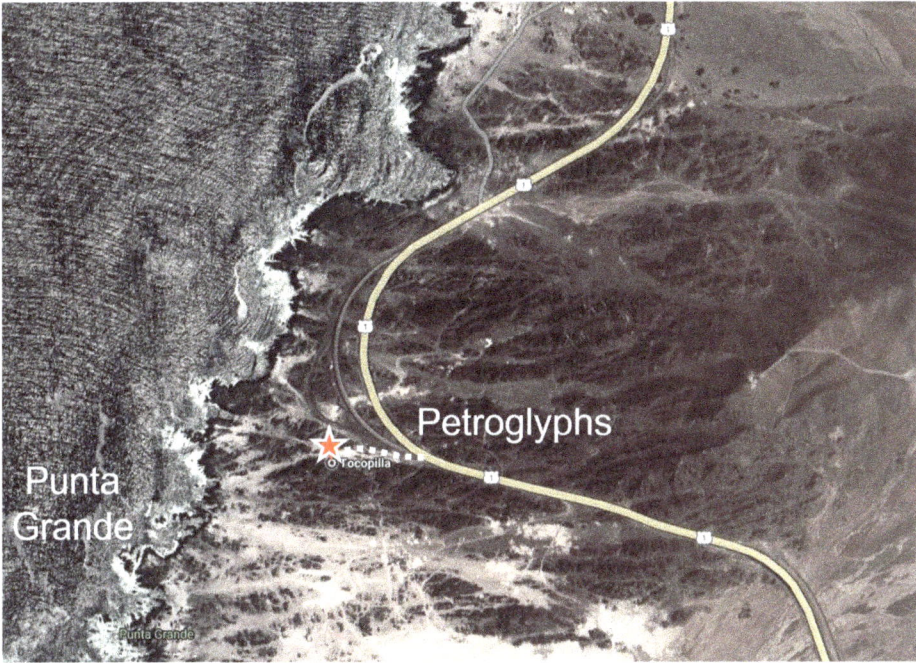

Google Maps satellite image showing the location of the petroglyphs at Punta Grande. Image copyright 2023 Maxar Technologies; CNES/Airbus.

Petroglyphs at Punta Grande.

But this is not just a geologic stop. The petroglyphs at this stop are on a large stone block adjacent to dozens of open graves and piles of sediment resulting from the looting of tombs. The drawings are of humans, llamas or vicuñas, and possibly fish, scratched with some kind of implement into the weathered surface of the rock. The different styles suggest different artists contributed to the panel over a span of time. As well, the subject matter suggests contributions by artists both from the interior and coast regions (Ballester, 2018).

Continuing south on the way to Playa Norte Hornitos, you pass by the Carolina de Michilla copper district in the Coastal Cordillera. Information on the mines is provided, although the road to the mines is closed to the general public.

Carolina de Michilla Stratiform Copper District

The interplay between the converging oceanic lithosphere and the South American Plate since Upper Triassic time along the western margin of South America led to a thick, Early to Late Jurassic volcano-sedimentary sequence that was intruded by largely dioritic Jurassic plutons and dikes. The Jurassic volcanics of the La Negra Formation in this area are mostly 35–40° northwest-dipping andesitic to basaltic andesite lavas with minor tuff breccias, sandstone, and limestone. The formation has a total thickness of at least 7.4 km (24,000 ft) but is on the order of 1 km (3,200 ft) thick in the district. Absolute ages of 186–172 Ma were obtained for the La Negra Formation in the Carolina de Michilla district. Intrusive rocks range from gabbro to granodiorite, with absolute ages between 168 and 112 Ma (Wolf et al., 1990; Kojima et al., 2003; Tristà-Aguilera et al., 2005).

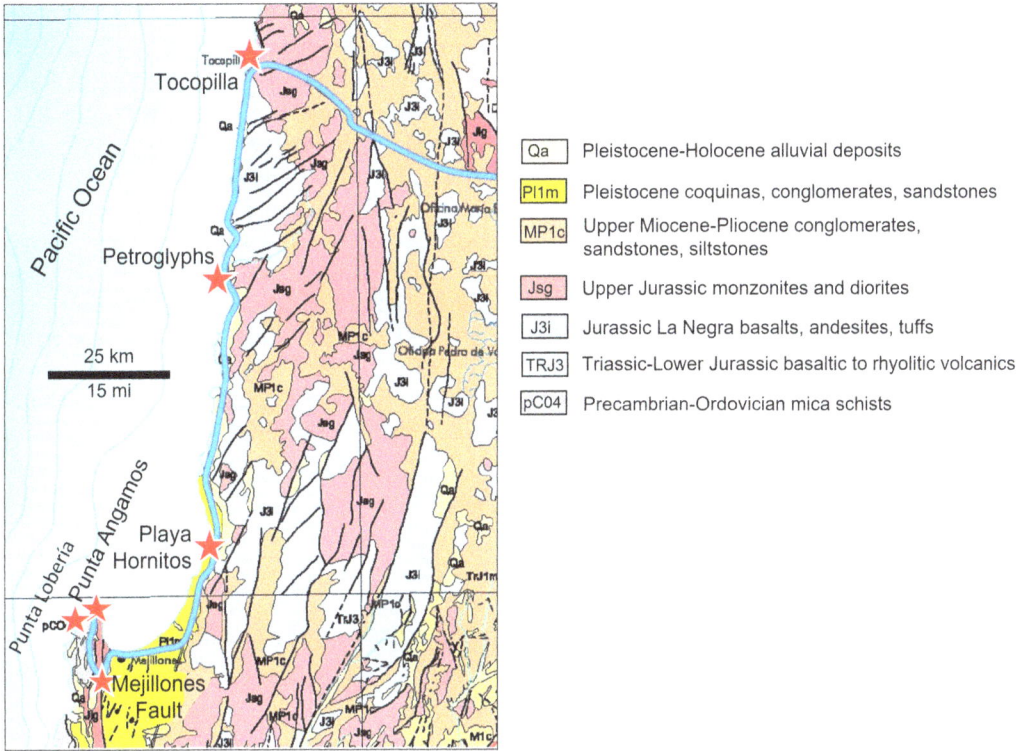

Generalized geologic map of the Coastal Cordillera from Tocopilla to Antofagasta. Modified after Sernageomin, 2003.

The Carolina de Michilla district is located about 120 km north of Antofagasta and is the largest mining area in the coastal Cordillera, producing more than 147,000 tonnes (162,000 tons) of copper ore per month with an average grade of 1.2% Cu (Kojima et al., 2003). This stratabound copper district is, after the porphyry copper deposits, the second most important source of copper in Chile. The first deposit was discovered in 1863, but industrial-scale production didn't begin till 1971 (The Diggings, Michilla Mine; The Diggings, Lince-Michilla Mine). The deposits are found along the trace of the Atacama Fault System as stratabound copper hosted in the La Negra Formation. Surface oxide ore is found over a transition zone, which is in turn over primary sulfide ore distributed mainly in breccias (Townley, 2007).

Primary copper mineralization is characterized by chalcocite + digenite + bornite ± chalcopyrite. Supergene (shallow) alteration produced chalcocite, covellite, atacamite, chrysocolla, and copper oxides. The source of the metals is still debated. Fluid-inclusions and carbon and oxygen isotopic compositions of calcite imply that the stratiform copper deposits are epigenetic (mineral deposit formed later than the enclosing rocks) and formed from non-magmatic, moderately oxidized hydrothermal solutions (Kojima et al., 2003). On the other hand, a magmatic origin is proposed for sulfur of the principal copper sulfides. The age of the main copper-sulfide mineralization event is 159 ± 16 Ma (Tristà-Aguilera et al., 2005).

Mines in the district include the underground Michilla Mine, the open pit at Lince-Michilla (the downfaulted part of the Susana deposit), the Susana/Rojo-Lince/Susana-Estefanía Mine, and the Buena Vista mine.

The Susana copper deposit is related to a mineralized hydrothermal breccia pipe with an associated set of lense-like to tabular orebodies ("mantos"). Mineralization is hosted by the La Negra basaltic to andesitic flows (Wolf et al., 1990).

Surface mining began on the Susana copper deposit in 1981. Discovery of a mineralized breccia pipe led to underground mining in 1983. By 1983, the deposit included 8 million metric tons (8.8 tons) of oxide/sulfide ore averaging 2% Cu at 0.5% cutoff, with probable reserves several million tonnes greater (Wolf et al., 1990). The Susana deposit, the largest of the stratiform type, produces more than 60,000 tonnes/month (66,000 tons/month) with an average grade of 0.6–1.0% Cu and 25 g/tonne (0.8 oz/ton) of silver. Brecciation associated with the gabbro-dioritic intrusion is a post-mineralization event. Following these events, the Susana-Lince area was strongly fractured by the dextral and normal movement along the Muelle Fault, which trends 65°NE and dips 55°SE here (Kojima et al., 2003).

The Buena Vista deposit, about 4 km (2.5 mi) southeast of the Susana deposit, is a small stratiform orebody producing 5,000 tonnes/mo (5,500 tons/mo) with an average grade of ~2.5% Cu and 7 g/tonne (0.22 oz/ton) Ag. This deposit is hosted in the andesitic La Negra Formation, the same as the host rocks at Susana. The copper-mineralized zone is composed of irregular-shaped mantos and, locally, hydrothermal breccias at depth (Kojima et al., 2003).

Google Earth oblique view west over the Carolina Michilla stratiform copper deposit.

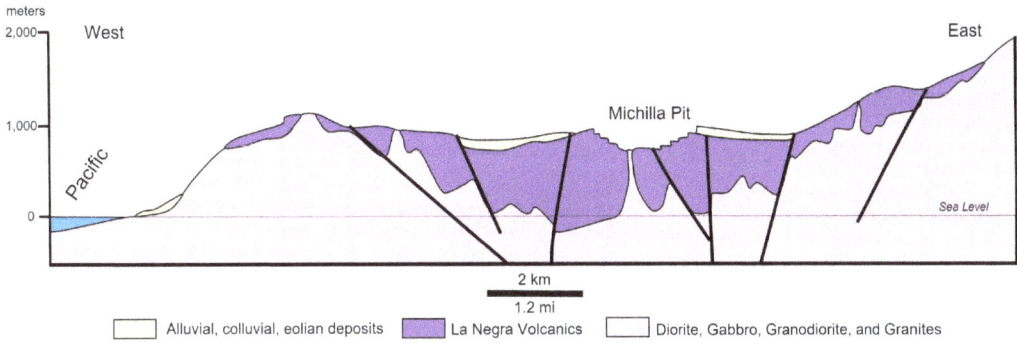

Cross-section of the Coastal Cordillera at Mina Michilla. Modified after Herrera et al., 2018.

Petroglyphs at Punta Grande to Uplifted Beach Terraces, Playa Norte Hornitos: *Drive south on R-1 to the sign for Playa Norte Hornitos 500 m on the right; turn right (west) on B-220 and drive to* **Stop 30.1, Uplifted Beach Terraces, Playa Norte Hornitos** *(−22.865278, −70.293056) for a total of 50.2 km (31.2 mi; 45 mins) and pull over on the right.*

Stop 30 Playa Norte Hornitos

One of the things that convinced Charles Darwin of the reality of earthquake-caused uplift was the raised beach terraces he found after the massive Concepción quake he witnessed in 1835. Playa Norte Hornitos has at least three uplifted beach terraces, although the ages are not well known. We will visit the most recent.

Stop 30.1 Uplifted Beach Terraces, Playa Norte Hornitos

One recent and at least two ancient wave-cut terraces (abandoned sea cliffs) in this area show that the coast is actively uplifting, and that the region is tectonically dynamic. Coastal uplift is related to subduction-driven earthquakes, many in the past 44,000 years (González-Alfaro et al., 2018).

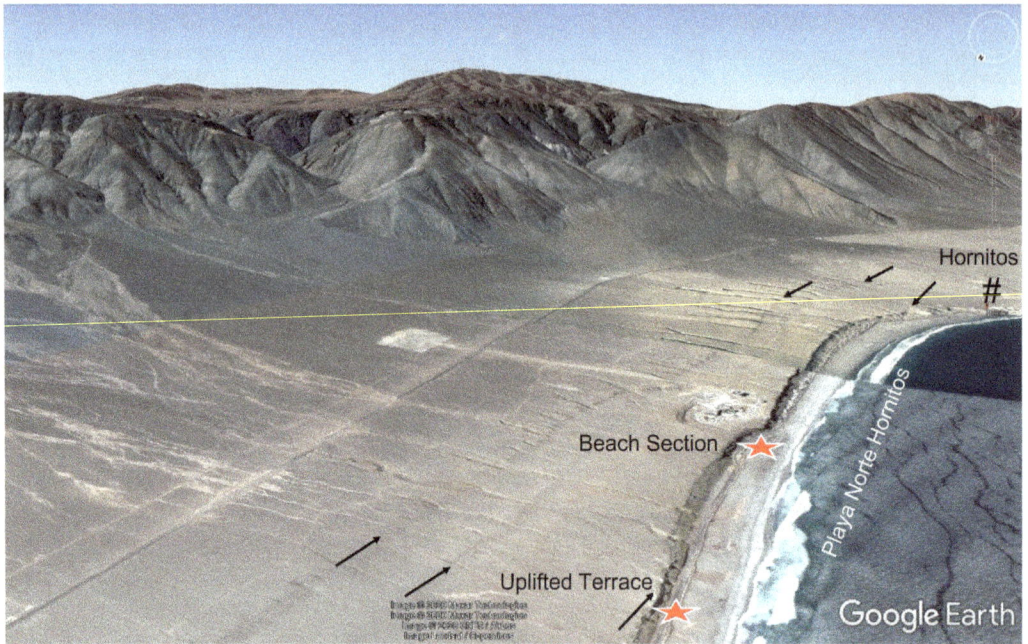

Playa Norte Hornitos, Google Earth oblique view south. The coastal plain is characterized by multiple wave-cut terraces (arrows) that indicate the coast is tectonically active and uplifting. Image copyright 2020 Maxar Technologies; CNES/Airbus; Landsat/Copernicus.

View north at the most recent raised beach terrace, Playa Norte Hornitos.

Uplifted Beach Terraces, Playa Norte Hornitos to Beach Section, Playa Norte Hornitos: *Continue driving south on B-220 for 3.2 km (2 mi; 6 min) to* **Stop 30.2, Beach Section** *(–22.893333, –70.287222) and pull into the parking area on the right.*

Stop 30.2 Beach Section, Playa Norte Hornitos

The unit exposed here is mapped as Pleistocene-Holocene poorly-consolidated marine-shoreline sandstones, conglomerates, and coquinas of the Mejillones Formation. The Mejillones forms a series of uplifted beach terraces, all less than 400,000 years old (Cortés et al., 2007).

The beach section consists of alternating layers of sand and gravel. Playa Norte Hornitos.

To avoid Side Trip 6:

Beach Section, Playa Norte Hornitos to Andrés Sabela Airport: *Continue south on B-220 to R-1; turn right (south) onto R-1 and drive to Accesso Aeropuerto Andrés Sabela; exit to the right and drive* to the endpoint, **Andrés Sabela Airport** *(–23.449412, –70.440415) for a total of 69.1 km (42.9 mi; 55 min).*

To go on Side Trip 6:

Beach Section, Playa Norte Hornitos to Mejillones Fault: *Continue driving south on B-220 to the village of Hornitos; continue on B-220 to the intersection with R-1; turn right (south) on R-1 and drive to B-262; turn right (west) on B-262 and drive to Mejillones; turn left (south) on Av Fertilizantes/B-268; drive to* **ST 6.1, Mejillones Fault** *(–23.130389, –70.498306) and pull over on the left for a total of 43.3 km (26.9 mi; 42 min).*

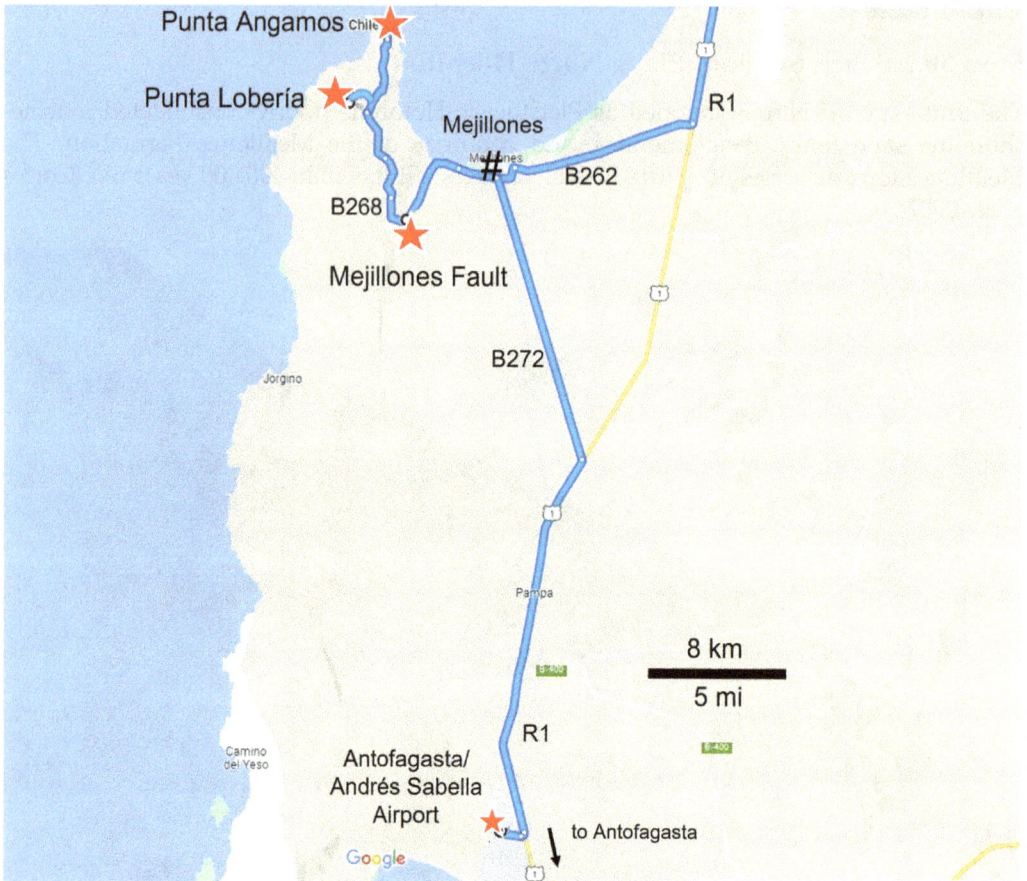

Side Trip 6 Mejillones Peninsula

The Mejillones Peninsula is a north-south uplifted range of hills cored by the Jurassic Mejillones Tonalite and Mejillones Metamorphic Complex. The metamorphic rocks have been dated as Cambrian (Cortés et al., 2007) and Precambrian-Ordovician (González-Alfaro et al., 2013). Regardless of which is correct, these are among the oldest rocks we have seen on this tour. The sea cliffs and headlands at the north end of the peninsula offer spectacular vistas of the Chilean Atacama coast. The peninsula is circumscribed by multi-level wave-cut terraces. It is thought to have been uplifted over the past 3 Ma at a rate of 20 to 50 cm/1,000 years (8 to 20 in/1,000 years; González-Alfaro et al., 2018).

Northern Mejillones Peninsula as seen looking west from Mejillones. Note at least three levels of uplifted wave-cut terraces.

Stop ST6.1 Mejillones Fault

The Mejillones Fault, a north-south-trending, down-to-the-east normal fault, bounds the eastern Mejillones Peninsula at La Rinconada. The peninsula appears to have been uplifted without much, if any, tilting. Jurassic San Luciano Gabbro on the west side of the fault is juxtaposed against Neogene sediments on the east. The scarp is characterized by triangular facets (inverted-V-shaped fault cut), suggesting it has relatively recent movement. We drive up the 500 m (1,640 ft) escarpment on the way to Punta Angamos. The upper surface of the peninsula is covered by a veneer of Neogene sediments (Delouis et al., 1998; Di Celma et al., 2013; González-Alfaro, 2013; González-Alfaro, 2018).

Trace of the Mejillones Fault (dotted line) just west of the village of Mejillones. View north.

Colorful Jurassic gabbro and tonalite on the uplifted Mejillones block. View west across the fault.

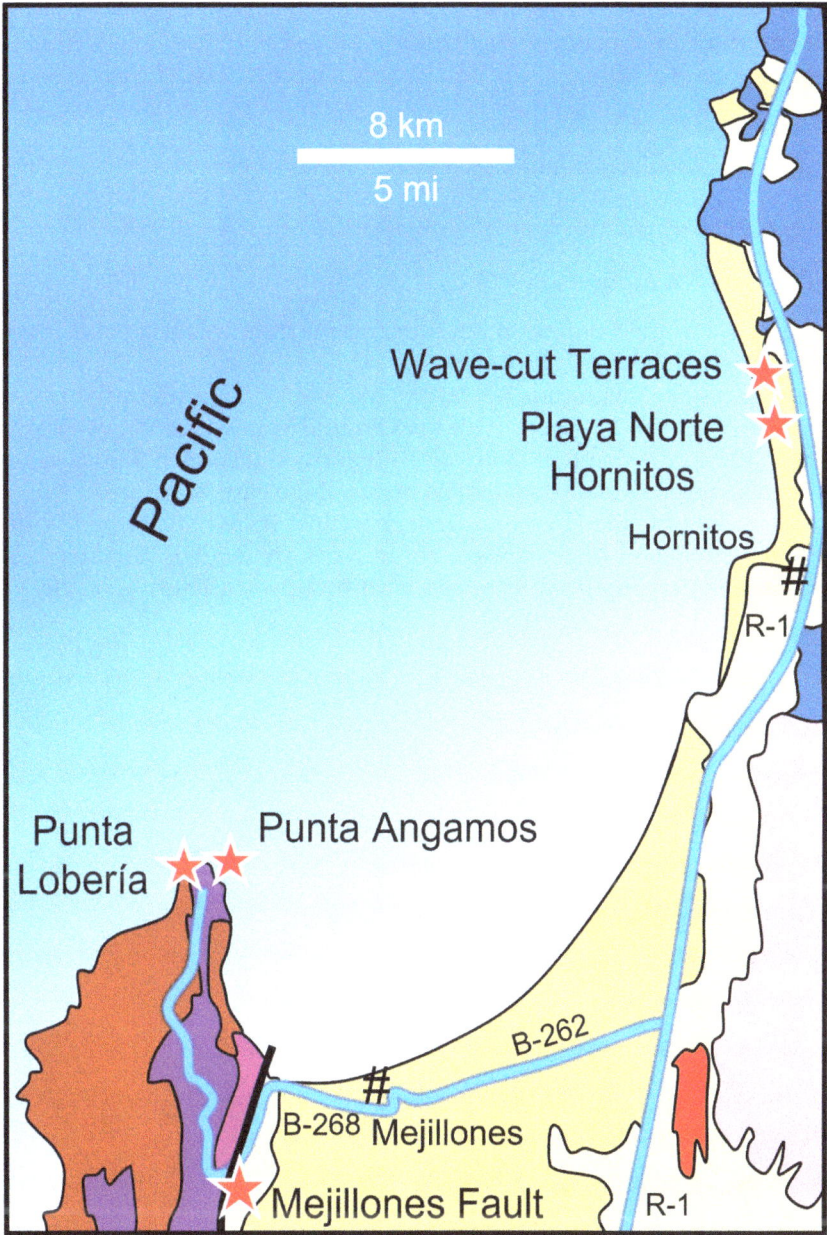

Geologic map of the Mejillones-Hornitos area. Modified after González-Alfaro, 2013.

*Mejillones Fault to Punta Angamos: Continue on B-268 up the escarpment and drive to the parking area at the gun emplacement at the north end of the road. This is **Stop ST 6.2, Punta Angamos** (−23.028511, −70.508828) for a total of 13.7 km (8.5 mi; 16 min).*

Stop ST6.2, Punta Angamos

The dazzling views of the Pacific and Atacama coasts from this headland inspire poets and geologists alike.

After experiencing the magnitude 8.5 earthquake of 1835, Charles Darwin spent weeks investigating the effects of the quake in the area around Valdivia and Concepción. He ultimately came to the conclusion that mountains are formed incrementally, as suggested by Charles Lyell, over eons by many small movements of the earth's surface.

Much of the coast of Chile is an emergent coastline. As you drive across the peninsula, several generations of uplifted wave-cut terraces are evident. Each terrace is a former beach that has been uplifted from sea level by upward faulting of the Mejillones Peninsula.

View north to Punta Angamos and uplifted wave-cut terrace.

Metamorphic core of the Mejillones uplifted block. View northwest toward Punta Lobería.

There is something unmistakably cool about a gun emplacement. View northeast from Punta Angamos.

The Battle of Angamos, during the War of the Pacific, took place off this point on October 8, 1879. Here the Chilean Navy encountered the Peruvian Navy and they slugged it out to determine who would control this region. The battle ended in victory for the Chileans. Led by Commodore Galvarino Riveros Cárdenas and Naval Captain Juan José Latorre, they captured the monitor Huáscar and killed the Peruvian Rear Admiral Miguel Grau Seminario (Wikipedia, Battle of Angamos). This battle led to Chilean dominance of the Pacific and, ultimately, to their victory in the war with Peru.

> ***Punta Angamos to Punta Lobería View***: *Return south on B-268 4.9 km (3.1 mi) to the sign for Punta Lobería; turn right (west) onto the graded dirt track and drive another 1.8 km (1.1 mi) to **Stop ST 6.3, Punta Lobería View** (–23.069167, –70.531390) for a total of 6.7 km (4.2 mi; 10 min) and pull over anywhere.*

Stop ST6.3 Punta Lobería View

In the ravine immediately ahead of you are outcrops of the Cambrian Mejillones Metamorphic Complex. The Complex consists of gneisses, amphibolites, and greenschists. On the point, you can see light-colored Quaternary beach deposits (and guano) as a veneer over the metamorphic rocks. This defines another uplifted beach terrace.

View west to Punta Lobería on the northwestern rim of the Mejillones Peninsula. A close-up of the lighthouse on the point is shown in the next figure.

Close-up of Punta Lobería and lighthouse, with uplifted Quaternary beach terrace deposits over metamorphics.

Punta Lobería to Andrés Sabela Airport: *Return to the road and take B-268 south and east to Mejillones; in Mejillones turn right (south) on B-272 and drive to R-1; turn right (south) on R-1 and drive to the endpoint, **Andrés Sabela Airport** (–23.449412, –70.440415) for a total of 60.6 km (37.6 mi; 55 min).*

The Mejillones graben (downdropped block), looking southeast across the Pampa Mejillones from the uplifted peninsula block.

As we return to Andrés Sabela airport, we pass through a broad, low plain, the Pampa Mejillones. This is a north-south-elongated downdropped block that separates the Mejillones Peninsula from the Coastal Cordillera proper. The horsts and grabens here indicate that this area is currently affected by east-west extension. We can also see that west of us the Mejillones Peninsula is segmented by faults into three uplifted blocks separated by at least one graben. The faults are normal faults trending north to northwest. The northern sector, the Morro Mejillones Block, as we have seen, consists of Jurassic gabbros and tonalites above a metamorphic basement. The central sector consists of the Morro Jorgino Uplifted Block (Early Jurassic Mejillones metamorphic complex) and Caleta Herradura Basin separated by the Jorgino Fault. In the southern sector, the Cerro Moreno Uplifted Block (Early Jurassic Morro Moreno Igneous Complex) is separated from the central block by the Bandurria Fault and from the Pampa Caleta Herradura Basin by the La Rinconada Fault. The Pampa Caleta Herradura Basin is a small, tilted half-graben.

Andrés Sabela airport, located in the barren, broad, and flat valley floor of the Mejillones graben, uses the local geology to its advantage.

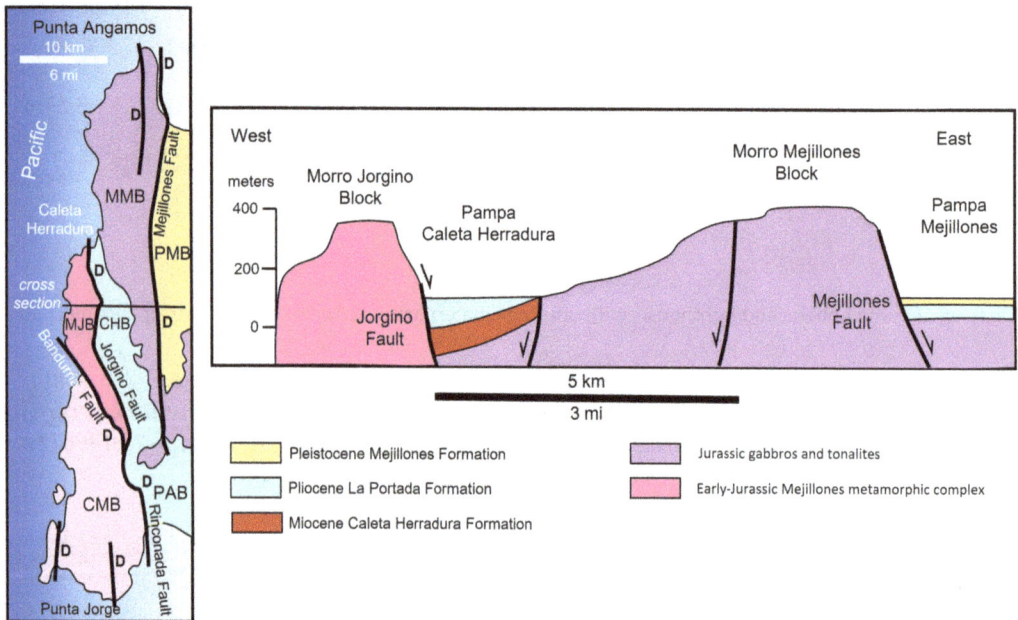

Geologic map and cross section through the Mejillones Peninsula. PMB = Pampa Mejillones Basin; MMB = Morro Mejillones Block; CHB = Caleta Herradura Basin; MJB = Morro Jorgino Block; CMB = Cerro Moreno Block; PAB = Pampa del Aeropuerto Basin. Modified after Di Celma et al., 2013; Mpodozis et al., 2014.

And so, on the verge between the arid desert and the deep blue sea, we end our journey through the Atacama Desert of north-central Chile. We have traversed the stark beauty of one of the driest places on Earth, passed by massive copper, nitrate, and lithium mines, and seen classic stratovolcanoes, the world's highest geyser field, and exquisite emergent coastlines. We have been to the otherworldly Valley of the Moon, to the drier than dry Salar de Atacama, and seen vicuñas and flamingos at springs and lagoons in the foothills of the Andes. We have seen active fumaroles on dizzyingly high Andean peaks, and if our timing was right, steam eruptions from the archetypal volcanic arc at a convergent margin. From here we move south and cross the Andes between the world-class wine regions of Argentina and Chile.

References

Allmendinger, R.W., T.E. Jordan, S.M. Kay, and B.L. Isacks. 1997. The evolution of the Altiplano-Puna Plateau of the central Andes. *Annual Review of Earth and Planetary Sciences* v. 25, p. 139–174.

Ambrus, J. 1977. Geology of the El Abra porphyry copper deposit, Chile. *Economic Geology* v. 72 no. 6, p. 1062–1085.

Amilibia, A., F. Sàbat, K.R. McClay, J.A. Muñoz, E. Roca, and G. Chong. 2008. The role of inherited tectono-sedimentary architecture in the development of the central Andean mountain belt: Insights from the Cordillera de Domeyko. *Journal of Structural Geology*, p. 1–20, DOI: 10.1016/j.jsg.2008.08.005.

Ariagada, C., P.R. Cobbold, and P. Roperch. 2006. Salar de Atacama basin: A record of compressional tectonics in the central Andes since mid-Cretaceous. *Tectonics* v. 25, p. 1–19, DOI: 10.1029/2004TC001770.

Atlas Obscura. El Tatio Geysers. Accessed 15 August 2020, https://www.atlasobscura.com/places/el-tatio-geysers.

Ballester, B. 2018. Revisita a los petroglifos de Gatico, Tocopilla. Preprint, Boletín de la Sociedad Chilena de Arqueología, no. 48, 6 p.

Barra, F., H. Alcota, S. Rivera, V. Valencia, F. Munizaga, and V. Maksaev. 2013. Timing and formation of porphyry Cu-Mo mineralization in the Chuquicamata district, northern Chile: New constraints from the Toki cluster. *Miner Deposita* v. 48, p. 629–651.

Barrera, P. Top copper production by country. Accessed 24 August 2020, https://investingnews.com/daily/resource-investing/base-metals-investing/copper-investing/copper-production-country/.

Boric, R., J. Díaz, H. Becerra, and M. Zentilli. 2009. Geology of the Ministro Hales Mine (MMH), Chuquicamata District, Chile. XII Congreso Geológico Chileno, Santiago, 22-26 Noviembre, 2009. S11-055, 4 p.

Britannica. Atacameño people. Accessed 17 August 2020, https://www.britannica.com/topic/Atacama-people.

Calder, E.S., R.S.J. Sparks, and M.C. Gardeweg. 2000. Erosion, transport and segregation of pumice and lithic clasts in pyroclastic flows inferred from ignimbrite at Lascar Volcano, Chile. *Journal of Volcanology and Geothermal Research* v. 104, p. 201–235.

Codelco. 2016. Annual report, mineral resources and reserves, p. 39–44. Accessed 16 August 2020, https://www.codelco.com/memoria2016/en/pdf/annual-report/annual-report-2016.pdf.

Correa, K.J., O.M. Rabbia, L.B. Hernández, D. Selby, and M. Astengo. 2016. The timing of magmatism and ore formation in the El Abra porphyry copper deposit, northern Chile: Implications for long-lived multiple-event magmatic-hydrothermal porphyry systems. *Economic Geology* v. 111 no. 1, p. 1–28.

Cortés, J., C. Marquardt, G. González, H.G. Wilke, and N. Marinovic. 2007. Cartas Mejillones y Península de Mejillones, Región de Antofagasta, nos. 103 and 104. Subdirección Nacional de Geología, Santiago. 1:100,000.

Cunningham, C.G., E.O. Zappettini, W. Vivallo S., C.M. Celada, J. Quispe, D.A. Singer, J.A. Briskey, D.M. Sutphin, M. Gajardo M., A. Diaz, C. Portigliati, V.I. Berger, R. Carrasco, and K.J. Schulz. 2008. Quantitative mineral resource assessment of copper, molybdenum, silver, and gold in undiscovered porphyry copper deposits in the Andes Mountains of South America. US Geological Survey Open-File Report 2008-1253, version 1.0, Reston, 290 p.

Delouis, B., H. Philip, L. Dorbath, and A. Cisternas. 1998. Recent crustal deformation in the Antofagasta region (northern Chile) and the subduction process. *Geophysical Journal International*, v. 132, p. 302–338.

de Silva, S., A.K. Schmidtt, D. Burns, C. Tierney, S. Grocke, and R. Peckyno. 2009. Virtual Altiplano, volcano world, Oregon State University. Accessed 12 August 2020, http://volcano.oregonstate.edu/book/export/html/43.

de Waele, J., V. Picotti, L. Zini, F. Cucchi, and P. Forti. 2009. Karst phenomena in the Cordillera de la Sal (Atacama, Chile). *In* Rossi, P.L. (Ed.), *Geological Constraints on the Onset and Evolution of an Extreme Environment: The Atacama Area*. GeoActa Special Publication 2, p. 113–127.

Di Celma, C., P.P. Pierantoni, and G. Cantalamessa. 2013. Geological map of the Miocene-Pleistocene successions of the Mejillones Peninsula, northern Chile. *Journal of Maps*, 14 p. Accessed 16 August 2020, https://www.tandfonline.com/doi/full/10.1080/17445647.2013.867419.

Ericksen, G.E. 1981. Geology and origin of the Chilean nitrate deposits. US Geological Survey Professional Paper 1188, 37 p.

Ewiak, O., P. Victor, and O. Oncken. 2015. Investigating multiple fault rupture at the Salar del Carmen segment of the Atacama Fault System (northern Chile): Fault scarp morphology and knickpoint analysis. *Tectonics*, v. 34, p. 187–212.

Galland, O., and C. Sassier. 2018. Geological treasures of the Puna Plateau. *GEO ExPro* v. 15 no. 4, 13 p. Accessed 22 February 2020, https://www.geoexpro.com/articles/2018/09/geological-treasures-of-the-puna-plateau.

Garcés Millas, I., and P. López Julián. 2012. Antecedentes hidroquímicos del Salar de Aguas Calientes I (Chile). Rev. Fac. Ing. Univ. Antioquia No. 62, Medellín – Colombia, p. 91–102.

Garside, M. Major countries in copper mine production worldwide from 2010 to 2019. Accessed 25 August 2020, https://www.statista.com/statistics/264626/copper-production-by-country/#:~:text=The%20ten%20leading%20countries%20in,Mexico%2C%20Russia%2C%20and%20Kazakhstan.&text=Chile%2C%20the%20world's%20leading%20copper,tons%20of%20copper%20in%202019.

Geen, A. 2015. The Atacama Fault System (AFS) in Chile. Accessed 12 October 2019, http://structuralgeologyof.weebly.com/extensional/the-atacama-fault-system-afs-in-chile.

Gerwe, J.E., J.J. Latorre, and L.F. Barrett. 2003. El Abra porphyry copper deposit, northern Chile – Update. Univ. de Concepción, 10th Congreso Geológico Chileno, Concepción, Chile, 1 p.

Giambiagi, L.B., V.A. Ramos, E. Godoy, and P.P. Alvarez. 2003. Cenozoic deformation and tectonic style of the Andes, between 33° and 34° South Latitude. *Tectonics* v. 22 no. 4, p. 1041, DOI: 10.1029/2001TC001354.

Glencore. 2015. Lomas Bayas mine. Accessed 13 August 2020, https://www.glencore.com/dam/jcr:06cf0e12-38d9-4e98-b90d-9698a899efb5/201510120800-Proposed-Sale-of-Cobar-and-or-Lomas-Bayas-Copper-Mines.pdf.

Glennon, J.A., and R.M. Pfaff. 2003. The extraordinary thermal activity of El Tatio Geyser Field, Antofagasta Region, Chile. *The GOSA Transactions* v. VIII, A Special Report, p. 31–77.

GoChile.com. 2020. Pukara de Lasana. Accessed 15 August 2020, https://www.gochile.cl/es/tours/tours-pukara-de-lasana.htm.

González, G., J. Cembrano, F. Aron, E.E. Veloso, and J.B.H. Shyu. 2009. Coeval compressional deformation and volcanism in the central Andes, case studies from northern Chile (23S–24S). *Tectonics* v. 28, p. 1–18, TC6003, DOI: 10.1029/2009TC002538.

González-Alfaro, J.A. 2013. Geología y estructura submarina de la Bahía de Mejillones: su vinculación con la deformación activa en la plataforma emergida a los 23°S. Memoria para optar al título de geólogo, Universidad Católica del Norte, Departamento de Ciencias Geológicas, Antofagasta, 137 p.

González-Alfaro, J., G. Vargas, L. Ortlieb, G. González, S. Ruiz, J.C. Báez, M. Mandeng-Yogo, S. Caquineau, G. Álvarez, F. del Campo, and I. del Río. 2018. Abrupt increase in the coastal uplift and earthquake rate since ~40 ka at the northern Chile seismic gap in the central Andes. *Earth and Planetary Science Letters* v. 502, p. 32–45.

Hartley, A.J., and G. May. 1998. Miocene gypcretes from the Calama Basin, northern Chile. *Sedimentology* v. 45, p. 351–364.

Hartley, A.J., and C.M. Rice. 2005. Controls on supergene enrichment of porphyry copper deposits in the Central Andes: A review and discussion. *Mineralium Deposita* v. 40, p. 515–525.

Hernández Prat, L.T. 2017. Evolucion geoquimica del domo Cerro la Torta, El Tatio, a traves de inclu-sions vitreas. Memoria para optar al titulo e geologa, Universidad de Chile, Departamento de Geologia, Santiago, 171 p.

Herrera, C., C. Gamboa , E. Custodio, T. Jordan, L. Godfrey, J. Jódar, J.A. Luque, J. Vargas, and A. Sáez. 2018. Groundwater origin and recharge in the hyperarid Cordillera de la Costa, Atacama Desert, northern Chile. *Science of The Total Environment* v. 624, p. 114–132.

Ide, F., and I.A. Kunasz. 1989. Origin of lithium in Salar de Atacama, northern Chile. Chapter 11 *in* Ericksen, G.E., M.T. Canas Pinochet, and J.A. Reinemund (eds.), Geology of the Andes and its relation to hydrocarbon and mineral resources: Houston, Texas, Circum-Pacific Council for Energy and Mineral Resources Earth Science Series, v. 11.

Jamasmie, C. 2019. Chilean miner Mantos Copper secures $250m to finance concentrator expansion – MINING.COM. Accessed 17 August 2020, https://www.mining.com/chilean-miner-mantos -copper-secures-250m-to-finance-concentrator-expansion/.

Jordan, T.E., C. Mpodozis, N. Muñoz, N. Blanco, P. Pananont, and M. Gardeweg. 2007. Cenozoic subsurface stratigraphy and structure of the Salar de Atacama Basin, northern Chile. *Journal of South American Earth Sciences* v. 23, p. 122–146.

Jordan, T., C. Herrera Lameli, N. Kirk-Lawlor, and L. Godfrey. 2015. Architecture of the aquifers of the Calama Basin, Loa catchment basin, northern Chile. *Geosphere* v. 11 no. 5, p. 1438–1474, DOI: 10.1130/GES01176.1.

Kay, S.M., B.L. Coira, P.J. Caffe, and C.H. Chen. 2010. Regional chemical diversity, crustal and mantle sources and evolution of central Andean Puna plateau ignimbrites. *Journal of Volcanology and Geothermal Research* v. 198, p. 81–111.

Kojima, S., D. Trista-Aguilera, and K. Hayashi. 1998. Genetic Aspects of the Manto-type Copper Deposits Based on Geochemical Studies of North Chilean Deposits. *Resource Geology* v. 59 no. 1, p. 87–98.

Kojima, S., J. Astudillo, J. Rojo, D. Trista´, and K. Hayashi. 2003. Ore mineralogy, fluid inclusion, and stable isotopic characteristics of stratiform copper deposits in the coastal Cordillera of northern Chile. *Mineralium Deposita* v. 38, p. 208–216, DOI: 10.1007/s00126-002-0304-5.

Leadbeater, C. 2017. The stark beauty of Chile's Atacama Desert. *The Week*, National Geographic Traveler. Accessed 17 August 2020, https://theweek.com/articles/721455/stark-beauty-chiles -atacama-desert.

Legends of the Atacama. Licancabur. Accessed 13 August 2020, https://www.explore-atacama.com /eng/atacama-guides/legends.htm.

Lehman, J. 2019. What's So Special About the Atacama Desert? Live Science, Accessed 5 October 2023, https://www.livescience.com/64752-atacama-desert.html

Lucchi, F., C.A. Tranne, P.L. Rossi, C. Gallardo, G. De Astis, and G.A. Pini. 2009. Volcanic and tectonic history of the El Tatio area (central Andes, northern Chile): Explanatory notes to the 1:50,000 scale geological map. *GeoActa*, Special Publication 2, p. 1–29.

Mining Data Solutions. 2020. Lomas Bayas Mine. Accessed 13 August 2020, https://miningdataon-line.com/property/140/Lomas-Bayas-Mine.aspx.

Mining-Technology.com. Mantos Blancos. Accessed 12 August 2020, https://www.mining-technol-ogy.com/projects/mantos-blancos/.

Moreno, T., and W. Gibbons. 2007. *The Geology of Chile*. The Geological Society, London, 395 p.

Motagh, M., B. Schurr, J. Anderssohn, B. Cailleau, T.R. Walter, R. Wang, and J.-P. Villotte. 2010. Subduction earthquake deformation associated with 14 November 2007, Mw 7.8 Tocopilla earthquake in Chile: Results from InSAR and aftershocks. *Tectonophysics* v. 490, p. 60–68.

Mpodozis, C., C. Arriagada, M. Basso, P. Roperch, P. Cobbold, and M. Reich. 2004. Late Mesozoic to Paleogene stratigraphy of the Salar de Atacama Basin, Antofagasta, Northern Chile: Implications for the tectonic evolution of the central Andes. *Tectonophysics* v. 399, 30 p. DOI: 10.1016/j.tecto.2004.12.019.

Mpodozis, C., and P. Cornejo. 2012. Chapter 14: Cenozoic tectonics and porphyry copper systems of the Chilean Andes. *Society of Economic Geologists*, Special Publication 16, pp. 329–360.

Mpodozis, C., P. Cornejo, and R. Mora. 2014. Geología y Metalogénesis de la Cordillera de la Costa entre Tocopilla y Antofagasta: 1) Geología Regional y Evolución Tectónica. Antofagasta Minerals, Santiago. 36 p.

Muñoz-Pedreros, A., P. de los Ríos, and P. Möller. 2013. Zooplankton in Laguna Lejía, a high-altitude Andean shallow lake of the Puna in northern Chile. *Crustaceana* v. 86 no. 13–14, p. 1634–1643.

NASA Earth Observatory. Licancabur. Accessed 13 August 2020, https://earthobservatory.nasa.gov /images/3999/volcano-licancabur.

Niemeyer, H., and C. Urrutia. 2009. Transcurrencia a lo largo de la Falla Sierra de Varas (Sistema de fallas de la Cordillera de Domeyko), norte de Chile. *Andean Geology* v. 36 no. 1, p. 37–49.

Oregon State University. 2020., Licancabur. Accessed 11 April 2020, http://volcano.oregonstate.edu /oldroot/CVZ/licancabur/index.html

Ossandón C., G., R. Fréraut C., L.B. Gustafson, D.D. Lindsay, and M. Zentilli. 2001. Geology of the Chuquicamata mine: A progress report. *Economic Geology* v. 96, p. 249–270.

Oyarzun, R., A. Márquez, J. Lillo, I. López, and S. Rivera. 2001. Giant versus small porphyry copper deposits of Cenozoic age in northern Chile: Adakitic versus normal calc-alkaline magmatism. *Mineralium Deposita* v. 36, p. 794–798.

Pananont, P., C. Mpodozis, N. Blanco, T.E. Jordan, and L.D. Brown. 2004. Cenozoic evolution of the northwestern Salar de Atacama Basin, northern Chile. *Tectonics* v. 23, TC6007, 28 p, DOI: 10.10289/2003TC001595.

Perello, J., and R.H. Sillitoe. 2004. Metallogenic aspects of giant porphyry systems of the Andes. Society of Economic Geologists, Predictive Mineral Discovery Under Cover, p. 45–50.

Ramirez, L.E., C. Palacios, B. Townley, M.A. Parada, A.N. Sial, J.L. Fernandez-Turiel, D. Gimeno, M. Garcia-Valles, and B. Lehmann. 2006. The Mantos Blancos copper deposit: An upper Jurassic breccia-style hydrothermal system in the Coastal Range of Northern Chile. *Mineralium Deposita* v. 41, p. 246–258.

Reutter, K.-J., R. Charrier, H.-J. Götze, B. Schurr, P. Wigger, E. Scheuber, P. Giese, C.-D. Reuther, S. Schmidt, A. Rietbrock, G. Chong, and A. Belmonte-Pool. 2005. Chapter 14: The Salar de Atacama Basin: A subsiding block within the western edge of the Altiplano-Puna Plateau. *In* Oncken, O., G. Chong, G. Franz, P. Giese, H.-J. Götze, V.A. Ramos, M.R. Strecker, and P. Wigger (Eds.), Springer Science & Business Media, 2006-Science, 569 p.

Rubilar Contreras, J.F.S. 2015. Arquitectura interna y desarrollo Oligoceno-Neógeno de la Cuenca del Salar de Atacama, Andes centrales del norte de Chile. Memoria para optar al Titulo de Geologo. Departamento de Geología, Universidad de Chile, Santiago, 66 p.

Samco Technologies. 2018. What is Lithium extraction? Accessed 13 August 2020, https://www.sam-cotech.com/what-is-lithium-extraction-and-how-does-it-work/.

Schurr, B., G. Asch, M. Rosenau, R. Wang, O. Oncken, S. Barrientos, P. Salazar, and J.-P. Vilotte. 2012. The 2007 M7.7 Tocopilla northern Chile earthquake sequence: Implications for along-strike and downdip rupture segmentation and megathrust frictional behavior. *Journal of Geophysical Research* v. 117, 19 p., B05305, DOI: 10.1029/2011JB009030.

Sernageomin.cl. Online geologic map of Chile, 1:1,000,000. https://portalgeomin.sernageomin.cl/.

Seymour, N.M., J.S. Singleton, R. Gomila, S.P. Mavor, G. Heuser, G. Arancibia, S. Williams & D.F. Stockli. 2021. *Journal of the Geological Society preprint*, 68 p.

Somoza, R., A.J. Tomlinson, P.J. Caffe, and J.F. Vilas. 2012. Paleomagnetic evidence of earliest Paleocene deformation in Calama (~22°S), northern Chile: Andean-type or ridge-collision tectonics? *Journal of South American Earth Sciences* v. 37, p. 208–213.

Somoza, R., A.J. Tomlinson, C.B. Zaffarana, S.E. Singer, C.G. Puigdomenech Negre, M.I.B. Raposo c, and J.H. Dilles. 2015. Tectonic rotations and internal structure of Eocene plutons in Chuquicamata, northern Chile. *Tectonophysics* v. 654, p. 113–130.

Stoertz, G.E., and G.E. Ericksen. 1974. Geology of Salars in northern Chile. US Geological Survey Professional Paper 811, 65 p.

The Diggings. El Abra Mine. Accessed 15 August 2020, https://thediggings.com/mines/25174.

The Diggings. Lomas Bayas Mine. Accessed 13 August 2020, https://thediggings.com/mines/27154.

The Diggings. Linz-Michilla Mine. Accessed 15 August 2020, https://thediggings.com/mines/24721.

The Diggings. Michila Mine. Accessed 15 August 2020, https://thediggings.com/mines/19031.

Tomlinson, A.J., N. Blanco, J.H. Dilles, V. Maksaev, and M. Ladino. 2018. Carta Calama, región de Antofagasta. Servicio Nacional de Geología y Minería, Carta Geológica de Chile, Serie Geología Básica No. 199, 1:100.000, 1 CD con anexos. Santiago.

Townley, B., P. Roperch, V. Oliveros, A. Tassara, and C. Arriagada. 2007. Hydrothermal alteration and magnetic properties of rocks in the Carolina de Michilla stratabound copper district, northern Chile. *Mineralium Deposita* v. 42, p. 771–789.

Tristà-Aguilera, D., J. Ruiz, F. Barra, D. Morata, O. Talavera Mendoza, S. Kojima, and F. Ferraris. 2005. Origin and age of Cu-stratabound ore deposits: Michilla district, Northern Chile. 6th International Symposium on Andean Geodynamics (ISAG 2005, Barcelona), Extended Abstracts. p. 742–745.

US Geological Survey. Lithium Mining in Salar de Atacama. Accessed 13 August 2020, https://www.usgs.gov/media/before-after/lithium-mining-salar-de-atacama-chile.

US Geological Survey. El Abra Porphyry copper deposit. Accessed 15 August 2020, https://mrdata.usgs.gov/sir20105090z/show-sir20105090z.php?gmrap_id=284.

Volcano Discovery. 2020Licancabur. Accessed 13 August 2020, https://www.volcanodiscovery.com/licancabur.html.

Wikipedia. 1960 Valdivia earthquake. Accessed 15 August 2020, https://en.wikipedia.org/wiki/1960_Valdivia_earthquake.

Wikipedia. 1995 Antofagasta earthquake. Accessed 28 August 2020, https://en.wikipedia.org/wiki/1995_Antofagasta_earthquake.

Wikipedia. Antofagasta. Accessed 28 August 2020, https://en.wikipedia.org/wiki/Antofagasta.

Wikipedia. Atacama desert. Accessed 28 August 2020, https://en.wikipedia.org/wiki/Atacama_Desert.

Wikipedia. Battle of Angamos. Accessed 15 August 2020, https://en.wikipedia.org/wiki/Battle_of_Angamos.

Wikipedia. Calama Chile. Accessed 15 August 2020, https://en.wikipedia.org/wiki/Calama,_Chile.

Wikipedia. Chuquicamata. Accessed 15 August 2020, https://en.wikipedia.org/wiki/Chuquicamata.

Wikipedia. Codelco. Accessed 15 August 2020, https://en.wikipedia.org/wiki/Codelco.

Wikipedia. La Portada. Accessed 15 August 2020, https://en.wikipedia.org/wiki/La_Portada.

Wikipedia. Laguna en Salar de Aguas Calientes. Accessed 13 August 2020, https://es.wikipedia.org/wiki/Salar_Aguas_Calientes_I.

Wikipedia. Lejía Lake. Accessed 13 August 2020, https://en.wikipedia.org/wiki/Lej%C3%ADa_Lake.

Wikipedia. Licancabur. Accessed 13 August 2020, https://en.wikipedia.org/wiki/Licancabur.

Wikipedia. Maria Elena. Accessed 15 August 2020, https://en.wikipedia.org/wiki/Mar%C3%ADa_Elena,_Chile.

Wikipedia. Pukara de Lasana. Accessed 15 August 2020, https://en.wikipedia.org/wiki/Lasana.

Wikipedia. Pukara de Quitor. Accessed 14 August 2020, https://en.wikipedia.org/wiki/Pukar%C3%A1_de_Quitor.

Wikipedia. Rio Loa. Accessed 15 August 2020, https://en.wikipedia.org/wiki/Loa_River.

Wikipedia. San Pedro Chile Volcano. Accessed 15 August 2020, https://en.wikipedia.org/wiki/San_Pedro_(Chile_volcano).

Wikipedia. Tocopilla. Accessed 15 August 2020, https://en.wikipedia.org/wiki/Tocopilla.

Wolf, F.B., L. Fontboté, and G.C. Amstutz. 1990. The Susana copper (-silver) deposit in Northern Chile - Hydrothermal mineralization associated with a Jurassic volcanic arc. *In* Fontboté, L., G.C. Amstutz, M. Cardozo, E. Cedillo, and J. Frutos (Eds.), *Stratabound Ore Deposits in the Andes*. Springer-Verlag, Berlin and Heidelberg, p. 319–320.

Zúñiga Bilbao, V.A. 2012. Ocurrencia y distribución de molibdenita, esfalerita y galena en el sistema pórfido cuprífero del Yacimiento Mina Ministro Hales. Geology thesis, Departamento de Geología, Universidad de Chile (Santiago), 132 p.

Additional Materials

Ferraris, B., Fernando, F. Di Biase, and Francisco. 1978. Hoja Antofagasta: region de Antofagasta, Servicio Nacional de Geología y Minería, Carta Geológica de Chile, 1:250.000.

Henríquez, S., J. Becerra, and C. Arriagada. 2014. Geología del área San Pedro de Atacama, Región de Antofagasta. Servicio Nacional de Geología y Minería, Carta Geológica de Chile, Serie Geología Básica, No. 171, 1:100.000. Santiago.

Medina, E., H. Niemeyer, H.W. Wilke, J. Cembrano, M. García, R. Riquelme, S. Espinoza, and G. Chong. 2012. Cartas Tocopilla y María Elena, Región de Antofagasta. Servicio Nacional de Geología y Minería, Carta Geológica de Chile, Serie Geología Básica 141 y 142 p., 1:100.000, Santiago.

4

Across the Andes in Darwin's Footsteps: Mendoza, Argentina, to Valparaíso and the Colchagua Wine Region, Chile

Aconcagua, the highest peak in all of the Americas. View from Mirador Aconcagua.

Overview

On this trip, we cross the Andes from east to west following in the footsteps of Darwin's exploration of 1835–1836. We start in the wine-growing region of Mendoza, Argentina, pause in the Casa Blanca and Maipo wine regions, and end in the wine country of the Colchagua Valley. We discuss the geology that makes the wines from here world-famous. But it's not *all* about wine. We encounter units deposited during the uplift and erosion of the Andes; volcanoes and intrusives related to subduction of the Nazca Plate beneath South America, see evidence of the dominant east-directed thrusting and deformation in the Aconcagua Fold-Thrust Belt, and drop into the Central Valley of Chile, the gap between

DOI: 10.1201/9781351168281-4

the Andes and Coast Ranges. In the Coast Ranges we find the Coast Range Batholith, the uplifted and eroded roots of a Jurassic volcanic arc, in Valparaíso and Viña del Mar. We venture into the Andes once again in Maipo Canyon southeast of Santiago, where we come upon deformed sedimentary and volcanic units, gypsum mines, volcanoes, and hot springs. We end our journey in volcanic-sedimentary terrane that, like the Napa Valley in the Northern Hemisphere, has just the right combination of soils, climate, and slopes to produce world-class wines.

Tour map from Mendoza, Argentina, over the Andes to Valparaíso, Chile, on the Pacific coast and wine regions near Santiago.

Itinerary

Start Mendoza International Airport, Argentina

Stop 1 Luján de Cuyo and Volcán Tupungato

Stop 2 Precordillera and Cordillera Frontal

Stop 3 Neogene Synorogenic Sediments

Stop 4 Early-Middle Triassic Continental Units at Potrerillos

Side Trip 1 Fossil Forest and Darwin Plaque

Stop 5 Uspallata Group Volcanics and Pediment

Stop 6 Altered Volcanics near Los Penitentes

Stop 7 Puente del Inca

Stop 8 Mirador Aconcagua, Aconcagua Provincial Park

Stop 9 Las Cuevas Slide

Stop 10 Aduana Chilena and West-Dipping Andesites

Stop 11 Laguna del Inca

Stop 12 Curve 17, Los Caracoles

Stop 13 Salto de Soldado and the Cordillera Principal/Río Aconcagua Valley

Stop 14 San Felipe

Stop 15 Coast Range Granodiorite, Chagres Roadcut

Stop 16 Roca Oceanica, Granite of the Coast Range Batholith

Stop 17 Sheraton Miramar, Valparaíso Granodiorite and Gneiss

Stop 18 Mirador Faro Punta Angeles Granodiorite

Stop 19 Sauce Diorite, Curauma

Stop 20 Casablanca Valley Vineyards

Stop 21 Cerro Santa Lucía, Santiago

Stop 22 Cerro San Cristóbal, Santiago

Stop 23 Maipo Valley and Viña Concha y Toro

Side Trip 2 Maipo Canyon

    ST 2.1 El Toyo Roadcut

    ST 2.2 Paso Angosto El Tinoco

    ST 2.3 Observatorio

    ST2.4 Tierra Amarilla

    ST2.5 Mina Lo Valdés

    ST2.6 Placa Verde

    ST2.7 Termas Valle de Colina

Stop 24 Diorite, Hotel Monticello Roadcut

Stop 25 Centinela Roadcut

Stop 26 Ignimbrite, Río Tinguiririca Roadcut

Stop 27 Santa Cruz and the Colchagua Valley

Stop 27.1 Viña Viu Manent Colchagua Museum

Stop 27.2 The Colchagua Museum

End Santa Cruz, Chile

Darwin's Trek

In 1831, Charles Darwin, then 22 years old, set off on what would be a five-year voyage of discovery. Among the expedition objectives was the mapping of the east and west coasts of South America. With him, he carried a recently published book, Charles Lyell's *Principles of Geology* (1830). Darwin devoured the book and determined to find supporting evidence for many of the new concepts proposed in the text. Concepts like the immensity of geologic time, the slow and gradual uplift of mountains, the existence of a former ice age, and the extinction of species.

In 1834, while his ship was undergoing repairs in Valparaíso, Darwin crossed the Andes twice, going from Santiago to Mendoza via Portillo Pass, and returning to Santiago over Uspallata Pass (the route of the modern highway). On reaching Santiago, he wrote:

> My object in coming here was to see the great beds of shells, which stand some yards above the level of the sea, and are burnt for lime. The proofs of the elevation of this whole line of coast are unequivocal: at the height of a few hundred feet old-looking shells are numerous, and I found some at 1300 feet.

While in Santiago, Darwin stayed with Alexander Caldcleugh. Caldcleugh was a Fellow of the Royal Society and private secretary to the British minister at Rio de Janeiro. He was also the author of what Darwin described as "some bad travels in South America" which contained a simple yet beautiful geologic map of this part of South America. Caldcleugh was probably a great help to Darwin (Chancellor and van Wyhe, Introduction to Darwin's *St. Fe Notebook*).

Darwin admired the fertile land around the Maipo River, famous today for its prolific vineyards. He noted that the terraces that border the valley consisted of stones uplifted from the sea that are now being eroded. This, he thought, supported the concept of the enormity of geological time in that the rocks needed to first be uplifted, then worn away. He wrote that "whole races of animals have passed away from the face of the earth" as these geological forces reshaped the environment (Course Hero, The Voyage of the Beagle).

At one point Darwin writes, "Valley very curious higher up," followed by a diagram and a description of a huge mass of "alluvium enormous angular fragments" separating two valleys near the Valle del Yeso. He was puzzled by this mass which he could not believe had been deposited by rivers: "I hardly dare affirm these hills are alluvium." He described the 250-meter-thick (800 ft) pile of rubble in his journal, but it was not until after the voyage that he realized it was a glacial moraine. He then cited the moraine as evidence for a global ice age, which was not widely accepted until the 1850s (Chancellor and van Wyhe, Introduction to Darwin's *St. Fe Notebook*).

Darwin followed the Maipo River until he arrived "at the foot of the ridge, that separates the waters flowing into the Pacific and Atlantic Oceans." He observed the rocks along the Cordillera and determined that they are mostly volcanic porphyry, sandstone, and conglomerate. Areas of the clay-slate pass into "prodigious beds of gypsum." At Peuquenes Pass he found black calcareous shales containing fossil Gryphaea (oysters), gastropods (snails), Terebratula brachiopods, and "a piece of an ammonite as thick as my arm" that was later dated as Early Cretaceous (about 120 Ma). He was amazed by the fact that marine fossils are now 14,000 feet above sea level. He concluded that the rocks must have subsided several thousand feet before being uplifted.

At the continental divide, he waxed poetic:

> When we reached the crest and looked backwards, a glorious view was presented. The atmosphere resplendently clear; the sky an intense blue; the profound valleys; the wild broken forms: the heaps of ruins, piled up during the lapse of ages; the bright-colored rocks, contrasted with the quiet mountains of snow, all these together produced a scene no one could have imagined. Neither plant nor bird, excepting a few condors wheeling around the higher pinnacles, distracted my attention from the inanimate mass. I felt glad that I was alone: it was like watching a thunderstorm, or hearing in full orchestra a chorus of the Messiah.

Along the way, Darwin observed and collected plants and animals. He found that species on the Atlantic side of the mountains were different from similar species on the Pacific side. He concluded that the Andes formed a barrier to the movement of plants and animals, isolating species and forcing them to adapt to different environments on either side of the mountains. Darwin was coming to believe that species rose out of geographic separation and isolation.

His observations also led him to conclude that the Cordillera was uplifted slowly, not abruptly as some other geologists claimed (Course Hero, The Voyage of the Beagle).

Upon reaching the eastern plains Darwin turned north towards Mendoza. He was not impressed: He describes the "sad drunken raggermuffins" of Luján and mentions that Mendoza was not worth describing, other than the "very fine grapes," which are today the basis of Argentina's wine industry.

Turning west he encountered a fossil forest and described the process whereby minerals replace the original wood. He described the Punte del Inca and adjacent hot springs and reveled in the multihued porphyries, granites, slates, and sandstones at Aconcagua before crossing Uspallata Pass and descending once more to Santiago.

Thus ended the notes in which Darwin recorded the first complete traverse by a geologist across the Andes. His notes were eventually published in expanded form in his *Geological Observations on South America* (1846). His notes also provided the raw material for his paper linking earthquakes, volcanoes, and the vertical movements of the earth's crust.

Darwin's notebooks provide the first consistent observations on the geology of the Andes and are the foundation on which all subsequent work has been built. As he stated, "What a history of changes of level, and of wear and tear, all since the age of the latter Secondary formations of Europe, does the structure of this one great mountain-chain reveal!" (Chancellor and van Wyhe, Introduction to Darwin's *St. Fe Notebook*).

Evolution and Tectonics of the Central Andes

The Main Cordillera of the Andes, or Cordillera Principal, consists primarily of 3,300 m (10,800 ft) of Late Cretaceous-Miocene basalts, basaltic andesites, and volcaniclastics of the Abanico and Farellones formations and their associated intrusions. These rocks are typical of a continental volcanic arc and represent the main Andean Orogeny. These units unconformably overlie Mesozoic marine and continental deposits. Volcanism of this age is related to the subduction of the Nazca/Farallon oceanic plate beneath the South American tectonic plate (Muñoz et al., 2006).

During Early Cretaceous, at the latitude of Aconcagua (32.5° S) in the present-day Cordillera Principal, the continental crust was thin (< 33 km; 20 mi) below a Mesozoic marine basin. In Late Cretaceous times, shortening and compression affected the Coast Range and western Cordillera Principal, but crustal thickness in the eastern Cordillera Principal remained normal at around 35 km (21 mi). Mild horizontal extension took place in the Main Andes during late Eocene-early Miocene at the same time that the Coast Range (Coastal Cordillera) was undergoing uplift. The Miocene-Present Andean contraction started about 21–18 million years ago in the western part of the Cordillera Principal (Giambiagi et al., 2023).

Andean deformation migrated eastward and began in early to mid-Miocene (18–20 Ma) in the Cordillera Principal. It is recognized by a wedge of syntectonic sediment of this age, that is, sediments shed from the active uplift. Early Miocene intrusives suggest uplift and an extensional basin in the western Cordillera Principal is dated as early Eocene to early Miocene (Yáñez et al., 2015). Compression linked to an increase in plate convergence velocity caused a partial tectonic inversion of north-south normal fault systems created during the preceding extensional episode. Associated volcanic activity produced the Miocene Farellones Formation and the emplacement of plutons.

At 18 Ma, the Aconcagua Fold-Thrust Belt started to develop thin-skinned thrusting in the eastern Cordillera Principal. Some thick-skinned basement uplifts developed in the west, with the inversion of pre-existing normal faults of the Abanico Basin. At this time, the volcanic arc migrated from the Farellones Volcanic Arc (23–17 Ma) in the western Cordillera Principal, to the Aconcagua Arc (15-8 Ma) in the eastern Cordillera Principal. The Aconcagua Fold-Thrust Belt was formed by both basement-involved and thin-skinned thrusting. Thrusts were generally older in the west and younger in the east, but there was some out-of-sequence thrusting. Uplift of the Frontal Cordillera (Cordillera Frontal) also took place during this time (~17 Ma). Both the Cordillera Frontal and Precordillera (easternmost uplifts near Mendoza) were elevated between 12 and 9 Ma. Between 6 and 3 Ma, horizontal shortening migrated to the present thrust front in the easternmost Precordillera. Far to the east, the Pampean Ranges formed in a broken foreland where opposite-directed faults, inherited from older tectonic episodes, controlled Quaternary uplifts and basins (Giambiagi et al., 2003; Giambiagi et al., 2023).

A reconstructed cross-section across the Andes near Santiago (Ramos et al., 2004) indicates between 5.5 and 7.7 mm/yr (0.2 to 0.3 in/yr) of west-east shortening in this central part of Chile. The magnitude of crustal shortening and resulting topographic uplift decrease progressively from north to south through Chile.

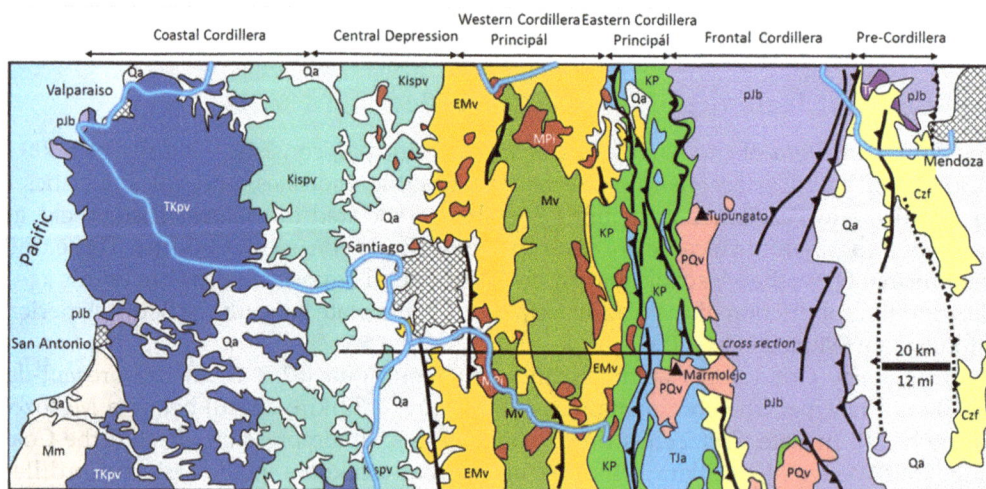

Generalized geologic map of the Andes between Valparaíso, Chile, and Mendoza, Argentina. Qa = Quaternary alluvium; Czf = Upper Cenozoic foreland deposits; PQv = Pliocene-Quaternary volcanic arc deposits; MPi =Miocene-Pliocene intrusives; Mm = Miocene marine deposits; Mv = Miocene volcanic arc rocks; EMv = Eocene-lower Miocene volcanics; KP = Cretaceous-Paleogene sediments; Kispv = Lower-Upper Cretaceous plutonic and volcanic rocks; TKpv = Triassic-Cretaceous plutonic and volcanic rocks; TJa = Triassic-Jurassic back-arc basin rocks; Tr = Triassic continental rift units; pJb = pre-Jurassic basement. Modified after Giambiagi et al., 2017.

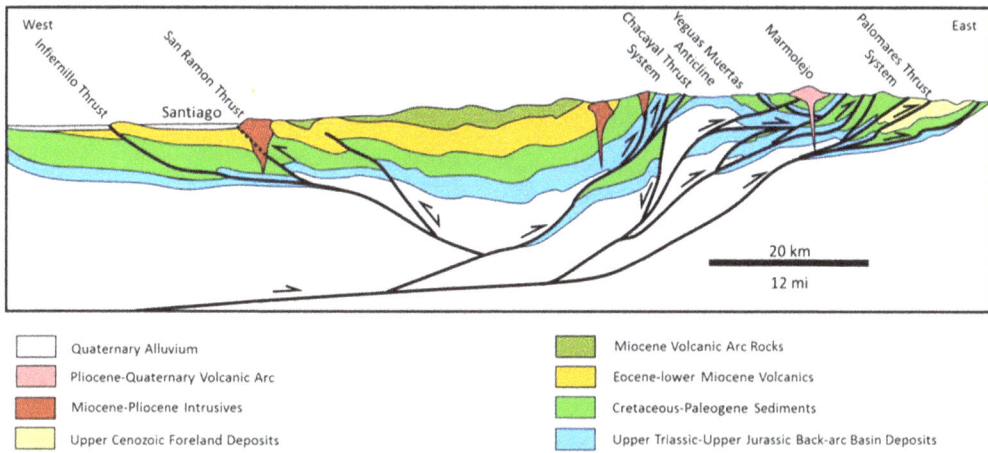

West-east cross-section through the Santiago area. Modified after Giambiagi et al., 2017.

Mendoza, Argentina

Geologically, Mendoza sits on the western margin of the Neuquén Basin and at the eastern foothills of the Andes.

The original inhabitants of this area were native groups known as the Huarpes and Puelches. They invented a system of irrigation for their crops of corn, beans, squash, and quinoa. While not fully part of the Inca Empire, they were influenced by Inca culture and adopted llama ranching and the Quechua language. Those not taken as European slaves were annihilated by plagues and epidemics in the late 1700s.

Mendoza was founded in 1561 by Pedro del Castillo, who named it for the governor of Chile, Don García Hurtado de Mendoza. After about 1600, the area began to prosper because of slave labor and the presence of Jesuit priests. Expanding on the system of irrigation ditches used by the natives, the settlers increased agriculture to the point where they were trading foodstuffs as far as Buenos Aires and Santiago. In 1813 the local governor, José de San Martín, organized the Army of the Andes to liberate Chile and Argentina from Spain. By 1818 he had won several battles and Argentina had won its independence. Chile became independent several years later.

After a massive earthquake in 1861 that killed over 5,000 people, Mendoza was rebuilt with wide streets and large plazas, as well as buildings that could better tolerate quakes.

Today Mendoza is the capital of Mendoza Province. In 2010, the city had 115,000 residents, with a total in the metropolitan area of 1,055,000. The main industries are olive oil, wine, energy (oil, gas, uranium), and tourism.

Start: El Plumerillo International Airport, Mendoza (−32.828000, −68.798845)

***Mendoza International Airport to Luján de Cuyo and Volcán Tupungato**: Exit the airport and turn right (north) onto RN-40; bear right toward Jorge Newberry; turn left at Acceso Norte/Autopista Mendoza/RN-40; turn left (south) onto Alvarez Condarco; turn left (east) onto Constitucíon; merge onto RN-40 heading south; turn sharp left to stay on Av. Ricardo Videla; use the right two lanes to turn right (east) onto Acceso Este; use the right lane to take the ramp to RN-40 to Uspallata/Las Cuevas/Chile; use the right lane to*

take the ramp to Uspallata/Las Cuevas/Santiago and merge onto RN-7 going west; drive to
Stop 1, Luján de Cuyo and Volcán Tupungato *(−33.094226, −68.928148) for a total of*
40.4 km (25.4 mi; 45 min) and pull over on the right.

Stop 1 Luján de Cuyo and Volcán Tupungato

We start our journey across the Andes in the flats and foothills west of Mendoza. The irrigated flats of Luján de Cuyo are known world-wide for their wines. The Cuyo district is the best-known wine region in Argentina, and Mendoza is the wine capital of the Cuyo district. Roughly 300,000 ha (741,000 ac) of grapes are under cultivation at altitudes ranging from 430–2,000 m (1,400–6,500 ft).

The *terroir* of the Mendoza district is described in Chapter 2. Suffice to say that the region's arid climate, well-drained soils, and clear glacial water work together to make great wines. The dry air means few insects and makes grape diseases, fungus, and mold uncommon (Darwin commented on both the lack of insects and the fine grapes grown here). Thus, there is little need for insecticides and fungicides. The region has four distinct seasons, with an average of 320 sunny days per year. Frost is rare in this high desert region, where annual rainfall rarely exceeds 250 mm (10 in) per year, and summer daytime temperatures can be upwards of 40°C (104°F). Irrigation is necessary due to the arid climate.

Soils are well-drained sandy alluvium with some gravel and limestone and a clay substrate. The lack of nutrients tends to stress the vines, concentrating the flavors.

Several grape varieties are grown in the Mendoza Region, but Malbec is the region's leading wine. The grape does well in the intense sun, warm days, and cool nights around Mendoza. After harvesting the grape, wine growers convert the tart malic acid in the freshly crushed juice to lactic acid by means of malolactic fermentation, a process that reduces the acidity and creates a smoother and creamier taste.

Other grapes grown in the district include Bonarda, Cabernet Sauvignon, Chardonnay, Syrah, Tempranillo, and Torrontes. The high-altitude vineyards of Tupungato, in the Uco Valley southwest of Mendoza, are gaining popularity for their Chardonnays, while Cabernet Sauvignon is doing well in the cooler climate of the Maipú region.

In 1993, Luján de Cuyo, the wine-growing district on the southern outskirts of Mendoza, was the first Argentine appellation recorded by the International Organization of Vine and Wine. Many of the vineyards are clustered around the valley of the Río Mendoza at elevations near 1,000 m (3,000 ft).

The vineyards, especially the larger ones, have been favorably compared to those of the Napa Valley in California when it comes to a wine-tasting experience. The architecture of the wineries is impressive, and the selection wineries to visit is large. Indeed, in parts of Mendoza you can walk to several wine tastings in a matter of a few hours. The scenery, however, is incomparable. Magnificent snow-capped peaks sprawl across the western horizon. You can just show up, but many vineyards require reservations. Arranged tours are a popular option because you don't have to worry about a designated driver.

Vineyard in the Luján de Cuyo district south of RN-7. View is southwest to Volcán Tupungato (right) and Cerro Marmolejo (left).

From this stop, you can see the impressive snow-covered peak of Volcán Tupungato. Tupungato is a 6,570 m (21,555 ft) stratovolcano straddling the Chile-Argentina border 110 km (62 mi) southwest of Mendoza. The fact that you can see this far is a stunning tribute to the clear, dry air in this region. The massive, classic cone-shaped volcano consists of andesitic and rhyolitic lavas and pyroclastic material that last erupted during the Pleistocene roughly 800,000 years ago. The name comes from the Quechua language and refers to "land that flourishes." Immediately to the south, the Tupungatito (small Tupungato) volcano last erupted in 1987.

Tupungato Provincial Park and Nature Preserve was created in 1983 to protect 150,000 ha (371,000 ac) of this unique mountain environment.

Luján de Cuyo and Volcán Tupungato to Precordillera and Cordillera Frontal: *Continue west on RN-7 for 12.6 km (7.8 mi; 9 min) to* **Stop 2, Precordillera and Cordillera Frontal** *(−33.078930, −69.060367) and pull over on the right.*

Stop 2 Precordillera and Cordillera Frontal

The Precordillera, a north-south-trending mountain range, is part of the east-directed Andean Fold-Thrust Belt and the youngest Andean uplift in this area. It comprises Silurian-Devonian marine units intruded by Permian and Early Triassic igneous rocks. The western sector is characterized by high-angle backthrusts (west-directed thrusts) and the greatest amount of uplift. A central zone contains both high-angle strike-slip faults and high-angle reverse faults with east and west vergence. The eastern domain is characterized by only east-directed thrusts. The double vergence of the Precordillera in this area has been related to the reactivation of pre-Andean structures .

There is evidence for four deformational events in the Precordillera: (1) an early Paleozoic compressional event, (2) a late Paleozoic compressional event, (3) a Triassic extensional event, and (4) a Cenozoic Andean compressional event. The Cenozoic deformation occurred in three stages: 1) Paleozoic east-dipping faults were reactivated as west-verging reverse faults; 2) reactivation of Paleozoic and Permo-Triassic structures and the generation of reverse faults; 3) generation of east-directed thrusts (Seia et al., 2023).

The Cordillera Frontal lies between the Main Cordillera (Cordillera Principal) to the west and the Precordillera to the east. It consists of pre-Carboniferous basement, mainly Devonian Vallecito Formation quartzites along with other sedimentary, metamorphic, and igneous rocks that have been strongly deformed by east-directed thrusting during the Gondwanan Orogeny. They were later intruded by late Paleozoic granitic rocks. The Upper Carboniferous (Pennsylvanian) El Plata Formation (conglomerate-sandstone-shale) and Permian Río Blanco conglomerates were deposited on the active margin of Gondwana (Heredia et al., 2012).

Permo-Triassic volcanics of the Choiyoi Group were deposited unconformably on the preceding units in an extensional setting prior to the Andean Orogeny. The Choiyoi Group and Cenozoic syntectonic units, mainly sandstone, conglomerate, and volcaniclastic rocks, are intruded by Mesozoic and Cenozoic granitic plutons. The entire sequence was deformed again during the Cenozoic Andean Orogeny. The Andean Orogeny uplifted the Cordillera Frontal during Neogene-Quaternary time and led to the deposition of thick continental units on the flanks of the uplift.

Charles Darwin, during his 1835 crossing of the Andes, identified the metamorphic basement, granitic intrusions, and volcanic sequences of late Paleozoic to Triassic ages in the Cordillera Frontal. He determined that the marine successions cropping out along the eastern Cordillera Principal were Cretaceous in age, and examined the conglomerates associated with the uplift of the Cordillera in the Alto Tunuyán Basin. Based on the composition of the synorogenic deposits (units deposited during active uplift), Darwin recognized that the Cordillera Frontal was uplifted later than the Cordillera Principal. Present work has confirmed these observations. Thus, Darwin was the first to recognize that the deformation proceeded from hinterland to foreland in these deformed belts (Giambiagi et al., 2009; Heredia et al., 2012).

Dark pre-Carboniferous units of the Precordillera west of Mendoza. View west from RN-7.

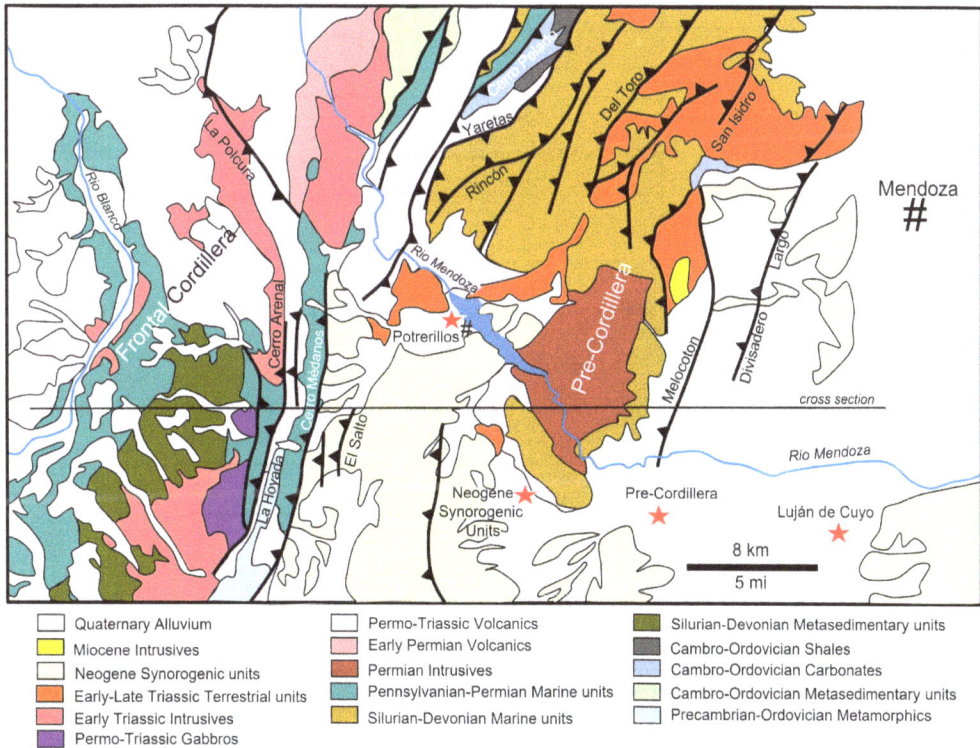

Geologic map of the Mendoza – Potrerillos area. Modified after Giambiagi et al., 2014.

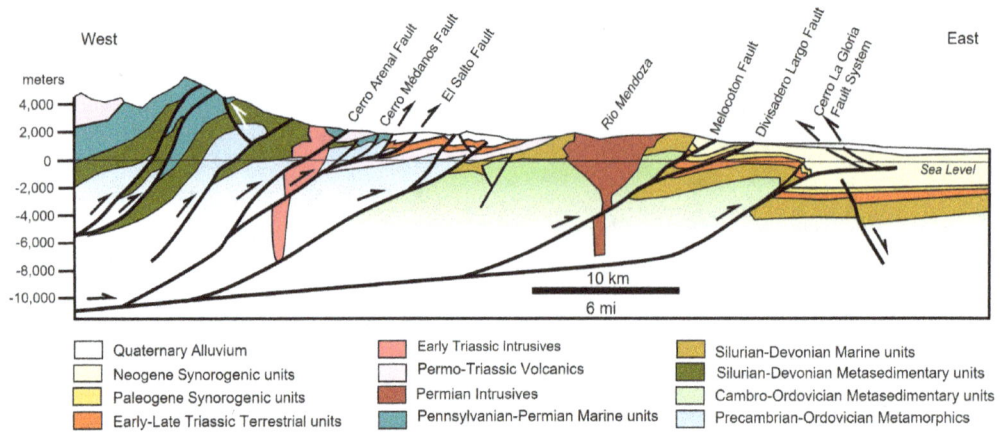

Cross-section through the Andean foothills west of Mendoza. Modified after Giambiagi et al., 2014.

Stratigraphy in the Andes west of Mendoza. Information taken from Martos et al., 2022 and others.

Cordillera Frontal to Neogene Synorogenic Sediments: Continue west on RN-7 for 9.9 km (6.1 mi; 8 min) to Stop 3, Neogene Synorogenic Sediments (−33.063103, −69.154215) and pull over on the right.

Stop 3 Neogene Synorogenic Sediments

In the foothills of the Cordillera Frontal, Neogene conglomerates, sandstones, and shales shed off the growing Andes rest unconformably on the Permo-Triassic rocks described earlier. They are synorogenic alluvial fans and river-channel deposits of the Andean Orogeny (Heredia et al., 2012). For those visitors from North America, these strata are the Neogene equivalent to the Pennsylvanian Fountain Formation redbeds west of Denver and Maroon Formation redbeds near Vail, Colorado.

Near the community of Potrerillos, and alongside RN-7, is a large reservoir, the Embalse Potrerillos. The reservoir, which impounds the Río Mendoza, is used for flood control, recreation, and to generate hydroelectric power.

Neogene synorogenic redbeds in the Cordillera Frontal west of Mendoza. Cordillera Principal is on the skyline.

Neogene Synorogenic Sediments to Early-Middle Triassic Continental Units at Potrerillos: Continue west on RN-7 for 15.7 km (9.8 mi; 12 min) to Stop 4, Early-Middle Triassic Continental Units at Potrerillos (−32.952914, −69.213501) and pull over on the right.

Stop 4 Early-Middle Triassic Continental Units at Potrerillos

At this stop, you can examine Triassic terrestrial sandstones and conglomerates of the Rio Blanco Formation along the highway. This unit was deposited in the Cuyo rift basin on the margin of Gondwana, which had begun to separate from northern Pangea (Laurasia) during the Triassic.

South-dipping Triassic sediments just west of Portrerillos.

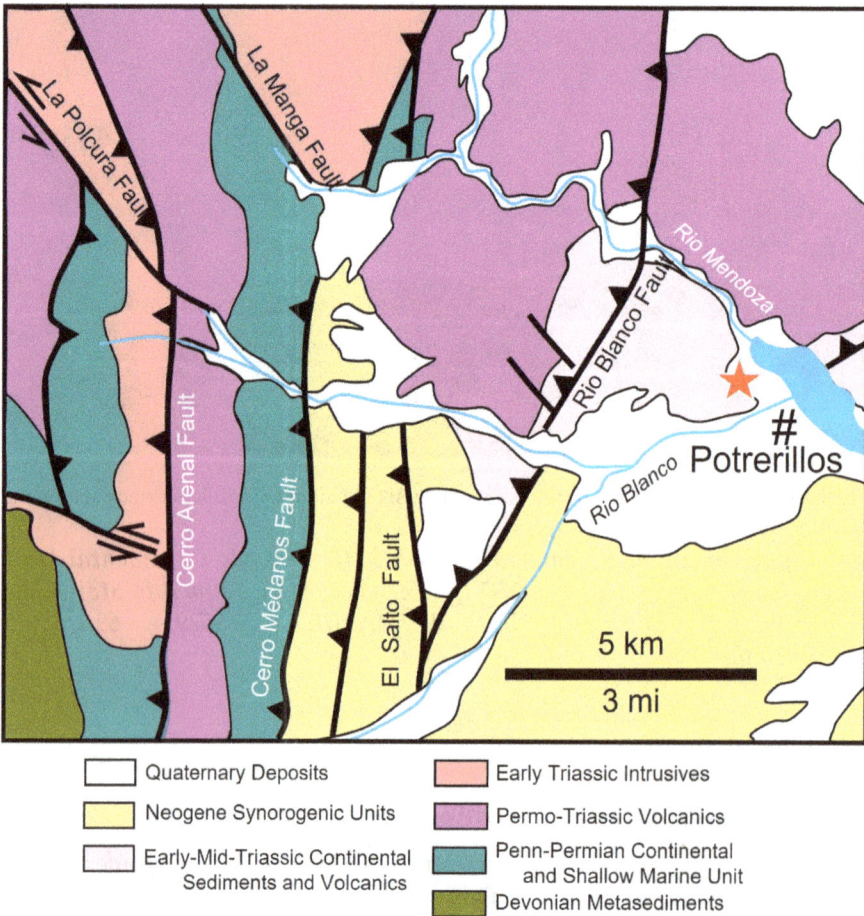

Geological map of the Cordillera Frontal west of Potrerillos Reservoir. Modified after Heredia et al., 2012.

To Avoid Side Trip 1:

Early-Middle Triassic Continental Units at Potrerillos to Uspallata Group Volcanics and Pediment: *Continue driving west on RN-7 for 60.2 km (37.4 mi; 1 hr 8 min) and pull over on the right. This is* **Stop 5, Uspallata Group Volcanics and Pediment** *(−32.617812, −69.438025).*

To go on Side Trip 1:

Side Trip 1, Early-Middle Triassic Continental Units at Potrerillos to Fossil Forest and Darwin Plaque: *Continue west on RN-7 to Uspallata; continue straight (north) on RN-149 to the north end of town; continue straight (north, then east) on RP-52 (graded gravel road) to* **Stop ST 1, Fossil Forest** *(−32.477008, −69.152608) on the left for a total of 76.0 km (47.2 mi; 1 hr 19 min).*

Side Trip 1 Fossil Forest and Darwin Plaque

Driving to the Darwin Plaque, we pass through the town of Uspallata. This farming community was originally a stop on the Trans-Andean Railroad and served local gold and silver mines.

On the northern outskirts of Uspallata just off RN-149 is Las Bóvedas de Uspallata, a small museum and monument to the Army of the Andes and General San Martín. This is one of the routes used by San Martín to cross the Andes in 1817 to attack the Spanish forces in Chile.

North and east of Uspallata we are driving through Triassic basalts and volcanic ash, tuff, and pyroclastic flows of the Río Mendoza, Las Cabras, Potrerillos, and Río Blanco formations. These were deposited in the Triassic Cuyo Basin, formed during the rifting and split-up of Gondwana. The present north-northwest elongated shape of this basin is the result of Andean deformation.

The rather modest-looking Fossil Forest stop does, in fact, have significant historical interest. We are in interbedded volcanic sediments and basalt flows of the Paramillo Formation, or lower section the Potrerillos Formation. These sediments were deposited as fluvial channel and floodplain deposits in a Triassic rift. The southwest-dipping layers contain a petrified forest of Araucarian conifers, trees related to today's Norfolk Island Pines (Brea et al, 2009). These trees are the same age (about 245 Ma) and family as those found at Petrified Forest National Park in Arizona. This stop was described by Darwin as he crossed the Andes in 1835. At the time he counted at least 52 fossil trees. One of the trunks, replaced by calcite, reminded him of the Biblical Lot's wife who had been turned into a pillar of salt.

Darwin correctly identified the local rocks as derived mainly from "various kinds of submarine lava," and was "gratified" to find the petrified forest at Agua del Zoro. "In an escarpment of compact greenish Sand-stone I found a small wood of petrified trees in a vertical position...." Darwin states that Robert Brown identified the fossil trees as coniferous, and believed the trees proved a complex history of uplift and re-emergence of the Andes (Chancellor and van Wyhe, Introduction to Darwin's *St. Fe Notebook*).

A previous monument at this site had been vandalized by local Creationists. It was replaced by a new, stronger plaque.

Darwin drew a cross-section of the Cordillera Principal through this site and the Puente del Inca. This and another cross-section through the Portillo Pass south of here are considered classics of modern geology.

When Darwin crossed the Andes, the Cordillera Principal was known as the Peuquenes Range. It and the Cordillera Frontal are separated by a broad valley, the Depresión Intermontana del Alto Tunuyán (intermountain basin of the Tunuyán High), filled with Neogene sediments. The eastern part of the Cordillera Principal contains deformed Mesozoic sedimentary rocks of the Aconcagua Fold-Thrust Belt. While observing these rocks, Darwin noted "shells which were once crawling on the bottom of the sea, [are] now standing nearly 14,000 feet above its level" (Darwin, 1845, p. 320).

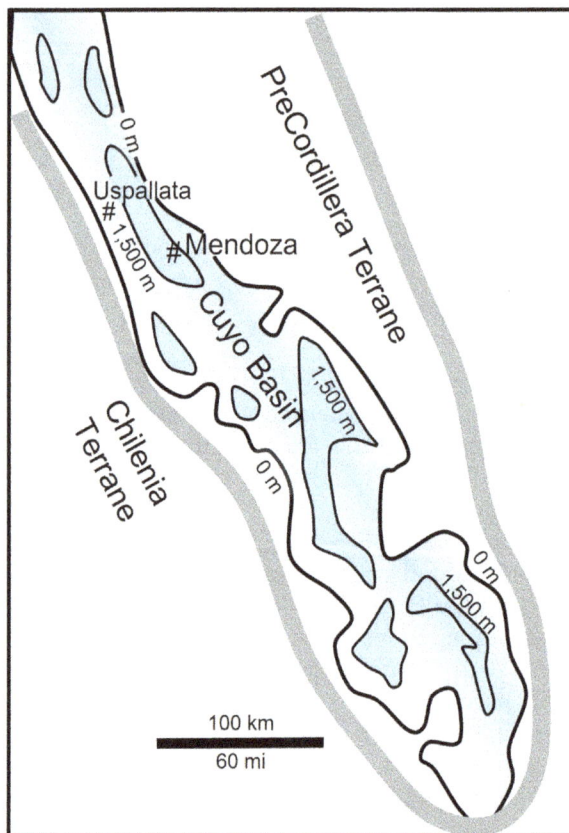

Thickness of the Triassic Cuyo Basin fill. Modified after Ramos and Kay, 1991.

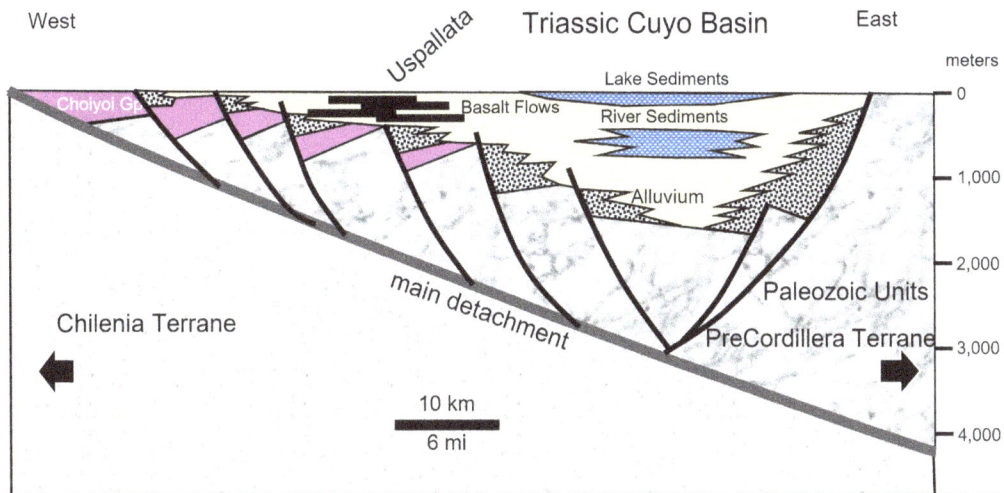

Schematic cross-section through the rifted Triassic Cuyo Basin in the area of Uspallata. Modified after Ramos and Kay, 1991.

Petrified Triassic Araucarian conifers, related to Norfolk Island Pines, in Darwin's Forest. Photo courtesy of Mcamilamenendezm, https://commons.wikimedia.org/wiki/File:Los_Colorados.jpg.

Fossil Forest and Darwin plaque, RP-52. looking southeast.

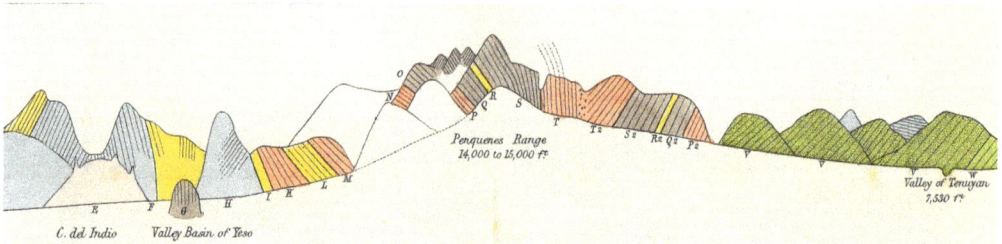

(I) SKETCH - SECTION OF THE PEUQUENES OR PORTILLO PASS OF THE CORDILLERA.

Darwin's 1846 cross-section through Portillo Pass.

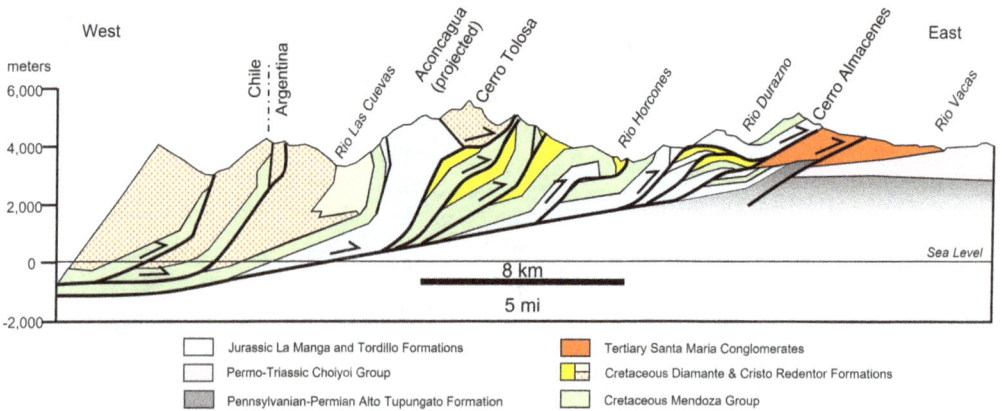

Cross-section through the Andes approximately 12.5 km (8 mi) south of Aconcagua (22°45'S). Modified after Giambiagi et al., 2017.

Side Trip 1, Fossil Forest and Darwin Plaque to Uspallata Group Volcanics and Pediment: *Return west and south on RP-52 and RN-149 to Uspallata; turn right (west) on RN-7 and drive to* **Stop 5, Uspallata Group Volcanics and Pediment** *(–32.617812, –69.438025) for a total of 35.3 km (21.9 mi; 40 min).*

Stop 5 Uspallata Group Volcanics and Pediment

The colorful rocks southwest of Uspallata are volcanics deposited in the Triassic rift-related Cuyo Basin. The volcanic units are younger than, but related to, the Choiyoi granite-rhyolite terrane in that both were deposited in an extensional regime (Ramos and Kay, 1991). The mid-Triassic rift basins formed east of a volcanic arc that reached its peak in Middle Jurassic time.

Most of the volcanics are rhyolitic in composition. The garish colors, yellow and ocher, that make this stop worthwhile are due to hydrothermal alteration that converted existing minerals to iron oxides and clay and introduced vast amounts of silica.

The eroded pediment surface south of the highway shows a paleo-valley floor 40 m (130 ft) above the current level of the river. A pediment is a gently sloping bedrock surface with a veneer of alluvial material at the foot of a mountain. The abrupt erosional escarpment indicates that active uplift and erosion continue apace. The valley fill consists of glacial till or drift, alluvium, and sediments from ancient lakes that were temporarily dammed by landslides. Up to four glacial episodes have been postulated for these valleys (Moreiras, 2010).

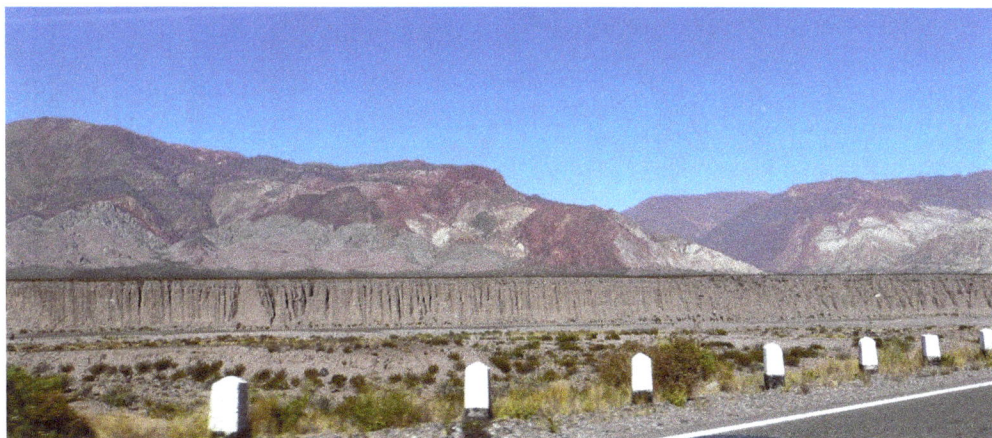

Strikingly colorful Permo-Triassic Choiyoi Group volcanics rise above the Quaternary valley floor and eroded pediment. View southwest.

Uspallata Group Volcanics and Pediment to Altered Volcanics near Los Penitentes: *Continue west on RN-7 for 38.9 km (24.2 mi; 30 min) to* **Stop 6, Altered Volcanics near Los Penitentes** *(–32.832454, –69.717401) and pull over on the right.*

Stop 6 Altered Volcanics near Los Penitentes

The Neogene Penitentes Basin in this area contains the earliest synorogenic infill of the southern Central Andes, the Penitentes Conglomerate, which outcrops in the Aconcagua Fold-Thrust Belt of the Principal Cordillera. The Penitentes Conglomerate has a maximum age of 15 Ma. Clast composition near the base of the sequence reflects the uplift and erosion of the Aconcagua's Fold-Thrust belt to the west of the basin, whereas the top of the sequence contains sediment derived from the eastern Cordillera Frontal by middle Miocene (~14 Ma). This again demonstrates the eastward migration of deformation during the Andean Orogeny. Structural analyses indicate a minimum shortening of 62% for the Aconcagua Fold-Thrust belt, and a late, out-of-sequence deformational phase during middle to late Miocene, after the uplift of the Cordillera Frontal (Martos et al., 2022).

The Los Penitentes Ski Area a few kilometers west of this stop has a dramatic backdrop. The mountains go pretty much straight up all around. You will have noticed by now the lack of forested hillsides that allow the rocks to be seen in all their glory. This is due partly to altitude, and mostly to the arid climate.

A lateral moraine lies on the north side of the valley beneath the ski runs of Penitentes Resort. The surrounding mountains are composed of the vivid, hydrothermally altered Choiyoi Group volcanics.

The El Abuelo-Mario Ardito rock avalanche northeast of the Penitentes Resort generated a debris flow that extended into the Río Mendoza valley. It is dated between 10,620 and 11,820 years ago (Moreiras, 2010).

Hydrothermal alteration makes these brightly colored rocks in the Choiyoi Group rhyolitic volcanics. View west.

Altered Volcanics near Los Penitentes to Puente del Inca: *Continue west on RN-7 for 21.2 km (13.2 mi; 22 min) to* **Stop 7, Puente del Inca** *(–32.825848, –69.910257) and pull over on the left.*

Stop 7 Puente del Inca

This bridge had nothing to do with Incas, although their influence was felt in the area.

Puente del Inca (Inca Bridge) is a natural bridge made up of travertine-cemented landslide debris. The travertine comes from thermal springs similar to those scattered along the axes of the Cordillera Principal. The bridge and adjacent travertine mound were formed by mineral precipitation from a 37°C (99°F) hot spring that emerges along the Penitentes Fault. The vibrant colors are due to algae, iron oxide, and calcium carbonate (travertine). The drop in temperature and release of carbon dioxide gas from the naturally carbonated water leads to the precipitation of dissolved minerals.

Darwin remarked on this feature when he crossed the Andes in 1835:

> At this place, there are some hot and cold springs, the warmest having a temperature, according to Lieut. Brand …, of 91°; they emit much gas. According to Mr. Brande, of the Royal Institution, ten cubical inches contain forty-five grains of solid matter, consisting chiefly of salt, gypsum, carbonate of lime, and oxide of iron. The water is charged with carbonic acid.
>
> **(Darwin, 1846)**

Darwin considered the Puente del Inca overrated but recorded the stratigraphic section there in great detail. He sketched the bridge and described it as a "crust of stratified shingle, cemented together by the deposits of the neighbouring hot springs" (Chancellor and van Wyhe, Introduction to Darwin's *St. Fe Notebook*).

In 1917 an English company, the Compañia Hotelera Sudamericana, began building a luxury hotel and spa around the hot springs. There were nine tiled pools connected to the hotel and it was claimed that the water could heal everything from chronic rheumatism to syphilis. I'm not sure why anyone would want to "take the waters" alongside a syphilitic, but that is what they advertised. The abundant carbon dioxide made good ventilation necessary to prevent suffocation. In 1965, a large landslide damaged the hotel buildings and the company abandoned the site.

Declining water flow over the bridge caused erosion and cracking that led the provincial government to restrict access. Ongoing studies are examining how to return the flow of mineral-rich water that might repair the bridge.

Puente del Inca, a travertine bridge over the Río Mendoza.

*Puente del Inca to Mirador Aconcagua, Aconcagua Provincial Park: Continue driving west on RN-7 for 3.7 km (2.3 mi; 7 min) to the turnoff to Aconcagua Provincial Park on the right; turn right (north) and park at the Visitor Center. This is **Stop 8, Mirador Aconcagua, Aconcagua Provincial Park** (–32.823658, –69.942280).*

Stop 8 Mirador Aconcagua, Aconcagua Provincial Park

Aconcagua is the tallest peak in all of the Americas, standing 6,959 m (22,831 ft) above sea level. There are glaciers on the south face and in the upper reaches of the Horcones Valley. The name is thought to derive from the Quechua (Inca) words for stone (ackon) and viewer (cahuak), meaning "stone sentry" or "lookout."

The massif lies in the Cordillera Principal and is part of the thin-skinned Aconcagua Fold-Thrust Belt. The lower slopes consist of Jurassic-Cretaceous carbonates, clastics, and volcanic sediments thrust to the east. Marine invertebrate fossils are common. The upper slopes consist of roughly 15 Ma (Neogene) dacitic to andesitic volcanics.

Deformation in the area began with gradual uplift between 80 and 120 Ma and culminated in the Andean Orogeny about 22 Ma. Uplift continues, to some extent, today.

Quaternary deposits are mostly alluvial and include river deposits, glacial till and moraines, rock glaciers, and landslide deposits. Lateral moraines have been dated to 14,000 to 16,500 years old (Hermanns et al., 2014).

View north to Aconcagua, highest peak in the Americas.

West-dipping Jurassic-Cretaceous units looking south from the Aconcagua overlook.

Visit

A number of trails go up the Horcones Valley, ranging from 2 km (1.2 mi; 1 hr) and gentle, to hikes that take 7 to 10 hours one-way. Check the park website or call for current weather conditions.

First climbed by Swiss mountaineer Matthias Zurbriggen in 1897, the peak is climbed rather routinely today. Trekking and climbing is limited to the season between November 15 and March 15.

There is a small fee for hiking beyond the roadside overlook. Check the website for the current amount.

Hours: 8 am to 6 pm.

Phone: 0261-425-8751.

Website: http://areasnaturales.mendoza.gov.ar/parque-aconcagua.html.

*Mirador Aconcagua, Aconcagua Provincial Park to Las Cuevas Slide: Continue west on RN-7 for 9.2 km (5.7 mi; 10 min) to **Stop 9, Las Cuevas Slide** (−32.81550, −70.03253) and pull over on the right.*

Stop 9 Las Cuevas Slide

The Las Cuevas landslide, technically known as the Tolosa rock avalanche, is visible below the town of Las Cuevas. The slide was the result of bedding-plane slip on steeply dipping Jurassic Tordillo Formation on the west side of Tolosa Peak. That is, the surface between the layers of rock acted as a plane of weakness that allowed the overlying rocks to slip catastrophically down the mountainside.

The slide is at least 50 m (165 ft) thick and has a maximum length of 1.5 km (0.9 mi). It shows no sign of having been eroded by Pleistocene glaciers. It contains blocks the size of a house, as large as 28 m (92 ft) in diameter. The slide is dated between 11,380 and 15,500 years old (Moreiras, 2010; Moreiras and Sepulveda, 2015). This is just one of many large slides in the Río Mendoza Valley. The slide is thought to have been triggered by earthquakes, or possibly by water lubricating the slip surface during a warmer, wetter climate.

View west toward Las Cuevas slide. There are no caves here. There is, however, a massive slide that came off the north side of the valley due at least in part to south-dipping bedding planes in the Cretaceous Diamante Formation. The slide probably occurred between 11,000 and 15,000 years ago.

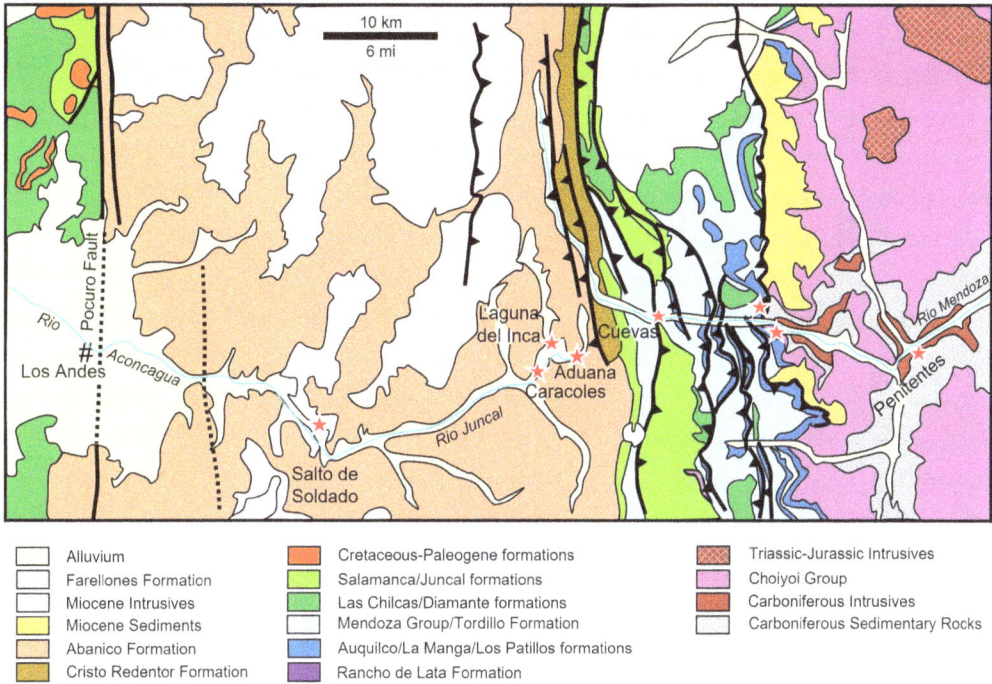

Geologic map of the Penitentes to Los Andes area. Modified after Mackaman-Lofland et al., 2018.

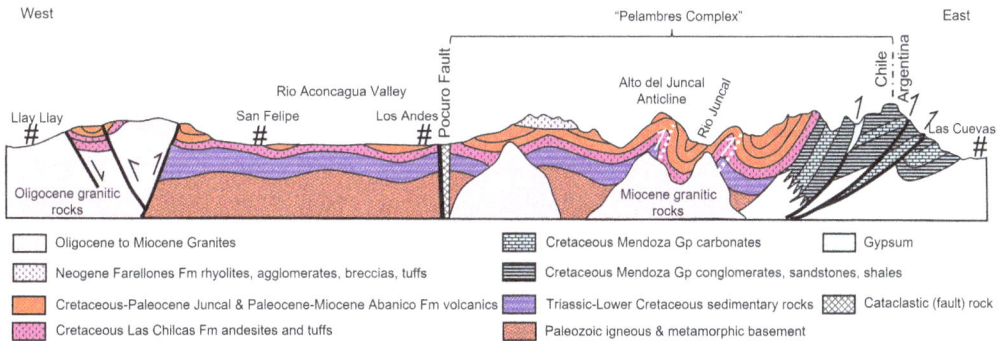

Cross-section from Llay Llay, Chile, to the Argentine border. 1 = gypsum; 2 = carbonates; 3 = sandstone and conglomerate; 4 = volcaniclastics; 5 = andesite and tuff; 6 = rhyolite; 7 = granites; 8 = cataclastic (broken) rock. Modified after Ramos, 2008.

Las Cuevas Slide to Aduana Chilena and West-Dipping Andesites: *Continue driving west on RN-7 to the border, where the highway changes name to Chile R-60; continue west on R-60 to **Stop 10, Aduana Chilena and West-Dipping Andesites** (−32.844945, −70.106667) for a total of 10.8 km (6.7 mi; 15 to 35 min). Time in transit can change depending on the backup at the Chilean Customs entry point. Rather than crossing the border in a tunnel, an alternative with a better view is to take the graded gravel road to the continental divide at Cristo Redentor (21.6 km; 1 hr 20 min).*

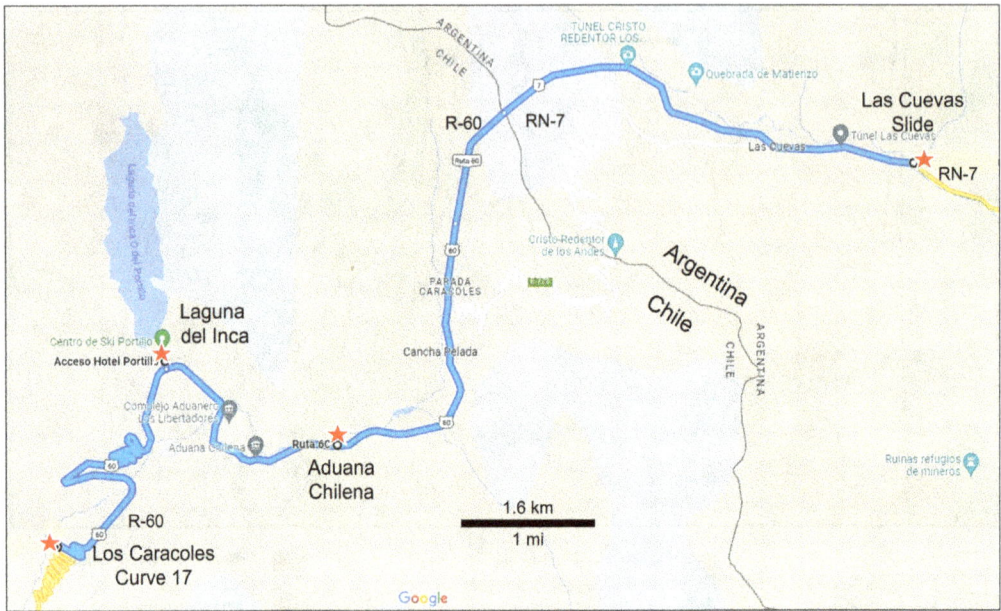

Stop 10 Aduana Chilena and West-Dipping Andesites

These rocks were mapped by the Chilean Geological Survey (Servicio Nacional de Geologia y Mineria) as Lower Cretaceous Los Pelambres Formation andesitic lavas, tuffs, volcanic sandstones, and conglomerates interbedded with fossiliferous calcareous sandstones and some marine carbonates (Rivano-G. et al., 1993). The stratigraphy is a bit unsettled in this area, but the Los Pelambres is time-equivalent to the Diamante Formation igneous complex (Mackaman-Lofland et al, 2018).

A cross-section by Ramos (2008) shows these units as Cretaceous Mendoza Group carbonates, conglomerates, sandstones, and shales overlain by Upper Cretaceous Las Chilcas Formation andesites and tuffs. And yet, his geologic map shows the units are Latest Cretaceous to Eocene Abanico and Juncal formations volcanic flows, tuffs, breccias, and conglomerates.

They were most recently mapped as Oligocene-Miocene Abanico Formation volcanic breccias, tuffs, and flows, and volcaniclastic conglomerates and sandstones (Mackaman-Loffland et al., 2018). Suffice it to say that they are volcanic units of uncertain age.

Ramos and Mackaman-Loffland both show these units to be carried on east-directed thrusts. They have been deformed into east-verging folds with a dominant west dip on their west flanks.

Lower Cretaceous Los Pelambres andesites. View north from just east of the Aduana Chilena, the Chilean border and customs station.

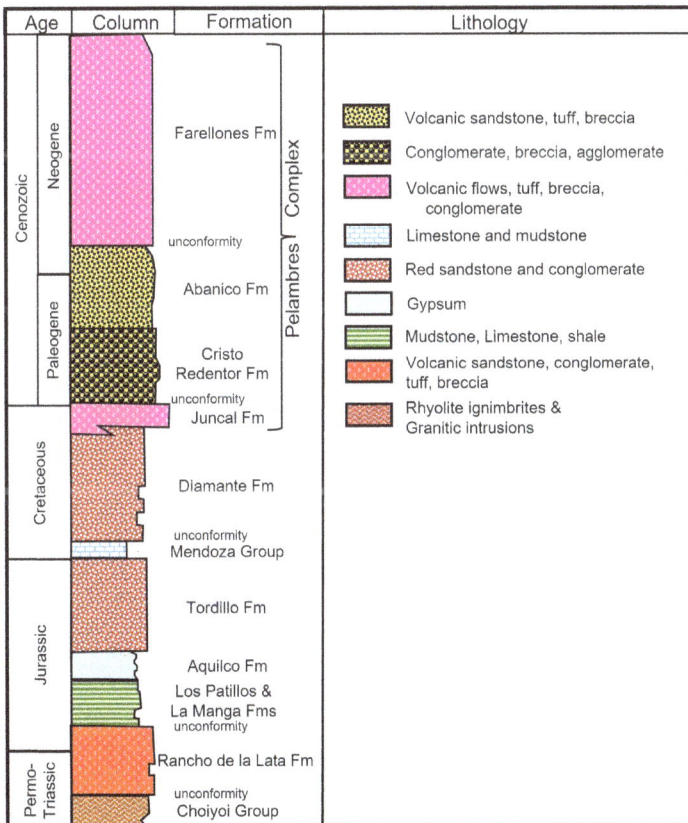

Stratigraphic column for the Chilean and Argentine Andes. Information drawn from Mackaman-Lofland et al., 2018.

Aduana Chilena and West-Dipping Andesites to Laguna del Inca*: Continue driving west on R-60 for 3.0 km (1.9 mi; 5 min) and pull into the hotel parking area on the right. This is* **Stop 11, Laguna del Inca** *(−32.836180, −70.128936). Walk 300 m (1,000 ft) north to the lake.*

Along the way, you will pass through Chilean Customs (Aduana Chilena). Expect to spend up to an hour waiting in line and going through paperwork, possibly having luggage x-rayed, and your vehicle inspected at Chilean Customs. You may be asked about your COVID vaccination status. Have your rental vehicle import/export and insurance papers available.

Stop 11 Laguna del Inca

Look north to steeply dipping Cretaceous Juncal Formation volcanics. The lake appears to be in a large syncline, although mapping indicates that there are synclines on both sides of the lake (Rivano-G. et al., 1993). It is not obvious, but Early Cretaceous limestone is interbedded with the volcanic Pelambres Complex on the east side of the valley (Ramos, 2008). The Pelambres Complex, consisting (from base to top) of the Juncal, Cristo Redentor, Abanico, and Farellones formations, represents the Early Cretaceous to Miocene Lo Prado-Pelambres Volcanic Arc that lay between the Central Chilean Basin to the west and the Neuquén Basin to the east (Charrier et al., 2007).

On the slopes to the northwest are Miocene diorite dikes intruded into the Juncal Formation (Ramos, 2008).

If you visit this lake during a full moon you may hear strange cries across the crystal waters. Local legend says that the spirit of the Inca Illi Yunqui haunts the lagoon. His princess, Kora-Llé, during a royal feast, had fallen from a nearby cliff and perished. Believing that no earthly grave was good enough, he lowered the body of the princess into the waters of the lake. From that time on the waters of the lake were tinted the same emerald color as the eyes of the princess. Only the sad wailing of Illi Yunqui remains to remind us of this tragic story.

The Laguna del Inca appears to occupy a large syncline. View north.

> **Laguna del Inca to Curve 17, Los Caracoles**: *Continue driving west on R-60 to the Curve 17 pullout on the right. This is **Stop 12, Curve 17, Los Caracoles** (−32.855668, −70.143000) for a total of 6.2 km (3.9 mi; 10 min).*

Stop 12 Curve 17, Los Caracoles

This pullout is on the seventeenth curve of a series of 19 switchbacks that drop the highway 231 m (760 ft) in less than a kilometer. Look west to see the volcanic flows, tuffs, and breccias of the Cretaceous-Paleocene Juncal Formation intruded by dacite dikes and sills injected during Miocene thrusting (Ramos, 2008). The Juncal volcanics were deposited near their source in the Cretaceous Central Chilean Basin that lay between the eastern and western volcanic arcs.

Looking south from Curve 17 you can see the Alto de Juncal Anticline, a large east-verging fold developed in Early Cretaceous volcanic layers carried on east-directed thrusts of the Aconcagua Fold-Thrust Belt (Ramos, 2008).

The upper Río Juncal Valley here has the classic U-shape associated with glacial erosion.

Curve 17, Los Caracoles. View southwest at east-verging Alto de Juncal Anticline with small internal back-thrusts (as seen by white layer).

Curve 17, Los Caracoles to Salto de Soldado*: Continue driving west on R-60 for 28.6 km (17.7 mi; 33 min) and pull over on the left. This is **Stop 13, Salto de Soldado and the Cordillera Principal/Río Aconcagua Valley** (–32.90225, –70.364067).*

Stop 13 Salto de Soldado and the Cordillera Principal/Rio Aconcagua Valley

The main Andean range is known as the Cordillera Principal. It is the result of the Andean Orogeny that spans Late Cretaceous to Recent times. The rocks involved in the deformation are largely Jurassic and Cretaceous marine deposits that were folded and thrust from west to east as part of the Aconcagua Fold-Thrust Belt. These sedimentary rocks are capped by up to 6 km (19,700 ft) of volcanics derived from an older, western volcanic arc near the Coastal Cordillera and a younger, eastern Andean volcanic arc near the Chilean-Argentine border (Ramos, 2008).

The sequence of events that led to the Andes includes the development of a Jurassic, subduction-related volcanic arc in the Coastal Cordillera that unconformably overlies basement rocks consisting of Carboniferous deep-marine turbidites and Choiyoi volcanics. A back-arc basin, the Aconcagua Platform, developed east of the Coastal Cordillera. This was the setting for Early Cretaceous sedimentation, followed by Late Cretaceous-Tertiary magmatism, folding, and thrusting. The magmatic activity was largely complete by late Miocene time.

The Chilean slopes of the Andes consist of folded Cretaceous, Paleogene, and Neogene sedimentary and volcanic units. Folding is more intense to the east and becomes gentler to the west. These folds were intruded by Miocene granodiorites (Ramos, 2008).

The Río Juncal has cut a slot canyon 140 m (450 ft) deep through Miocene intrusives (granite and dacite porphyry). The porphyry has intruded the Late Cretaceous-early Miocene Abanico Formation (Ramos, 2008; Mackaman-Loffland et al., 2018) or Los Pelambres Formation (Rivano G et al., 1993) consisting of andesitic lavas and tuffs, volcanic sandstone, and breccia. The canyon is so narrow that the bottom is in almost perpetual shadow. The name 'Salto del Soldado' (Soldier's Leap) derives from a daring escape during the struggle for independence from Spain. A Chilean soldier being pursued by the Spanish army is said to have escaped by jumping his horse across the narrow gorge, which is supposedly only around 9 m (30 ft) wide at one point. I measured the narrowest section on Google Earth and got 83 m (272 ft) wide at the top. Apocryphal? You decide.

Salto de Soldado slot canyon cut into Miocene intrusives by the Río Juncal. View southeast.

Salto de Soldado to San Felipe: *Continue driving west on R-60; at the east side of San Felipe take the ramp for San Felipe; keep left at the fork and follow the signs for R-5; on the overpass follow the signs for Auco/Rinconada and merge onto E-85 heading south; turn right (west) onto Coronel Santiago Bueras and drive to* **Stop 14, San Felipe** *(−32.749722, −70.745833) and pull over on the right for a total of for 44.6 km (27.7 mi; 47 min).*

Stop 14 San Felipe

The outcrops here are Early Cretaceous Las Chilcas Formation andesites, tuffs, volcani-clastics, breccias, and conglomerates (Rivano-G et al., 1993; Ramos, 2008). This is our first encounter with the Las Chilcas Formation, an important unit in unraveling the structural history of the region. This unit records the initial phase of Andean uplift and compression occurring just to the west in Cretaceous time (105–83 Ma). Up to 3.5 km (11,500 ft) of synorogenic non-marine sandstone and conglomerate were deposited in an extensional retro-arc basin. The Las Chilcas sediments, eroded from the volcanic arc, were accumulating in a subsiding basin just east of the subduction-related arc (Charrier et al., 2007; Boyce et al., 2020).

A fault with slickensides (a polished surface with scratch marks) here strikes 314° and dips 70° to the northeast, indicating it is a normal fault down to the east.

Structures and associated syntectonic deposits suggest that the strongest deformation (folding, inversion, and thrusting) occurred between 100 and 95 Ma. The Las Chilcas Formation at this stop is being overthrusted by east-directed granodiorites that outcrop a few kilometers to the west (Rivano-G et al., 1993; Ramos, 2008).

Roadcut into Las Chilcas, west of San Felipe.

Fresh and weathered Las Chilcas andesite in San Felipe roadcut.

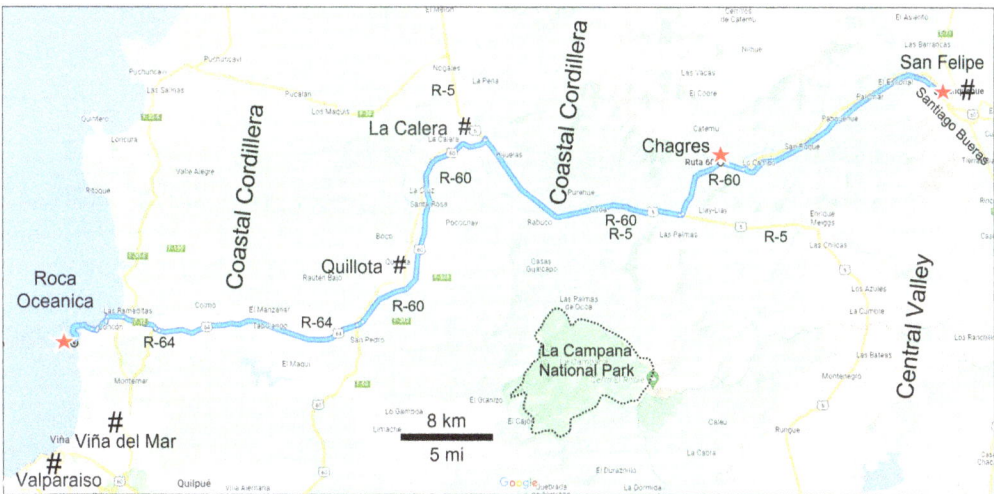

San Felipe to Coast Range Granodiorite, Chagres Roadcut: *Continue driving west on old R-60 for 22.3 km (13.9 mi; 25 min) to **Stop 15, Coast Range Granodiorite, Chagres Roadcut** (−32.804199, −70.949334) and pull over on the right by the shrine and just before the offramp. Carefully walk east on the path/shoulder ~100 m (325 ft) to examine the roadcut.*

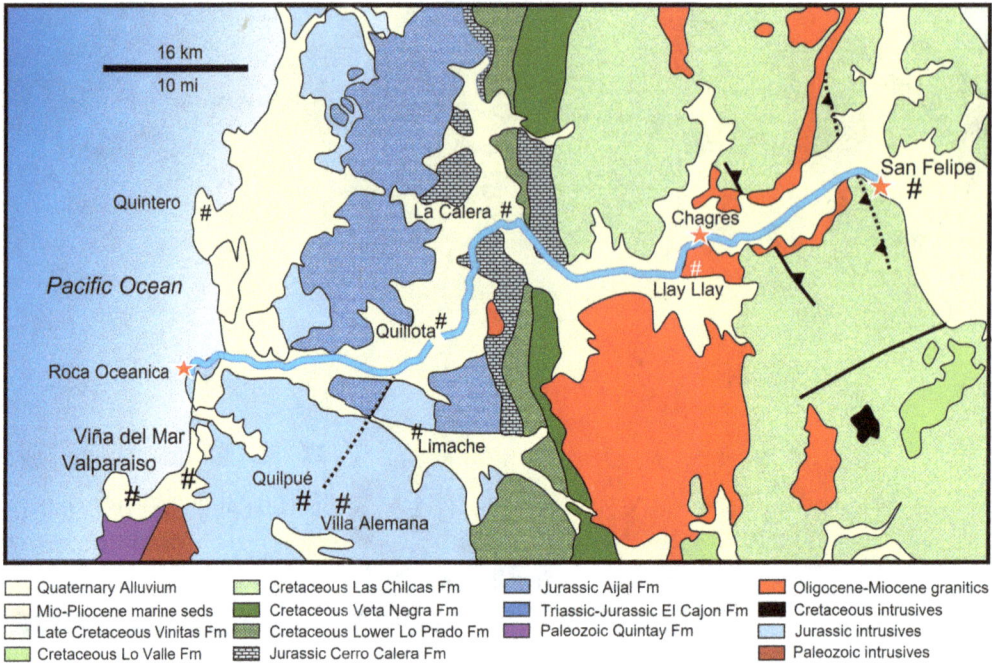

Geologic map of the Valparaíso to San Felipe segment of the transect. Information derived from Ramos, 2008; Boyce et al., 2020; Sernageomin.cl, online geologic map of Chile.

Stop 15 Coast Range Granodiorite, Chagres Roadcut

Excellent roadcuts at Chagres expose granites and granodiorites. These have been mapped as Lower Cretaceous Chilinga granodiorite and diorite (Rivano G et al., 1993), as Oligo-Miocene granitic rocks by Ramos (2008), and as mid-Cretaceous intrusives (Boyce et al., 2020). Regardless, they are intrusive rocks of the Coastal Cordillera. These rocks are a deep manifestation of the volcanic arc that existed in this area from Permian through mid-Cretaceous time.

Chagres roadcut in granitic rock of the Coastal Batholith. Street View east.

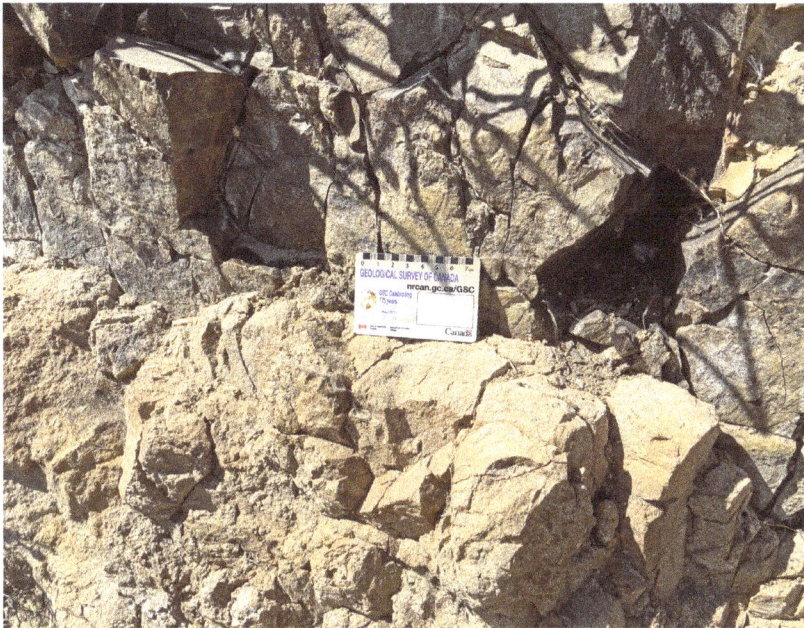

Orange-pink granite, possibly Cretaceous to Miocene in age, of the Coastal Batholith at Chagres.

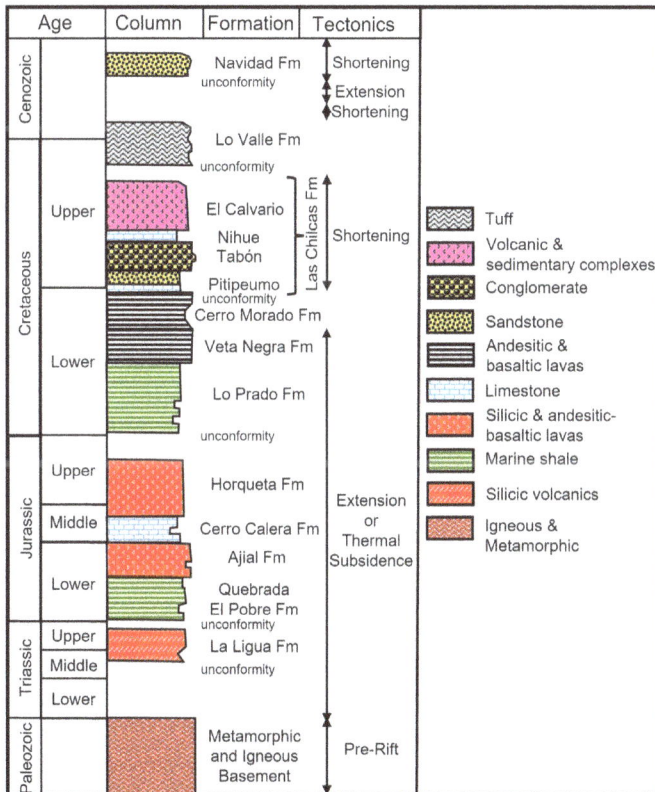

Stratigraphy of the Coastal Cordillera. Information drawn from Boyce et al., 2020.

La Campana National Park

Nine km (5.6 mi) south of R-60 between Llay Llay and La Calera is La Campana National Park. This park encompasses some of the most scenic and geologically historic peaks of the Coastal Cordillera. The park's name comes from Cerro La Campana peak, which apparently looks a bit like a bell ("campana" in Spanish) when viewed from the nearby Aconcagua Valley. The area, designated a National Park in 1984, is about 80 km² (31 mi²).

In August 1835, during the second voyage of the HMS Beagle (1831–1836), Charles Darwin climbed Cerro La Compana (1,880 m, or 6,168 ft), as part of his trans-Andean explorations (Wikipedia, La Campana National Park). His notes show that he both appreciated the beauty of the location and pondered the geological forces that went into making the landscape:

> ...in the morning we set out to ascend the Campana or Bell Mountain.... The paths are very bad, but both the geology and scenery amply repaid the trouble. We spent the day on the summit, and I never enjoyed one more thoroughly. Chile, bounded by the Andes and the Pacific, was seen as in a map. The pleasure from the scenery, in itself beautiful, was heightened by many reflections which arose from the mere view of the Campana range.... Who can avoid wondering at the force which has upheaved these mountains, and even more so at the countless ages which it must have required. (Rundel and Weisser, 1975)

Darwin's descriptions of the area's unique species were instrumental in establishing the park. The park is renowned for having some of the last forests of the endangered Chilean Wine Palm (*Jubaea chilensis*) in the Palmas de Ocoa, as well as sclerophyllous and Valdivian forests.

Geologically, the park is part of the Coastal Batholith, comprising granodiorite and gabbro of Late Cretaceous to Miocene age. These magmas intruded Early Cretaceous Lo Prado andesites and marine shales. The Lo Prado Formation has been altered and mineralized, and mines in the unit have produced gold, silver, and lead from Spanish times until 1994, when the last mine closed (Campana National Park).

The best time to visit is in the spring, when the waterfall is flowing and the wildflowers are in bloom. Summers can be hot and dry (Dare2Go, La Campana National Park).

Park Entrance Fee:	CLP 2,500 per adult (foreigners); CLP 2,000 for Chileans.
CLP 1,500 foreign children;	CLP 1,000 for Chilean children (2022).
Camping Fee:	CLP 6,000 per site/day.

Since 1984, the Parque Nacional La Campana has been part of the La Campana-Peñuelas UNESCO Biosphere Reserve.

Coast Range Granodiorite, Chagres Roadcut to Roca Oceanica*: Continue driving west on R-60; merge onto R-5/Panamerica Norte going west; on the east side of La Calera take the exit to Viña del Mar/Quillota and merge onto R-60 west; take the exit to San Pedro/ Concón and merge onto R-64 west; take the exit to Quintero/Reñaca/Concón and merge* *onto F-32 west; at the roundabout take the third[d] exit onto Maroto; turn left (south) on Barros; bear left (south) onto Calle Dos; turn right (west) onto Vergara; continue straight on Las Elenas; turn left (west) onto Av. Borgoño; pull into the parking area on the right at* **Stop 16, Roca Oceanica, Granite of the Coast Batholith** *(−32.941382, −71.552948) for a total of 77.4 km (48.1 mi; 1 hr 10 min).*

Stop 16 Roca Oceanica, Granite of the Coast Batholith

Roca Oceanica was declared a nature preserve in 1990 owing to its unique flora, fauna, and geology. The promontory and natural overlook looms 25 m (82 ft) over the Pacific. Gneiss, granodiorites, and gabbro dikes exposed in the cliffs reveal the nature of the Coast Batholith. Sand dunes above the road tell the story of a long-lost beach.

The Dunes of Concón consist of sand derived from erosion of the mostly granitic Coastal Cordillera. Carried by rivers to the sea, the sand was deposited on an ancient beach. Winds of the Pacific blew the sand to its present location high on the bluffs. At the end of the last ice age, roughly 10,000 years ago, sea level was at least 30 m (100 ft) lower here. Since then, melting glaciers and rising sea levels have drowned the beach and cut off the source of sand, while the relentless churning of the waves has eroded the cliffs along the coast.

The Coast Batholith is mainly Late Carboniferous and Permian (320–280 Ma) granitic rocks that cooled far below the surface. Then, around 160 Ma (Late Jurassic), dark diorite and gabbro dikes intruded the light-colored granites (Deckart et al., 2014). Since then, the coast has been uplifted and eroded several kilometers, exposing the granites and dikes.

The Coast Batholith, part of the Coastal Cordillera, is a deep remnant of the volcanic arc that erupted the volcanic Choiyoi Group that we encountered as basement rocks in the Argentine Andes and Neuquén Basin. The volcanic arc developed at the margin of the continent over a subducting oceanic plate. The Coast Batholith is emplaced into rocks of a Paleozoic accretionary complex. The complex consists of a great thickness of sediments, now metamorphosed, that accumulated in an oceanic trench/subduction zone and were later accreted or welded onto the continent (Wikipedia, Coastal Batholith). The metamorphosed sediments were buried between 8 to 10 km (4.8 to 6 mi), where they were intruded by granitic magmas. Today the metamorphic accretionary complex outcrops mainly west of the Coast Batholith. The accretionary complex is subdivided into a western segment deposited during the Carboniferous, and an eastern series deposited a bit later, in the Permian (Deckart et al., 2014).

Perched dune field above granite intruded by basalt dikes, Roca Oceanica. View south toward Viña del Mar.

Late Carboniferous (Pennsylvanian) granodiorite, Roca Oceanica.

Dune field, Roca Oceanica. View northeast.

As you drive south to Valparaíso you pass Castillo Reñaca. In high roadcuts, you can see Late Carboniferous tonalites (a coarse felsic intrusive that looks a lot like granodiorite) that were intruded by much younger Jurassic granitic rock during the formation of the Coast Range batholith.

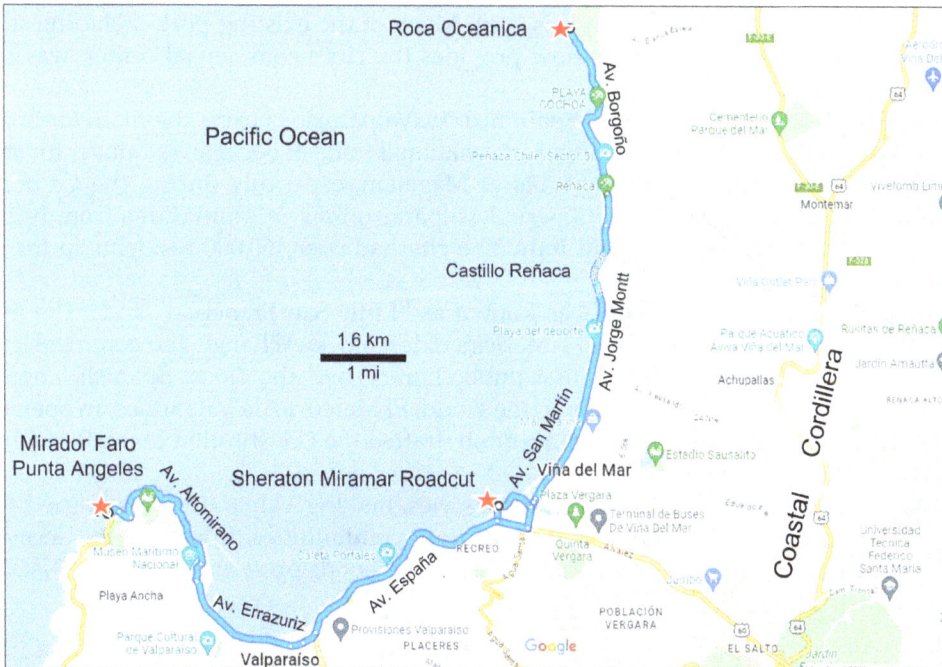

> *Roca Oceanica to Sheraton Miramar: Continue driving south on Av. Borgoño; at the roundabout continue straight on Av. San Martin; cross the Puente Casino over the estuary and turn right onto Von Schroeder; pull into the parking lot on the right just south of the Sheraton Miramar. This is* **Stop 17, Sheraton Miramar, Valparaíso Granodiorite and Gneiss** *(–33.021157, –71.567492) for a total of 11 km (6.9 mi; 20 min). Cross the street to the roadcut. Mind the traffic!*

Valparaíso

Valparaíso is a major city and seaport on the Pacific coast of Chile. The population (2012) was 284,000 in the city, and 930,000 in the metropolitan area. "Greater Valparaíso" is the second largest metropolitan area in the country, after Santiago. Valparaíso has a mild Mediterranean climate where the summer is essentially dry, and it rains mainly in the winter. Average monthly temperatures vary from a high of 17°C (63°F) in January to a low of 11.4°C (52.5°F) in July. The city is affected by fog from the cold Humboldt Current during most of the year.

History

The Bay of Valparaíso was originally inhabited by the Picunche people, known for their agriculture, and the Chango, who were nomadic fishermen. The first Europeans arrived in 1536, aboard the *Santiaguillo*, a supply ship under the command of Juan de Saavedra, who named the town after his native Valparaíso de Arriba in Spain.

In 1810, a wealthy merchant built the first pier in Chile. In that location today is the El Mercurio de Valparaíso newspaper building. Reclamation of land from the sea moved the coastline five blocks seaward from this spot. Much of the existing port, including much of the land reclamation work that now provides the city's commercial center, was built between 1810 and 1830.

After Chile's independence from Spain in 1818, Valparaíso became the main harbor for the Chilean navy, and the main port for international trade. It became a stopover for ships rounding South America via the Straits of Magellan, especially during the California Gold Rush (1848–1858). During this period Valparaíso took in immigrants from Britain, Germany, France, Switzerland, and Italy. The city had over 160,000 residents in the late 1800s.

During its golden age, the city was known as "Little San Francisco" and "The Jewel of the Pacific." Valparaíso has Latin America's oldest stock exchange, the continent's first volunteer fire department, Chile's first public library, and the oldest Spanish language newspaper in continuous publication in the world, El Mercurio de Valparaíso, in operation since 1827. In 1828, a constitutional convention drafted the Constitution of the Republic of Chile at the Church of San Francisco.

Buildings reflected a variety of European styles, making Valparaíso more varied than most other Chilean cities. The largest immigrant communities, from Britain, Germany, and Italy, each created their own hillside neighborhood, preserved today as National Historic Districts ("Zonas Típicas").

The opening of the Panama Canal in 1914 and the associated reduction in ship traffic dealt a blow to Valparaíso's port-based economy. Especially during the second half of the twentieth century, Valparaíso experienced a decline as wealthy families left the historic quarter and moved to Santiago or nearby Viña del Mar.

The city has been the seat of the Chilean National Congress since 1990. Starting in 1998, activists convinced the Chilean government to apply for UNESCO World Heritage status for the city. In 2003, the old quarter of Valparaíso was declared a UNESCO World Heritage Site on the basis of its historic buildings and culture.

Since the start of the twenty-first century, the city has staged a recovery, attracting artists and many thousands of tourists from around the world. Valparaíso is a major port for container ships, copper, wine and fruit exports, and cruise ships. The city is an educational center with four major universities and several vocational colleges. The city is also a cultural center.

Geology

Valparaíso lies on the west flank of the Coastal Cordillera, a mountain range that contains the Coastal Batholith, the core of an old volcanic arc. The city is underlain by granite, with some metamorphic rock along the coast to the south. Man-made fill added more land along the margin of the bay. But that is not what Valparaíso is known for, geologically speaking. It is known for earthquakes.

The city sits just onshore of the Peru-Chile Trench, an active subduction zone where the Nazca tectonic plate is diving beneath the South American Plate at an average rate of 6.6 cm (2.6 in) per year. Most of the time the plates are locked together by friction. Every so often the pressure builds to a breaking point and the subducting plate moves downward while the overriding plate jumps upward.

Valparaíso has a long record of destructive quakes. Earthquakes with an estimated magnitude greater than 8 at or near the city are recorded for the years 1570, 1657, 1730, 1751, 1802, 1822 (estimated magnitude 8.5), 1835 (estimated magnitude 8.5; recorded by Charles Darwin, among others), 1906 (estimated magnitude 8.2), 1985 (magnitude 8.0), 2010 (magnitude 8.8), and 2015 (magnitude 8.3). Numerous smaller earthquakes have been recorded. For comparison, the "big one" that destroyed San Francisco in 1906, is estimated at magnitude 7.7 to 7.9. The 1730 Valparaíso earthquake and tsunami are the largest ever recorded in Chile. Buildings were destroyed along over 1,000 km (600 mi) of coast, and the tsunami generated was large enough to cause damage in Japan. Tsunami heights and recorded uplift indicate a rupture on the order of 600 to 800 km (370 to 500 mi) with 10 to14 m (28 to 46 ft) of uplift, probably caused by a magnitude 9.1 to 9.3 quake. On November 19, 1822, Valparaíso experienced a violent earthquake that left the city in ruins. A major quake struck Valparaíso in August 1906. At that time, the city was the center of the Chilean economy. The damage was in the hundreds of millions of pesos at the time, and there were 3,000 dead and over 20,000 injured. Following the quake, the city was rebuilt, including the widening of streets, undergrounding streams to create Francia and Argentina Avenues, creating Pedro Montt, the main street, and building Plaza O'Higgins (Carvajal et al., 2017; Wikipedia, Valparaiso).

The 1835 earthquake had an enduring impact on the development of geologic concepts. It so happened that Charles Darwin was there to experience the earthquake and spent several weeks studying the aftereffects. He found recent seashells elevated above sea level. Santa Maria Island had been uplifted about 3 m (10 ft). He noted reports of three volcanoes erupting at about the same time. Traveling inland he saw seashells at several elevations above sea level, suggesting that the land had moved upward many times in small amounts. Eventually, he concluded that mountains grow incrementally rather than in one great cataclysm, that earthquakes are related to some as-yet-unknown mechanism in the Earth's interior, and that the surface, or crust, floats on a sea of molten rock (van Wyhe, 2010).

Stop 17 Sheraton Miramar, Valparaíso Granodiorite and Gneiss

Much of the area around Valparaíso consists of extensive outcrops of plutonic (deep magmatic) rocks dated at close to 160 Ma. These include a series of bodies that vary from dark gabbros to light-colored granites. These are medium to fine-grained igneous rocks with both gneissic and mylonitic metamorphic foliation. They are part of a composite pluton, less than 16 km (10 mi) long, restricted to the coast between Laguna Verde and Punta Gallo. They intrude Paleozoic gneisses and granitic rocks (Gana et al., 1996).

The rocks in this roadcut are granodiorite intruding older gneiss. You can also find zones of basic (dark) migmatite and pegmatite dikes in the granodiorite. All are part of the volcanic arc that existed in this area from Permian through Jurassic times.

Granodiorite intruding older gneiss at the Sheraton Miramar roadcut.

Sheraton Miramar to Mirador Faro Punta Angeles Granodiorite: *Continue driving south on Av. La Marina; turn right (west) onto Av. España; bear right to take the road to Puerto; continue straight onto Errázuriz; continue straight onto Antonio Varas; continue straight onto Altamirano; continue straight onto Camino Costero and pull into the overlook parking area on the right. This is* **Stop 18, Mirador Faro Punta Angeles Granodiorite** *(−33.022486, −71.648017) for a total of 13.1 km (8.1 mi; 22 min).*

Stop 18 Mirador Faro Punta Angeles Granodiorite

Punta Angeles has been mapped as a "zone of magmatic enclaves." These are zones below the Permian-Jurassic volcanic arc where the mixing of acid (light) and basic (dark) magmas occurs (Gana et al., 1996). The Paleozoic granodiorite, orthogneiss, granite, and migmatite, followed later by Jurassic gabbro south and east of Valparaíso, are found within the upper Paleozoic Valparaíso Metamorphic Complex. This complex consists of foliated metamorphic rocks derived mainly from rock that originally was a deep igneous intrusive. Metasediments also occur as bodies less than 2 km (1.2 mi) long within the coastal intrusive complex. The metamorphism has been dated between 330 Ma and 278 Ma (Upper Carboniferous to Permian), more-or-less the same timing as early emplacement of the regional magma chamber.

View south from Mirador Punta Angeles. This is granodiorite of the Coastal Batholith.

Granodiorite of the Coastal Batholith at Mirador Punta Angeles.

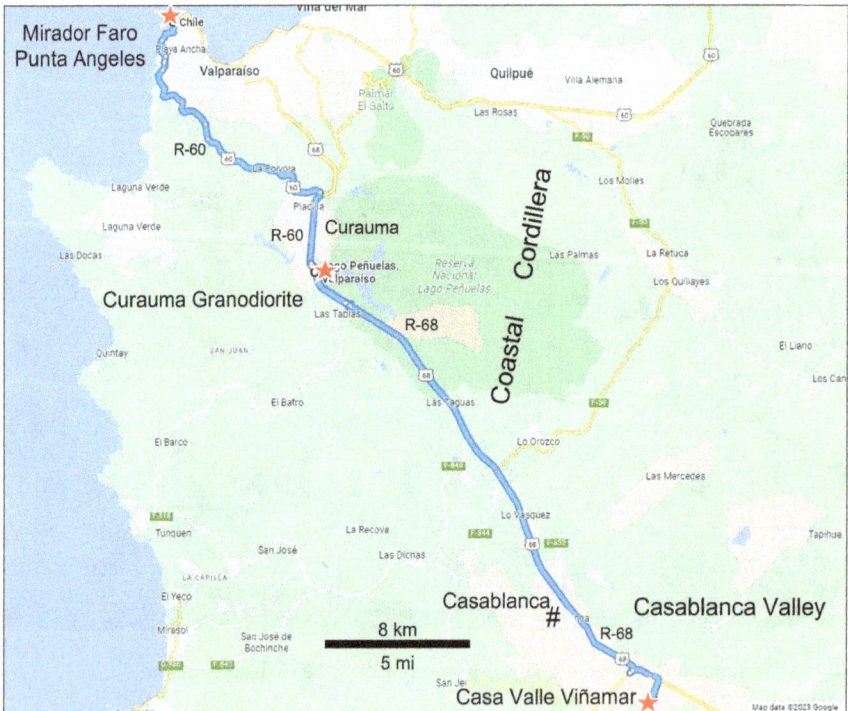

Mirador Faro Punta Angeles Granodiorite to Sauce Diorite, Curauma: Continue south on Camino Costero; continue straight onto Cam. La Pólvora; merge onto R-60 east; use the right lane to merge onto R-68 east/Camino a Santiago; drive 6.8 km (4.2 mi) on R-68 and exit at Retorno; cross the highway and merge back onto R-68 going northwest to Valparaíso; drive 2.3 km (1.4 mi) and exit on the right into the "Area de Control" parking area. This is Stop 19, Sauce Diorite, Curauma (−33.146540, −71.559674) for a total of 28.6 km (17.8 mi; 26 min). This is a parking area for vehicles wrecked in highway accidents.

Stop 19 Sauce Diorite, Curauma

The Late Jurassic Sauce Diorite of the Coastal Cordillera is exposed at this stop. It is part of a plutonic complex that contains diorites with tonalites and subordinate gabbros. In certain outcrops these units appear in the form of alternating bands 1–3 m (3–10 ft) thick. These magmas formed beneath volcanic arcs associated with plate subduction.

The rocks are dark to medium gray in color, and medium to fine-grained. The contact with other nearby Jurassic intrusives is gradual, whereas the unit is in fault contact with Paleozoic granites. The age is in the range 155–157 Ma (Gana et al., 1996).

Outcrop of Sauce Diorite at the Carauma stop parking area.

Sauce Diorite, Carauma stop.

> ***Sauce Diorite, Curauma to Casablanca Valley Vineyards***: *Merge onto R-62 northwest and immediately exit right to Peñuelas; turn left (west) onto Paso Inferior Lago Peñuelas and pass under R-62; turn right (north) on R-718 and drive to the sign for Santiago; turn right and immediately turn right again to merge onto R-68 east to Santiago; take the exit to Algarrobo/San Antonio; follow the signs to Viñamar and turn left (east) onto the frontage road; turn right (south) at the entrance to Casa Valle Viñamar and drive to the parking area at the villa. This is **Stop 20, Casablanca Valley Vineyards** (–33.358278, –71.353042) for a total of 33.8 km (21.0 mi; 25 min).*

Stop 20 Casablanca Valley Vineyards

Casablanca Valley is the first of three Chilean wine-growing regions we will visit on this transect. The other two are the Maipo Valley and Colchagua Valley.

Conquistadores and missionaries brought the first vines in the 1500s. Later, as Chilean aristocrats visited France, they returned with Cabernet Sauvignon, Sauvignon Blanc, Franc, Merlot, and Malbec grapes. Chile's vineyards benefitted from the phyloxera grape blight that ravaged Europe during the 1850s to 1870s, and yet a combination of bureaucracy and high taxes held back the wine industry in Chile during much of the twentieth century. The wine was considered low quality and mostly consumed in-country. A resurgence of wine-growing began in the 1980s and has continued ever since, supported largely by international sales. As of 2015, Chile was the fifth-largest exporter and seventh-largest producer of wines in the world. Concentrated in the Valle Central (Central Valley) around Santiago, the industry is favored by a dry climate, wide daily temperature variation, and low risk of frost. Most vineyards are irrigated by rivers originating in the Andes to the east. Overall, the climate is temperate, with summers generally in the range of 15–18ºC (59–64ºF) and highs no more than 30ºC (86ºF).

A unique combination of climate and soils has made the Casablanca Valley Chile's top white-wine region, and it has recently become a major wine-tour destination given its location between Santiago and Valparaíso. The valley extends 30 km (18 mi) miles east-west from near the Pacific to the interior of the Coastal Cordillera. At no point is the valley more than 32 km (20 mi) from the Pacific. This geography makes the wines distinctive. The valley benefits from the moderating effects of the Pacific. The cold Humboldt Current, which flows up the west coast of Chile, leads to cooling afternoon breezes that blow from the ocean toward the mountains, filling the vacuum created by warm air rising in the east. Cool fog rolls over the valley overnight. Frequent cloud cover mitigates the harsh summer temperatures. The average summer temperature is 25ºC (77ºF), and the annual average temperature is 14ºC (57ºF). It usually rains between May and October. Harvest extends from mid-March to the end of April. Grapes are able to ripen longer, maintaining a balance of sugars and acids, and resulting in more intense, complex, fruity flavors (Winesearcher, 2018; London Wine Competition, 2023; Great Wine Capitals, 2023).

The first vineyards were planted in the Casablanca Valley in the 1980s. It is now known for producing some of the finest white wines, including Sauvignon Blanc and Chardonnay, in the entire country. In recent years, winemakers have experimented with a range of crisp, aromatic whites, such as Viognier, Gewurztraminer, and Riesling. Pinot Noir, which responds well to the cooler climate in this valley, is also grown with some success (Winesearcher, 2018; London Wine Competition, 2023).

The Casablanca Valley's diverse soils give the valley the ability to accept new varieties such as Cabernet Franc and Malbec. It has recently begun producing sparkling wine thanks to the quality of its Chardonnay and Pinot Noir. The sandy clay soils, derived from erosion of the Coastal Cordillera, are well-drained and well-suited for vines (Winesearcher, 2018; Great Wine Capitals, 2023).

The expansion of vineyards around the industrial town of Casablanca followed the initial success of the industry here, and vines now dominate the landscape. Only a lack of water for irrigation (and restrictive local laws related to water use) have held back vineyard expansion (Winesearcher, 2018).

The Casablanca Valley, totaling about 22,000 ha (54,400 ac), has more than 5,000 ha (12,400 ac) under cultivation. It is crossed by Route 68, the highway that links Santiago to Valparaíso. The valley has 11 vineyards open to tourists and offers a culinary and tourism experience that includes wine museums, bike tours, picnics, and restaurants (Great Wine Capitals, 2023).

The choice of vineyards mentioned on this tour is not intended to endorse any particular winery. It is pure chance, in that these are the places that happened to be open when we drove by. In the Casablanca Valley during March 2023, it was Casa Valle Viñamar.

Casa Valle Viñamar

Casa Valle Viñamar was founded to develop a line of sparkling wines.

The winery is easily accessible, only 30 minutes from Valparaíso and less than an hour from Santiago. The excellent wines, active community-building, and focus on sustainable viticulture have earned the winery a coveted Great Wine Capital's "Global Best of Wine Tourism" Award.

Sparkling white wines are the specialty at Casa Valle. Each guest is welcomed with a complimentary glass on arrival. There are eight varieties of sparkling wines with different taste profiles, from Viñamar ICE, a sweet and refreshing wine, to the Extra Brut, a dry wine characterized by its balanced acidity. Visitors sample the selection inside a large, white Moroccan-style mansion with stunning views of the vines and hills beyond.

The winery's cool coastal climate and rich soils are perfect for growing the Chardonnay, Pinot Noir, and Sauvignon Blanc found in the 98-ha (242-ac) vineyard.

The best time to visit is probably October to April, when the grapes are growing and the warm weather makes tastings on the terrace a most pleasing experience.

Casa Valle offers tours of the winemaking process, wine tastings, and is available for private events. Spring is ideal for picnics, and thematic wine-tasting sessions vary each month. Casa Valle is available for private visits, corporate events, working days, and weddings.

Their store has all of the Viñamar sparkling wines. The store also acts as a showcase for local handicrafts and products created by the community of Casablanca. The winery collaborates with local entrepreneurs, providing spaces for them to sell their products and empowering them through sales and marketing workshops.

The winery also has Macerado Bistró, one of Casablanca's most renowned restaurants. The bistró is designed to showcase the wines as well as local food of the Casablanca Valley. The ultimate meal is the "Trip through Casablanca," a four-course lunch that pairs wines with elegant dishes made from fresh local produce.

Visit

Address: Camino interior Nuevo Mundo S/N Ruta 68, Km. 72, Casablanca.

Hours: Open Monday to Sunday.

Winter – 10:00 am to 17:30 pm.

Summer – 10:00 am to 18:30 pm.

Phone: +56 32 3313387 and +56 32 3313385.

Website: http://vinamar.cl.

Macerado Bistró Reservations: +56 9 7831 4823 or vinamar@macerado.cl.

View northwest over the Casablanca Valley from the entrance to Casa Valle Viñamar.

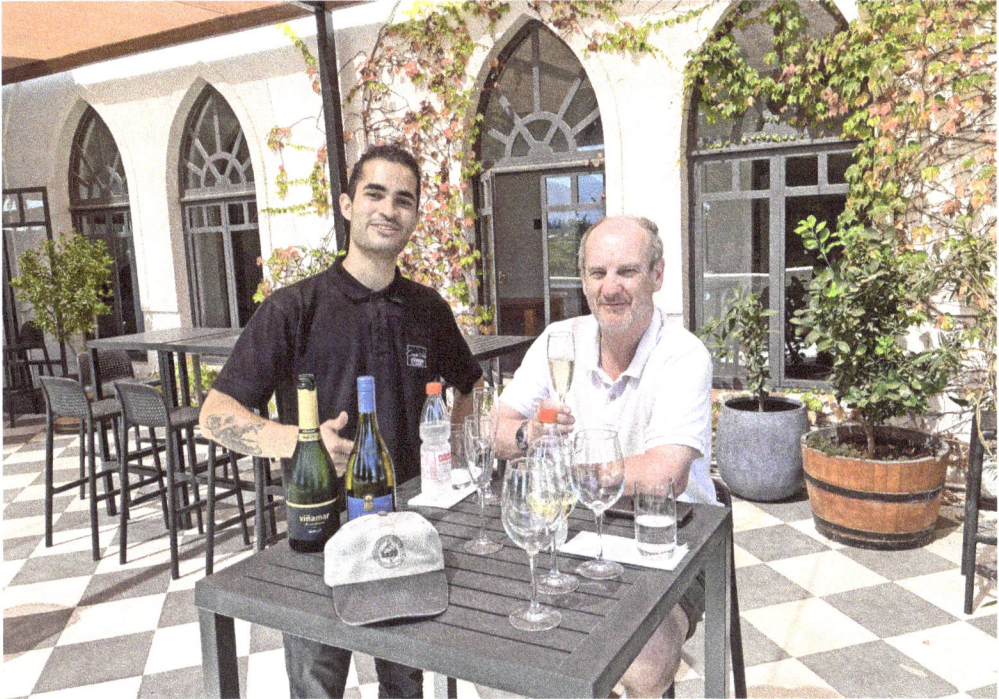

Casa Valle Viñamar: Enjoying a cool beverage on a warm summer's day.

Santiago

If you get a chance to stop in Santiago, do it. Do it for the urban culture. Do it for the geology. Do it for the first-rate wines that are produced in the nearby valleys.

Santiago is the largest city in Chile, with a population close to seven million, and is the industrial, agricultural, retail, cultural, and political center of the nation.

History

The original inhabitants of Chile were nomadic hunters. Shortly before Europeans arrived, the Incas made incursions into the area and established some strongholds, but fierce opposition by the native Araucanians, in particular the Mapuche ("people of the land") group, prevented a strong Inca influence.

The Araucanians, hunter-gatherers and small farmers, were the largest Native American group in Chile. The farmers among them lived in scattered small villages and traded with and made war on other indigenous groups. Those in central Chile were more settled and used irrigation to grow corn, potatoes, and beans, and raise llamas. Those living in the south combined hunting guanacos with slash-and-burn agriculture. With an estimated population of close to one million when the Spanish arrived, the Araucanians became famous for their stiff resistance to colonization. They quickly adapted to horses and added guns to their traditional clubs and bows and arrows. They raided Spanish settlements and managed to resist the Europeans until the late 1800s, when their numbers had dwindled to a few hundred thousand. Today they are considered the first heroes in Chile's national mythology.

The earliest Spaniards, a group sent south from Peru by Francisco Pizarro and led by Pedro de Valdivia, arrived in the Mapocho Valley in December 1540. In February 1541, they founded the city of Santiago del Nuevo Extremo along the Mapocho River. The first industry in the area was the raising of livestock. The city suffered a number of natural disasters, including earthquakes, a smallpox outbreak in 1575, and the flooding of the Mapocho River. In September 1810, Chile established its first non-colonial government, starting the war for independence, which lasted until the Army of the Andes, under General San Martín, defeated the Royalists at the Battle of Chabuco and the Battle of Maipú in 1818. Santiago became the capital of the new republic. The mid-1800s saw the establishment of the University of Chile, Museum of Fine Arts (now the Museum of Science and Technology), Museum of Natural History, Teatro Municipal, and O'Higggins Park, the city's equivalent to New York's Central Park with gardens, lakes, and trails. By the late 1800s, Santiago had replaced Valparaíso as the banking and industrial center of Chile.

Geology

We have left the Coastal Cordillera and entered the Central Valley. The Santiago Basin (part of the Central Chilean Basin) is a north-south-elongated alluvial valley bounded by the Andes (Cordillera Principal) on the east and by the Coastal Cordillera (Cordillera de la Costa) on the west.

Plio-Pleistocene alluvial and fluvial deposits of the Maipo and Mapocho rivers filled the Santiago Basin. Gravity data indicates that this is a relatively shallow basin with an average depth of 250 m (820 ft), and sub-basins with depths sometimes exceeding 500 m (1,640). Pyroclastic deposits (volcanic ash and rock fragments) of the Ignimbrita de Pudahuel occur locally in the western and central parts of the basin. Rocks beneath the western Santiago Basin comprise Jurassic to Upper Cretaceous volcanic and sedimentary sequences and Jurassic to Cretaceous intrusive rocks. Beneath the eastern part of the basin are Cenozoic volcanic and minor sedimentary units of the Cordillera Principal (Yáñez et al., 2015).

Santiago and the view east to the Andes from Cerro San Cristóbal.

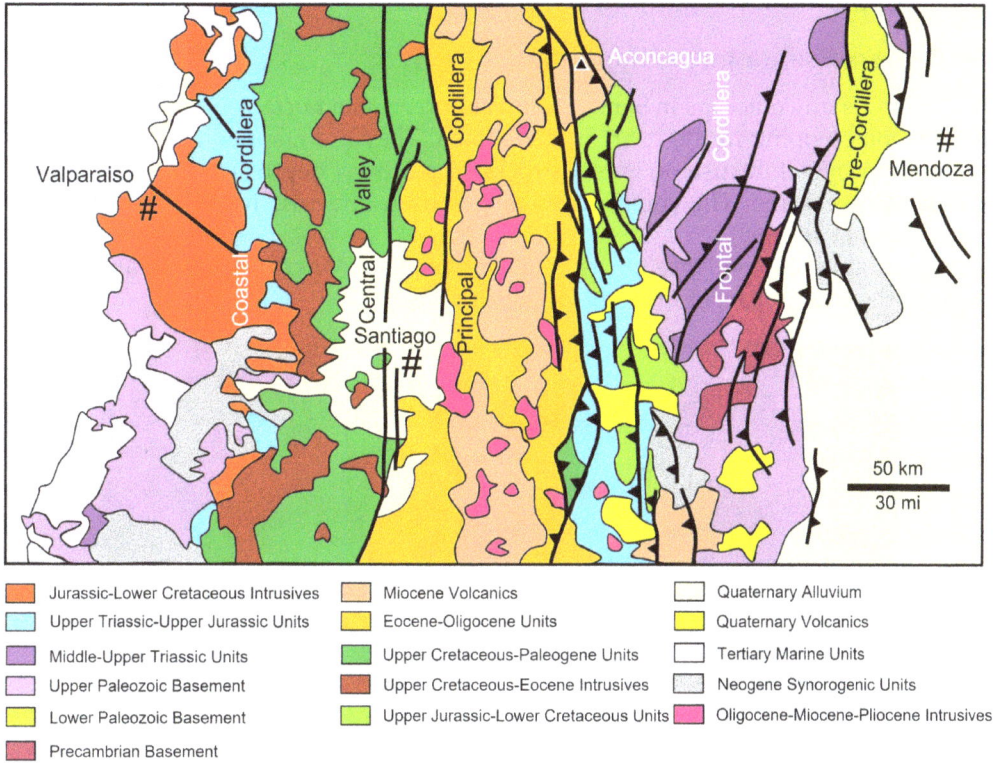

Generalized geologic map of Santiago and surrounding areas. Modified after Farias, 2007; Charrier et al., 2014.

Casablanca Valley to Cerro Santa Lucía, Santiago: *Return to the frontage road and turn right (southeast); in 1.2 km merge onto R-68 east; on the western edge of Santiago take the right two lanes to exit onto Costanera Norte al Oriente; exit onto Av. Cardenal Caro; turn left (south) toward Ismael Vergara; turn left (east) onto Ismael Vergara; turn right (south) onto Jose de la Barra; pull over on the right into Estacionamiento Santa Lucía and park. This is* **Stop 22, Cerro Santa Lucía, Santiago** *(−33.440011, −70.644278) for a total of 76 km (47.2 mi; 60 min).*

Stop 21 Cerro Santa Lucía, Santiago

There are not many outcrops in Santiago, but at Cerro Santa Lucía you can find dark igneous rocks outcropping in the heart of downtown. They are mapped as Eocene to early Miocene hypabyssal intrusives, igneous rocks emplaced at depths of less than 2 km (1.2 mi) beneath a volcano. The composition ranges from andesitic to basaltic andesites to dacites (Gana et al., 1996).

These volcanics are evidence of the Late Cretaceous to Miocene volcanic arc that once existed in the eastern part of the Central Chilean Basin.

Lastarria monument and outcrops, Cerro Santa Lucía. This hill is Eocene to early Miocene andesite and dacite. Lastarria (1817–1888) was Minister of Finance from 1862–1863.

Dacite outcrop near the Lastarria monument, Cerro Santa Lucía. 100 peso coin for scale.

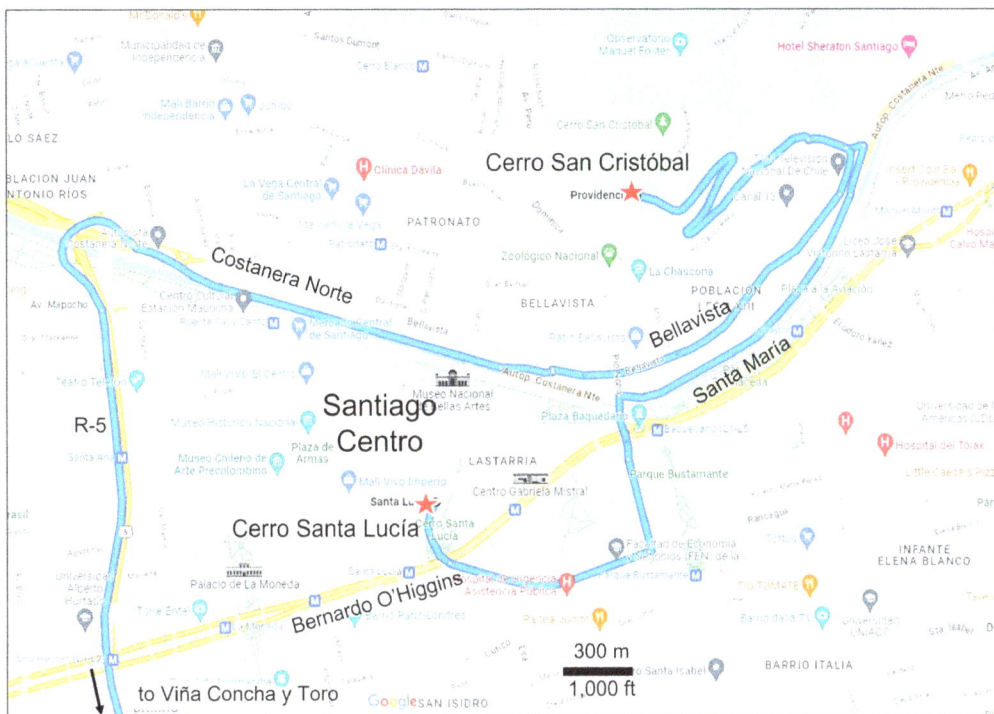

Cerro Santa Lucía, Santiago to Cerro San Cristóbal, Santiago*: Continue south on Santa Lucía; use the left lane to bear left (east) on Diagonal Paraguay; turn left (north) on Av. Vicuña Mackenna; use the right three lanes to stay on Av. Vicuña Mackenna and cross the Mapocho River; turn right (east) onto Av. Santa Maria; turn left (west) onto Av. Carlos Reed; park and take the path to the top. This is **Stop 22, Cerro San Cristóbal, Santiago** (–33.427770, –70.634307) for a total of 5.6 km (3.5 mi; 16 min).*

Alternatively, you could walk (5.5 km, 3.4 mi; about 1 hr 20 min) or take a taxi and the funicular (it is a nice walk).

Stop 22 Cerro San Cristóbal, Santiago

Cerro San Cristóbal is a hill and large park on the north side of central Santiago. It rises about 300 m (1,000 ft) above Santiago and provides sweeping views of the city and Andes. Cerro San Cristóbal was named by the Spanish conquistadors for St Christopher, patron saint of travelers, since it was used by travelers as a landmark. The Mills Observatory was built on the hill in 1903. Currently known as the Manuel Foster Observatory, it is a twin of the Lick Observatory of the University of California. The summit has a sanctuary dedicated to the Immaculate Conception, with a 22-meter-high (72 ft) statue of the Virgin Mary, an amphitheater, and a chapel. The statue is lit at night and can be seen from all of Santiago. Cerro San Cristóbal is Santiago's largest public park, the *Parque Metropolitano*. Climbing to the summit involves about a 45-minute walk, or you can drive via the road joining the Santiago Metropolitan Park, or you can take the Funicular de Santiago from the base next to the Zoo at the North end of Pio Nono in Barrio Bellavista (Wikipedia, San Cristóbal Hill).

The main peak is composed of the same Eocene to early Miocene andesites, basaltic andesites, and dacites we saw at Cerro Santa Lucía. Further northeast on the hill are outcrops containing the lower part of the Late Cretaceous-early Miocene Abanico Formation. The Abanico is primarily andesites and basalts with interbedded tuffs, breccias, and continental sediments (Gana et al., 1996). These units are part of the Late Cretaceous to Miocene volcanic arc that was primarily east of the Central Basin.

Cerro San Cristóbal, Santiago to Maipo Valley and Viña Concha y Toro**: Return to Pedro de Valdivia and drive south; turn right (southwest) on Los Conquistadores; use the right lane to take the ramp to Autopista Costanera Al Poniente (west) and merge onto the freeway; take Exit 18 to R-5 South; take the exit to Vespucio Sur/Lo Espejo; use the right lane to take the ramp to Av. Vespucio Al Oriente (east); drive east on R-70/Av. Vespucio Sur to Exit 36, Al Sur Rancagua; drive south on Acceso Sur; turn right (west) onto Arco Iris; turn right (south) onto Caletera Acceso Sur/La Serena; turn left (east) onto Sgto Menadier; turn right (south) onto Av. Salvador Allende; turn right (south) onto Av. Concha y Toro; turn right (west) onto Av. Virginia Subercaseaux; turn right into the vineyard, **Stop 23,

Maipo Valley and Viña Concha y Toro parking (−33.635722, −70.574153) for a total of 41.8 km (26.0 mi; 1 hr). You can also use public transit (metro and bus).

Stop 23 Maipo Valley and Viña Concha y Toro

Santiago lies near the epicenter of the Chilean wine-growing region. It is directly west of Argentina's well-known Mendoza wine-growing region. The region is bounded on the east by the Andes, and on the west by the Coastal Cordillera. The valley floor lies at 400 m (1,300 ft) in the west and rises to 800 m (2,600 ft) in the east.

The Maipo Valley (2,800 ha, or 7,000 ac) has been referred to as the "Bordeaux of South America." The valley is divided into three sub-regions: Alto Maipo, Central Maipo, and Pacific Maipo. Red wines produced in the Maipo valley include Syrah, Cabernet Sauvignon, Carmenère, Merlot, and Malbec. White varieties include Chardonnay and Sauvignon Blanc.

Alto Maipo is in the foothills of the Andes at altitudes between 400 and 800 m (1,300 and 2,600 ft). The region has a mountainous climate and extremely porous and rocky soil that results in a bold and elegant Cabernet Sauvignon with a fruity aroma and packed with flavor. Alto Maipo is also known for its red blends that are full-bodied and similar to the Cabernet Sauvignon of the Napa Valley. The red blends have notes of blackcurrant, fig, black cherries, and baking spices.

Central Maipo, located in the floodplain of the meandering Maipo River, extends into the southern suburbs of Santiago and is where the first vines were planted in the 1540s. Cabernet Sauvignon dominates here, with over 1,400 ha (3,500 ac) dedicated to this grape alone. But recently, Central Maipo has also begun producing Carmenère wines. Carmenère is a medium-bodied red wine similar to Merlot that is now grown primarily in Chile. Carmenère wines of the Maipo Valley have a deep red color and flavors of ripe fruits and coffee with notes of plum, spices, and blackberries. The climate is cool to cold at night and sunny and hot during the day. Alluvial soils are rocky, well-drained, and poor in nutrients, a factor known to concentrate the flavor in the grapes.

Pacific Maipo, along the banks of the Maipo River in the western parts of the Central Valley and into the Coast Range, benefits from the influence of the Pacific Ocean (fog, cool days, temperate nights) and alluvial soils. White-wine grapes are mainly grown in the coastal areas. Sauvignon Blanc from the Maipo Valley has a recognizable straw-yellow color. The wine hints at flavors of pears with tropical fruits and green apples. Chilled Chardonnay of this region is refreshing to drink, with notes of pineapple, oranges, lychees, and fresh herbs. The red wines produced in the Pacific Maipo are refreshing, with natural acidity.

Lastly, the relatively small (1,098 ha; 2,700 ac), arid Aconcagua Valley lies about 70 km (45 mi) north of Santiago and has cool, wet winters and hot, dry summers. This valley, which we passed through both east and west of San Felipe, is known for superior quality Carbernet Sauvignon, Merlot, and Syrah wines (Dunnell, 2017).

Viña Concha y Toro

Viña Concha y Toro is the largest producer and exporter of wines in Latin America, and among the ten largest wine companies in the world. Headquartered in Santiago, Chile, it is distributed in 135 countries. In 2016, the company had over 10,000 ha (2,470 ac) under cultivation, with vineyards in Chile, Argentina, and the United States (Wikipedia, Concha y Toro). It is in the Alto Maipo region.

Concha y Toro was founded by Don Melchor de Santiago Concha y Toro, VII Marquess of Casa Concha, and a prominent Chilean lawyer, politician, and businessman, along with his wife, Emiliana Subercaseaux, in 1883. He brought vines from the Bordeaux region in France, primarily Cabernet Sauvignon, Sauvignon Blanc, Semillon, Merlot, and Carmenère (Concha y Toro, Our History; Wikipedia, Concha y Toro).

The vineyard was incorporated in 1923. Concha y Toro began exporting wine to the port of Rotterdam in 1933. In 1950, the winery began to acquire more vineyards and also began adapting its business to new markets and meeting higher demand. In 1966, the company initiated *Casillero del Diablo*, its hallmark label, and in 1987, the company launched its first ultra-premium wine, *Don Melchor*. Viña Maipo, was acquired in 1968; Viña Cono Sur, was created in 1993; Trivento Bodegas in Argentina was started in 1996; a joint venture with Château Mouton Rothschild for the production of Viña Almaviva began in 1997. In 2011 Concha y Toro bought Fetzer Vineyards, a California winery, including subsidiaries such as Bonterra Vineyards, Five River Wines, Bel Arbor Winery, Jekel Vineyards, and Little Black Dress Wines.

Viña Concha y Toro is controlled by the Guilisasti and Larraín families. The company director is Mariano Fontecilla de Santiago-Concha, the present Marquess of Casa Concha and a Chilean diplomat and ambassador and great-grandson of the founder.

Concha y Toro produces several wine varieties. White wines include Chardonnay, Sauvignon Blanc, Semillon, and Gewürztraminer. Red wines include Cabernet Sauvignon, Merlot, and Carmenère. The company has won several international awards including *The World's Most Admired Wine Brand* in 2011, 2012, and 2013, and *The Americas' Most Admired Wine Brand* from 2011 to 2019 (Wikipedia, Concha y Toro).

Visit

No reservation is needed. Tours are given in Spanish and English and include a modest wine tasting. The tour takes you to the old mansion, around the park-like grounds, and to a section of the vineyard that has all of the varieties of grapes they grow. You even get to pick and taste the grapes right off the vine. More elaborate wine tastings are available. There is a restaurant and wine shop.

Address	Av. Virginia Subercaseaux 210, Pirque, Región Metropolitana, Chile.
Phone	56 224 765 100.
Hours	9:00 am to 5:00 pm.
Tour fee	from 30,000 p to 100,000 p.
Tour duration	70 minutes to 120 minutes.

Concha y Toro vineyard, Maipo Valley.

To avoid Side Trip 2:

> *Maipo Valley and Viña Concha y Toro to Diorite, Hotel Monticello Roadcut*: *Continue west on Virginia Subercaseaux; turn left (south) on G-45/Padre Hurtado; turn right (west) on G-51/El Arpa; turn left (south) onto the ramp to Rancagua via Acceso Sur; take the casino/Hotel Monticello exit and drive to the parking area. This is* **Stop 24, Hotel Monticello Roadcut** *(−33.920548, −70.724181), for a total of 41.9 km (26.0 mi; 40 min).*

To go on Side Trip 2:

> **−Side Trip 2, Maipo Valley and** *Concha y Toro to ST 2.1, El Toyo Roadcut*: *Exit the winery parking lot and turn left (east) on Av. Subercaseaux; continue straight onto Camino El Toyo; cross the Maipo River and turn left (west) onto Camino al Volcán; drive 580 m (1,900 ft) and pull over on the right. This is* **ST 2.1, El Toyo Roadcut** *(−33.673527, −70.345000) for a total of 34.1 km (21.2 mi; 37 min).*

Side Trip 2 Maipo Canyon

The scenic Maipo Canyon is a favorite of white-water rafters and is an easy getaway from the bustle of Santiago. It takes you into the heart of the High Andes. The scenery (and altitude) is literally breathtaking. Geologically, this valley provides a unique transect into the heart of the Aconcagua Fold-Thrust Belt in central Chile. The units range from Jurassic and Cretaceous to Miocene: The Mesozoic units are marine and continental with a sprinkling of volcanics; the Tertiary is primarily volcanic, reflecting subduction and the related Andean volcanic arc. There are a few scattered Miocene intrusives, and Recent volcanics from the San José Volcano are also found in the far eastern parts of the valley.

The oldest unit widely represented in Maipo Canyon is the Río Damas Formation. The uppermost Jurassic continental and volcanic deposits of the Río Damas Formation represent intense continental deposition in the Jurassic to Early Cretaceous dominantly marine Mendoza-Neuquén back-arc basin. Uplift of the volcanic arc led to a marine regression and deposition of continental sediments in the back-arc basin. Volcanism was associated with the extension in the back-arc basin. The volcanic arc at that time was located at the present-day Coastal Cordillera. The sedimentary rocks of the Río Damas Formation and laterally equivalent Tordillo Formation in Argentina represent Upper Jurassic continental back-arc deposits in the Cordillera Principal (Rossel et al., 2014).

Above the Río Damas Formation are the Late Jurassic-Early Cretaceous Baños Morales and Lo Valdés formations. The Baños Morales Formation is dominated by andesitic volcanics, whereas the Lo Valdés Formation is exclusively sandstones, shales, and limestones.

In 2017, the field course at the University of Chile mapped the Volcán River valley in detail. Their cross-section shows the Aconcagua Fold-Thrust Belt as a series of east-vergent, thin-skinned detachments that were deformed into anticlinal and synclinal folds. They propose that deformation progressed from west to east and that minimum shortening in this sector is 53%, or about 34 km (20.4 mi; Aranguiz et al., 2017).

The overlying Early Cretaceous Colimapu Formation consists of redbed conglomerates, sandstones, and mudstones (Sempere et al., 1994; Charrier et al., 2002). It represents the thickest uninterrupted sequence of continental sediments in the Andean portion of central Chile (Bowes et al., 1962).

Unconformably above the Colimapu is the BRCU (Brownish-Red Clastic Unit), mainly Late Cretaceous river deposits.

Above the BRCU is the Late Cretaceous to Miocene Abanico Formation. The Abanico Formation is at least 3,000 m (10,000 ft) thick and consists of interbedded pyroclastic and, locally, red to purple continental sediments, mostly eroded volcanics. In many areas, the Abanico was strongly deformed during the Andean Orogeny. The Abanico is interpreted as having been deposited in a volcanic arc setting (Sempere et al., 1994). The Abanico Formation's great thickness and the presence of repeated interbeds of river and lake deposits all suggest accumulation in a large, strongly subsiding, and probably north-south oriented basin. Deposition of the Abanico Formation is thought to be related to crustal extension, and its later deformation has been related to tectonic compression and structural inversion during the Andean Orogeny (Charrier et al., 2002).

The Farellones Formation lies above the Abanico and is often hard to distinguish from the older unit. However, it is dominantly Miocene, is less deformed, and has more volcanic ash than the Abanico Formation (Sempere et al., 1994; Charier et al., 2013).

The youngest rocks in the area are volcanics. Unconformably above the Farellones Formation in the upper Maipo/Río Colina valleys are Quaternary volcanics of the San José Volcanic Group. The andesitic lavas and pyroclastics are derived from 5,856 m high (19,213 ft) Volcán San José, a large stratovolcano on the Chile-Argentina border (Lopez-Escobar et al., 1985).

Geologic map of the Maipo Canyon showing stops. Information drawn from Thiele, 1980; sernageomin.cl, online geologic map of Chile.

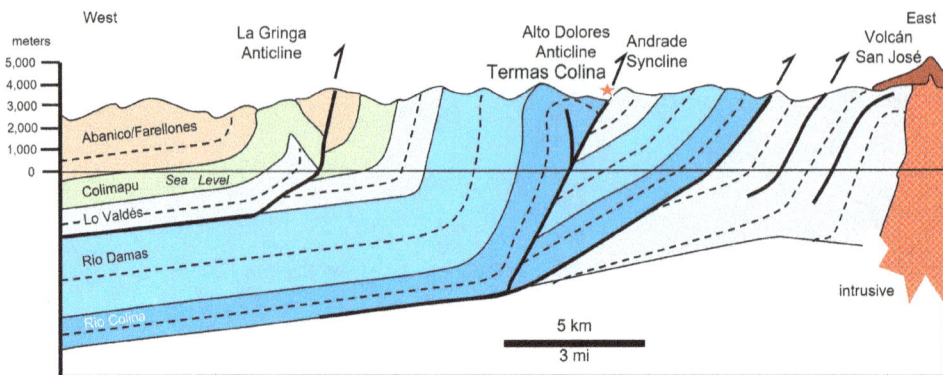

Cross-section through the upper Cajón Maipo/Río Colina valleys. Location is shown on the previous figure. Modified after Thiele, 1980; Aranguiz et al., 2017.

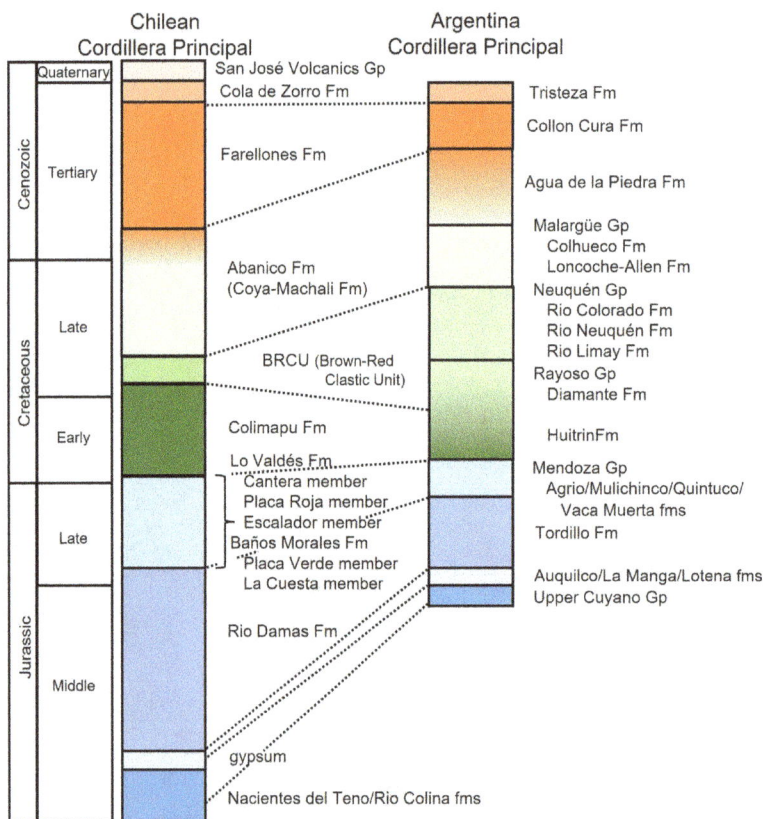

Stratigraphy of the Cajón de Maipo (Maipo Canyon) and the equivalent units on the Argentine side of the Andes. Information taken from Charrier et al., 2002; Salazar and Stinnesbeck, 2015.

Stop ST 2.1 El Toyo Roadcut

This is a great roadcut into gently southeast-dipping dark andesite. The rock has been mapped (Thiele, 1980) as Miocene Farellones andesites and rhyolites with interbedded continental sediments.

Farellones andesite at the El Toyo roadcut.

> **Side Trip 2, El Toyo to Paso Angosto El Tinoco**: *Continue driving south and east on Camino al Volcán for 10.2 km (6.3 mi; 12 min) to* **Stop ST 2.2, Paso Angosto EL Tinoco** *(–33.742670, –70.290845) and pull over on the right.*

Stop ST 2.2 Paso Angosto El Tinoco

Late Cretaceous to early Miocene Abanico Formation andesites, rhyolites, and interbedded continental sediments (Thiele, 1980) are folded into an east-verging, north-south-trending anticline. The Abanico lies unconformably below the Farellones Formation, seen near the top of the ridge.

Paso Angosto panorama, view to the southwest. Bedding is highlighted by the white dashed lines. This appears to be an east-verging anticline in the Abanico and Farellones formations.

*–Side Trip 2, Paso Angosto El Tinoco to Observatorio: Continue driving east on Camino al Volcán for 850 m (0.5 mi; 1 min) to **Stop ST 2.3, Observatorio** (–33.743499, –70.283879) and pull over on the left.*

Stop ST 2.3 Observatorio

A gentle syncline is visible on the south side of the Río Maipo. It is developed in purple andesites of the Abanico Formation.

A gentle syncline is developed in the Abanico Formation. View south from the Observatorio pullout.

Side Trip 2, Observatorio to Tierra Amarilla: *Continue driving east on Camino al Volcán for 25.3 km (15.7 mi; 33 min) to **Stop ST 2.4, Tierra Amarilla** (–33.828668, –70.073313) and pull over on the right.*

Stop ST 2.4 Tierra Amarilla

The hillside to the south is labeled "Tierra Amarilla" ("Yellow Earth") on the map. It is an area of hydrothermally altered Miocene granodiorite porphyry. The yellow colors are a result of the weathering of pyrite in the altered granodiorite. Pyrite weathers to iron oxide and releases sulfuric acid, which proceeds to convert feldspars in the rock to clay minerals, causing a bleached appearance. When prospectors tell you they are looking for "color," this is what they are talking about.

Until now the dominant dip of bedding has been to the west. Looking east you can see that the units are now dipping east. We are at or near the axis of a large, roughly north-south anticline.

View south to Tierra Amarilla.

> *Side Trip 2, Tierra Amarilla to Mina Lo Valdés: Continue driving east on Camino al Volcán for 2.2 km (1.3 mi; 4 min) to **Stop ST 2.5, Mina Lo Valdés** (−33.828466, −70.053154) and pull over on the right, away from truck traffic.*

Stop ST 2.5 Mina Lo Valdés

You are in the mining camp of Lo Valdés. This is a gypsum mine, one of several in the valley. Gypsum is mined for sheetrock, Plaster of Paris, as a soil additive, and has other uses as well. It is mapped as the Lower Cretaceous Colimapu Formation that dips steeply west (Bowes et al., 1962; Thiele, 1980). The unit is primarily interbedded marine shale, sandstone, conglomerate, and minor andesitic volcanics, but has lenses of gypsum in the upper part. Gypsum, an evaporite, indicates the drying of seawater in a shallow marine environment. We are near the time transition from marine to continental environments.

Looking down on the Mina Lo Valdés gypsum mining operation. View west.

*—Side Trip 2, Mina Lo Valdés to Placa Verde: Continue driving east on Camino al Volcán for 1.3 km (0.8 mi; 3 min) to **Stop ST 2.6, Placa Verde** (−33.828135, −70.043634) and pull over on the right.*

Stop ST 2.6 Placa Verde

Bedding here is now dipping steeply west. The green rocks make up the Placa Verde (literally "Green Plate," but a better translation might be "Green Slab") member of the Late Jurassic-Early Cretaceous Baños Morales Formation. This unit is 643 m (2,110 ft) of dominantly green and purple andesitic volcanics (Salazar and Stinnesbeck, 2015).

The green slab, or "placa verde," near-vertical beds of Late Jurassic Baños Morales Formation andesitic volcanics.

Looking west down the Río Volcán Valley toward Placa Verde (greenish beds) and Mina Lo Valdés gypsum mine (white spot on the valley floor).

About 2.1 km (1.3 mi) up the valley on the south side is a massive rockslide that obviously moved down a bedding plane that is inclined into the valley.

Rockslide that moved down bedding planes in the Lo Valdés Formation. View east.

*—Side Trip 2, Placa Verde to Termas Valle de Colina: Continue driving east on Camino al Volcán for 9.7 km (6.0 mi; 20 min) to **Stop ST 2.7, Termas Valle de Colina** (−33.854171, −69.981235) and pull into the parking area below the hot springs.*

Stop ST 2.7 Termas Valle de Colina

Finally, what we've come all this way for. These hot springs are a popular destination. At 2,484 m (8,150 ft) elevation, the sky is a deep blue, and the milky, steaming turquoise pools are surrounded by colorful and jagged peaks in all directions.

The springs issue from a probable north-south-oriented fault in the Upper Jurassic Río Colina Formation. The Río Colina Formation consists of fossiliferous marine limestone, shale, sandstone, and some gypsum. The hot springs are related to elevated heat flow associated with the still-active volcanoes in the area.

At Termas Valle de Colina, water moves upward through high-permeability structures, possibly local faults, that also allow groundwater from snowmelt to penetrate deep, warm zones. Temperatures over 40°C (104°F) have been measured in water flowing from the Termas Valle de Colina (Anselmo et al., 2017).

A view south up the valley of the Río Colina reveals bedding dipping both east and west. The rocks are of the Late Jurassic Río Damas Formation (reddish purple unit) and Late Jurassic-Early Cretaceous Lo Valdés Formation (light-colored unit to the west). These dips define the axis of the large, north-south-trending Alto Dolores Anticline (Aranguiz et al., 2017). The Río Damas Formation is dominantly continental and volcanic sediments deposited in the Mendoza-Neuquén back-arc basin during a period of extension and magmatism. The main volcanic arc at the time was in the Coastal Cordillera (Rossel et al., 2014).

The unit mapped as Lo Valdés Formation (Thiele, 1980) has been redefined as a lower Baños Morales Formation and upper Lo Valdés Formation. The Baños Morales Formation is up to 760 m (2,500 ft) thick and consists of the lower La Cuesta and upper Placa Verde members. The La Cuesta member is 117 m (384 ft) thick and consists of andesitic volcanics interbedded with thin limestone layers. The limestone and ammonites found in the interbeds suggest an offshore environment of deposition. The Placa Verde consists mainly of andesite.

The re-defined Lo Valdés Formation consists of siliciclastic and carbonate sedimentary rocks. It is up to 539 m (1,770 ft) thick and is interpreted on the basis of fossils as having been deposited, from base to top, in shoreline, offshore transition, and offshore environments (Salazar and Stinnesbeck, 2015).

Termas Valle de Colina pools, view west.

View south up the Río Colina Valley from Termas Valle de Colina. Rocks are of the Late Jurassic Río Damas Formation (reddish-purple, center) and Early Cretaceous Baños Morales/Lo Valdés Formation (light-colored, right). You are looking up the axis of a large anticline, Anticlinal Alto Dolores.

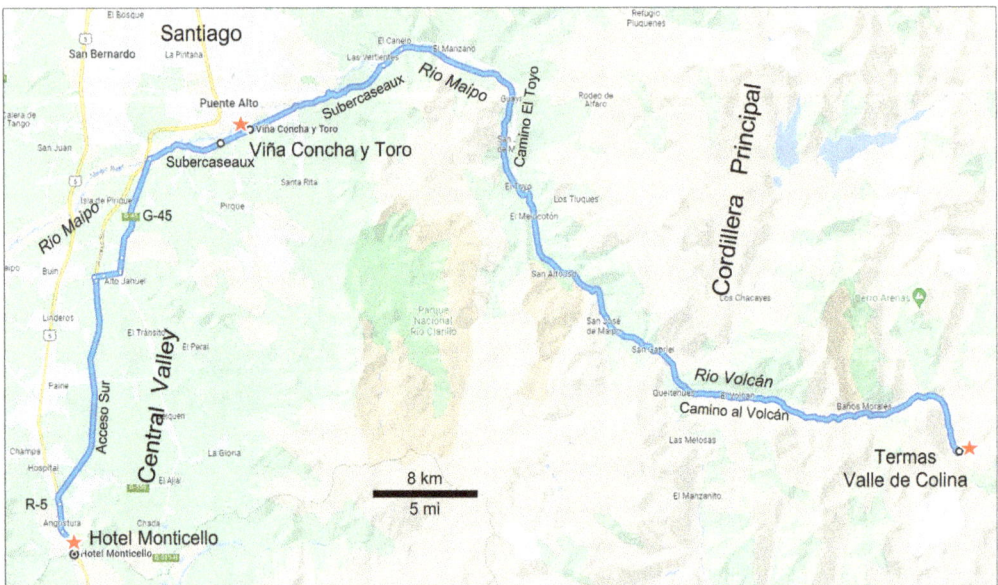

Visit

There is an 8,000 p fee to enter the hot springs area. There is ample parking and changing rooms are available. Bring a towel, bathing suit, and sunscreen. There are toilets and a nearby campground.

> *Side Trip 2, Termas Valle de Colina to Diorite, Hotel Monticello Roadcut: Return west on Camino al Volcán; turn left (west) onto Camino El Toyo; continue straight onto Subercaseaux; turn right (north) on El Toyo; turn left (west) onto Camino al Volcán; continue straight on Camino a San José del Maipo; continue straight onto Camino Henríquez; turn left (west) onto Av. San Carlos; turn left (south) on Nonato Coo; turn right (west) on Domingo Tocornal; use the left lane to turn left (south) onto Acceso Sur; merge onto R-5/Panamericana Sur; take the exit to the casino/Hotel Monticello and drive to the parking area. This is **Stop 24, Diorite, Hotel Monticello Roadcut** (−33.920548, −70.724181), for a total of 127 km (79 mi; 2 hrs 27 min). Walk back to the roadcut.*

Stop 24 Diorite, Hotel Monticello Roadcut

As you approach the Hotel Monticello on R-5/Panamericana Sur, you pass by some amazing roadcuts just north of the hotel. The cuts expose Abanico Formation that has been faulted and intruded by diorite dikes. There are no good pullouts, so glance at it carefully as you drive by.

The foundation for the Hotel Monticello was carved into an outcrop of Miocene diorite that intruded into the Late Cretaceous-early Miocene Abanico Formation (Thiele, 1980). This is our first chance to look close-up at a Tertiary intrusive.

Hillside of Miocene diorite at the Hotel Monticello.

Miocene diorite, Hotel Monticello roadcut.

Roadcut on R-5 just north of Hotel Monticello showing faults and diorite dikes and sills in the Abanico Formation. Dotted lines are faults. There is no good place to stop here, so watch for it. Street View east.

Diorite, Hotel Monticello Roadcut to Centinela Roadcut: *Return to R-5 south; take the exit for Santa Cruz/San Fernando; use the left lane to take the ramp to Pichilemu/Santa Cruz; merge onto R-90 south; right after the bus stop ("Parada") pull into the parking area on the right. This is* **Stop 25, Centinela Roadcut** *(–34.618561, –70.997110) for a total of 88.4 km (54.9 mi; 56 min).*

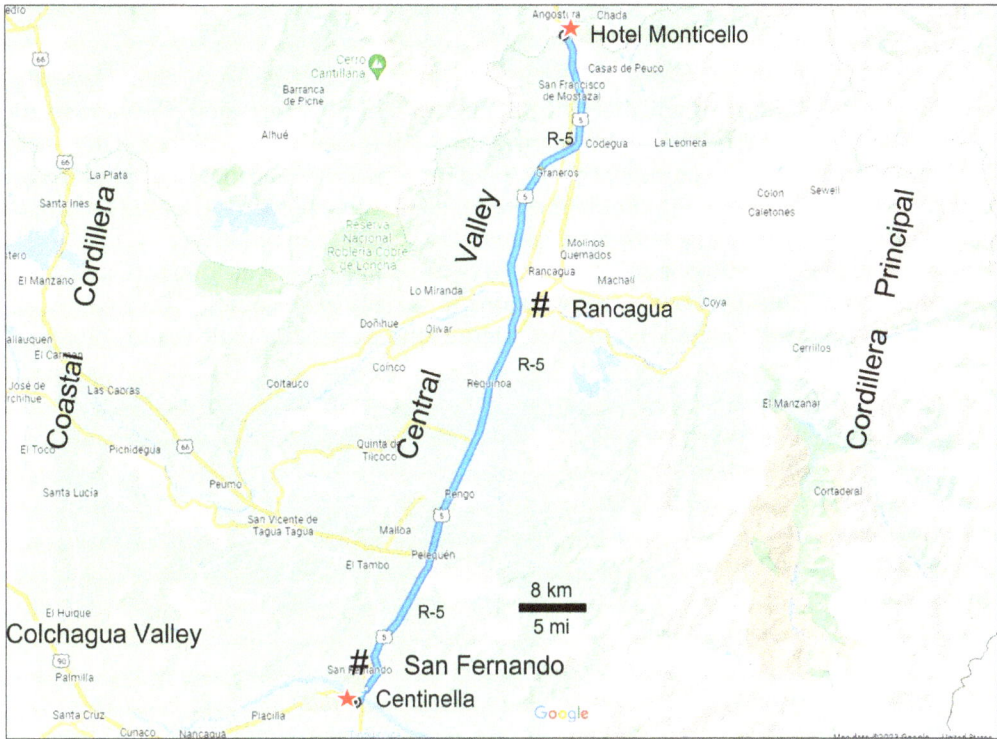

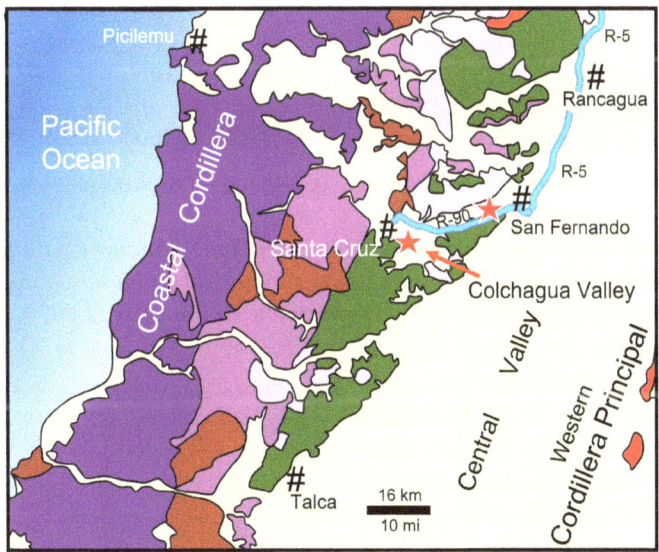

Generalized geologic map of the Colchagua Valley and surrounding areas. Modified after Mpodozis and Conejo, 2012; Muñoz et al., 2018; KLS Geological Tours, 2019.

Stop 25 Centinela Roadcut

We have been driving through the Central Valley/Central Depression, the north-south basin between the Andes/Cordillera Principal and the Coastal Cordillera. This valley, mostly covered in alluvial material, has a few outcrops of Cretaceous sedimentary and volcanic rocks and Upper Cretaceous intrusives. Generally speaking, the units along the western side of the valley are considered part of the Coastal Cordillera.

This stop provides a view of an andesitic dike intruding pink volcanic tuff. The tuff is considered Cretaceous, part of the Coastal Cordillera suite of volcanics.

There is no good way to walk up to this outcrop, and traffic along the road is heavy.

Centinela Roadcut looking south across R-90. A dark andesitic dike intrudes the pink volcanic tuff.

Centinela Roadcut to Ignimbrite, Río Tinguiririca Roadcut: Continue driving west on R-90 for 1.5 km (1 mi; 1 min) and pull over by the shrine on the old road on the right. This is **Stop 26, Ignimbrite, Río Tinguiririca Roadcut** *(−34.609847, −71.008723). Mind the traffic.*

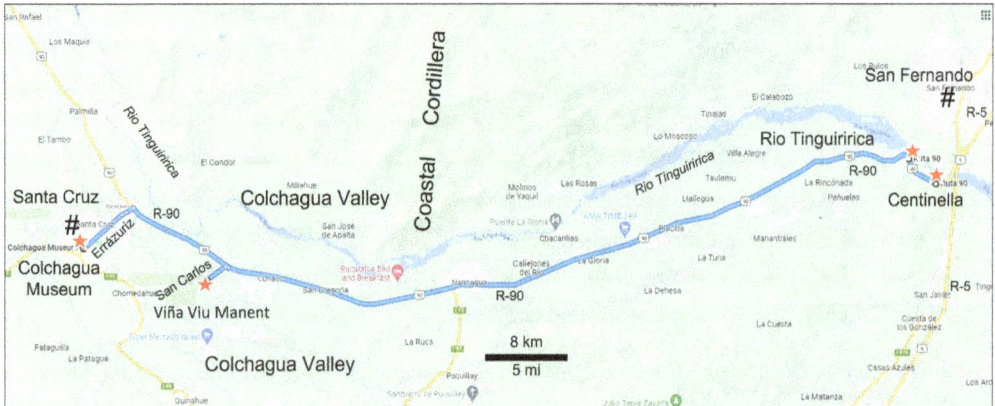

Stop 26 Ignimbrite, Río Tinguiririca Roadcut

Only generalized geologic maps exist for this area. The rocks here are considered "Cretaceous sedimentary and volcanic rocks." This excellent roadcut is in fact a pink tuff or ignimbrite. The angular fragments and poor sorting imply a pyroclastic flow rather than an airfall tuff for this unit.

Río Tinguiririca roadcut pullout on the right. Google Street View northwest.

Cretaceous ignimbrite, Río Tinguiririca roadcut. Note the angular fragments and poor sorting.

Ignimbrite, Río Tinguiririca Roadcut to Viña Viu Manent: *Continue driving west on R-90; after 28.3 km (17 mi) turn left (south) onto I-752/Agua Clara, then immediately turn right (west) toward San Carlos; turn left (southwest) on San Carlos and drive to* **Stop 27.1, Viña Viu Manent** *(−34.651260, −71.310052) on the left for a total of 29.3 km (18.2 mi; 28 min).*

Stop 27 Santa Cruz and the Colchagua Valley

The Colchagua Valley is located about 130 km (80 mi) south of Santiago. Over the past 20 years or so, this valley has developed into one of the largest and most active wine-producing regions in the country. Because the area was designed with wine tourism in mind, practically all wineries have impressive tasting rooms and cellars. The valley was named the "World's Best Wine Region" by Wine Enthusiast Magazine in 2005. It is best known for its red wines, introduced from Bordeaux in the late 1800s. The Colchagua Valley is known as Chile's "Napa Valley." Several of Chile's most prestigious wines come from the Colchagua Valley including, among others, Los Vascos winery in Peralillo, a joint venture between Santa Rita and the Rothschilds of Bordeaux.

The Valley stretches 120 km (75 mi) from the base of the Andes to the Pacific, with consistently warm, sunny days and cool, crisp nights. Roughly 34,400 ha (85,000 ac) of vines are cultivated. The region is characterized by steep slopes and well-drained granitic soils and is irrigated by the Tinguiririca River which brings clear snowmelt waters from the Andes.

Santa Cruz is an agricultural community of 32,000 (2002) in the heart of the Colchague Valley.

Colchagua Valley vineyards looking west from Viña El Condor. Hills in the background are Jurassic-Cretaceous volcano-sedimentary units.

Terroir

Steep north and west-facing slopes of the coastal mountains allow vineyards to take advantage of the sunlight, and the well-drained granitic soils tend to stress the vines, which leads to lower grape yields with highly concentrated flavor. Known as "granitic soils," they are derived from the Coastal Cordillera, are light brown, have a sandy, granular texture, have high permeability, and low fertility (few nutrients).

The hillsides tend to be slightly cooler than the valley floor and often have a greater daily temperature variation. This is said to develop grapes with an enhanced aromatic profile and an excellent balance of ripeness and acidity. Whereas the hillsides have granitic soils, in the valley bottoms the Tinguiririca River deposits alluvial silts and clays that are also ideal soils. Known as "Stoney Alluvial Soils," they are sandy clay loam of moderately fine texture. These dark-gray soils are composed of sediment carried by rivers and contain rock fragments as well as organic matter. They are generally deep soils of medium to high permeability and medium to high fertility.

The warmer eastern Colchagua Valley and Central Valley are renowned for their red wines, including Cabernet Sauvignon, Merlot, Carmenère, Malbec, Syrah, Carignan, País, Mourvèdre, Cabernet Franc, and Pinot Noir. In these hills and valleys, the afternoon heat is tempered by cool gusts from the Pacific and warm breezes from the interior mountains, creating dry and pleasant conditions conducive to various red grapes. Red wines from Colchagua Valley are powerful and bold with ripe tannins. The soils, also "granitic," are moderately fine-textured yellow-brown silty clay loam. The soil contains a large component of quartz fragments and grains and has medium to low permeability. Alongside this soil in some of the valley bottoms is a "sand-silt-ash soil" derived from the erosion of Jurassic and Cretaceous volcanic sandstones with small amounts of silt and clay.

Red wines of the warmer and drier east are replaced by white-wine grapes in the ocean-cooled western parts of the Colchagua Valley. The Costa, or "coastal" zone includes the up-and-coming Paredones subzone that grows cool-climate grapes, mainly Chardonnay and Sauvignon Blanc. Paredones vineyards have a mix of clay and quartz called *batolitocostero* (Coastal Batholith). Sauvignon Blanc has been the most-grown grape in Paredones. The area usually sees only five or six hours of sunlight each day due to morning and evening fog. It is said that the "minerality" in the Paredones wines is a result of salt in the sea air rather than the soil type (WineSearcher, 2022; WineTourism, 2022; Schachner, 2022).

Stop 27.1 Viña Viu Manent

We just happened into this amazing vineyard as we were driving by. You can reserve a wine tour, or just pop in for a tasting. We dropped in for a tasting and sampled seven wines for 15,000 p (about $19 US).

Viña Viu Manent was founded as Bodegas Viu in Santiago in 1935 by the Catalan immigrant Miguel Viu García, and his sons Agustín and Miguel Viu Manent (Wikipedia, Viña Viu Manent; WineTourism.com, Viña Viu Manent). In 1966, Miguel acquired the Hacienda San Carlos de Cunaco in the Colchagua Valley. The estate had 150 ha (370 ac) of vines on three plots of land planted with pre-phylloxera vines, as well as a winery and manor house (WineTourism.com, Viña Viu Manent). In 1993, Viu Manent produced the first Malbec made from 100% Chilean grapes. The vineyard teases out the nuances of the region's terroir by experimenting with single vineyard Carménère and working towards a Chardonnay made from purely Colchagua grapes (a small percentage is still sourced in the nearby Casablanca Valley; World's Best Vineyards, 2021).

The estate has been involved in the wine tourism scene since it opened to the public in 1995. Since 2000, the winery has been run by Jose Miguel Viu, grandson of the founder (WineTourism.com, Viña Viu Manent).

As of 2017, the company had 300 ha (740 ac) under cultivation and projected sales of 250,000 cases per year in 50 countries. The main grape varieties include Malbec, Syrah, Carmenère, Cabernet Sauvignon, and Merlot, as well as Chardonnay, Sauvignon Blanc, and Viognier.

Awards include *Vineyard of the Year*, in 2017, according to the Chilean Wines Association, and *Best Wine Tourism Center for Visitors*, in 2018, according to Drinks International. It was among the 60 best wineries in 2008, 2009, 2010, and 2011, according to the World Association of Journalists and Writers of Wines and Spirits (Wikipedia, Viña Viu Manent).

Visit

In addition to a café, the Rayuela Wine and Grill serves meat and fish prepared in a mud oven alongside refreshing salads, and for something extra special, there are pairing menus and master classes by celebrity chef Pilar Rodríguez. The 75-minute tour ambles through the Variety Garden before taking you through the vineyards in antique horse-drawn carriages and ends with a tasting session (World's Best Vineyards, 2021). There is a small wine museum, and bike tours are available. It is also a hotel.

Carretera Del Vino Km 39, Santa Cruz, Colchagua, Vi Región, 3130000, Chile.

Tourism Phone:	+56 22 8403180 or +56 22 8403181.
Winery Phone:	+56 72 2858350 or +56 72 2858751.
Website:	https://viumanent.cl/en/.

The author examines the *terroir* of wine, Viu Manent vineyard.

Viña Viu Manent to the Colchagua Museum: *Return to R-90 and turn left (northwest); continue on R-90; turn left (west) onto Cruce Ruta I-50 (Paniahue) – Santa Cruz – Bucale; continue straight on Av. Errázuriz into the town of Santa Cruz.* **Stop 27.2, the Colchagua Museum**, *Av. Errázuriz 145, Santa Cruz (–34.640382, –71.363955) parking is at the curve on the right for a total of 7.9 km (4.9 mi; 13 min).*

Stop 27.2 The Colchagua Museum

What better way to end this tour of the Andes and wine regions than to visit one of the best museums, in my opinion, in this part of the world? This private museum contains an eclectic series of collections that are as comprehensive and impressive as they are diverse. The Colchagua Museum addresses the history of the planet as well as that of humans in the Americas. Founded by businessman Carlos Cardoen, and run by his foundation, Cardoen has collected objects in the fields of paleontology, archaeology, and pre-Colombian art from Chile and throughout the Americas, as well as materials from the Spanish conquest, colonial times, and the Republican era in Chile.

The Colchagua Museum opened in 1995 and was renovated after the 2010 earthquake destroyed many of the displays. Behind its colonial façade, the museum contains over 7,000 pieces categorized by theme. There are rooms and displays dedicated to topics as diverse as the history and development of arms, traditional huaso (cowboy) gear, coaches and carriages, agricultural machinery, and railroads. There is also the Great Rescue Room, dedicated to the 2010 rescue of 33 miners trapped for two months in the San José mine in northern Chile, an event that captivated the world (Colchagua Wine Tours, Colchagua Museum). The museum has been described as the largest and most varied private collection in Latin America (Chile: Wine Route – Colchagua Museum).

Collections of interest to geologists include the Paleontology exhibit, which contains fossils that go back 500 million years and cover the origin and evolution of life. It includes collections of echinoderms, ammonites, trilobites, mollusks, fish, plants, insects, and vertebrates, among others. The mineral exhibit includes a magnificent collection of amber, native metals, copper minerals, and Chilean minerals in general. The Charles Darwin room is dedicated to the British naturalist who traveled across Chile between 1832 and 1835, making notes on the geology, botany, zoology, and anthropology of the region (Chile Travel & News, Museo de Colchagua). Since you are here, this museum is well worth spending a couple of hours in.

Visit

Hours: Winter: Monday to Sunday, 10 am to 6 pm.

Summer: Monday to Sunday, 10 am to 7 pm.

Address: Av. Presidente Errázuriz 145Santa Cruz, Valle de Colchagua, Chile.

Landline: 56 72 282 1050.

Entry: 8,000 p (about $10) adult; 4,000 p child.

Website: http://www.museocolchagua.cl/english/index.html.

The colonial façade of the Colchagua Museum.

Native copper. Part of the minerals display, Colchagua Museum.

Ammonite display, Colchagua Museum.

Colchagua Museum display showing the heroic rescue of Chilean miners in 2010. The 33 miners were rescued after spending 69 days almost a kilometer underground in the copper and gold mine.

Our geo-tour has taken us from the wine-growing region of Mendoza, Argentina, to the wine-growing region of Colchagua, Chile. In between we have seen the fold-thrust belt of the Andes, the Central Valley of Chile, and the Coastal Cordillera. Crossing the Andes in the footsteps of Darwin, we have seen petrified forests, the tallest mountain in the Americas, Alpine lakes, slot canyons, hot springs, gypsum mines, vineyards, and some of the great cities of Latin America: Mendoza, Santiago, and Valparaíso. This is where culture, history, and science converge. We have seen a remarkable blend of geology, geography, and human history that defines the beauty and diversity of this corner of the world.

References

Anselmo, A., D. Morata, M. Reich, and L. Daniele. 2017. Travertine genesis at Baños Morales and Baños Colina hot springs, Central Andes, Chile (33°23'-33°52'S). IGCP636 Annual Meeting 2017 – Santiago de Chile, 3 p.

Aranguiz, T., K. Kotthoff, and F. Olivares. 2017. Tectonic evolution and geological history of the Colina and Volcan River Valleys in the high Andes of Santiago, Chile. Geological Society of America 2017 Annual Meeting, Seattle. Poster, Accessed 14 May 2023, https://gsa.confex.com/gsa/2017AM/webprogram/Handout/Paper308051/POSTERFINAL.pdf.

Bowes, W.A., P.H. Knowles, A. Moraga B., E. Klohn H., and M. Serrano C. 1962. Reconnaissance for Uranium in the Colimapu formation of Central Chile. Instituto de Investigaciones Geologicas RME-4571 (Rev.), Santiago, 16 p.

Boyce, D., R. Charrier, and M. Farias. 2020. The first Andean compressive tectonic phase: Sedimentologic and structural analysis of mid-Cretaceous deposits in the Coastal Cordillera, Central Chile (32°50'S). *Tectonics* v. 39, 24 p. e2019TC005825, DOI: 10.1029/2019TC005825.

Brea, M., A.E. Artabe, and L.A. Spalletti. 2009. Darwin Forest at Agua de la Zorra: The first *in situ* forest discovered in South America by Darwin in 1835. *Revista de la Asociación Geológica Argentina* v. 64 no. 1, p. 21–31.

Carvajal, M., M. Cisternas, and P.A. Catalán. 2017. Source of the 1730 Chilean earthquake from historical records: Implications for the future tsunami hazard on the coast of Metropolitan Chile. *Journal Geophysical Research Solid Earth* v. 122, p. 3648–3660, DOI: 10.1002/2017JB014063.

Chancellor, G., and J. van Wyhe. Introduction to Darwin's St. Fe notebook in the complete work of Charles Darwin online. Accessed 6 August 2023, http://darwin-online.org.uk/.

Charrier, R., O. Baeza, S. Elgueta, J.J. Flynn, P. Gans, S.M. Kay, N. Muñoz, A.R. Wyss, and E. Zurita. 2002. Evidence for cenozoic extensional basin development and tectonic inversion south of the flat-slab segment, southern Central Andes, Chile (33º–36ºS.L.). *Journal of South American Earth Sciences*, 23 p.

Charrier, R., L. Pinto, and M. Rodriguez. 2007. Tectonostratigraphic evolution of the Andean Orogen in Chile. Chapter 3, Geological Society London Special Publications, p. 21–114.

Charrier, R., V.A. Ramos, F. Tapia, and L. Sagripanti. 2014. Tectono-stratigraphic evolution of the Andean Orogen between 31 and 37ºS (Chile and Western Argentina). *In* Sepúlveda, S.A., L.B. Giambiagi, S.M. Moreiras, L. Pinto, M. Tunik, G.D. Hoke, and M. Farías (Eds.), *Geodynamic Processes in the Andes of Central Chile and Argentina*. Geological Society, London, Special Publications, v. 399, 51 p., DOI: 10.1144/SP399.20.

Chile Travel & News. Museo de Colchagua. Accessed 15 May 2023, https://www.chile-travel-and-news.com/2015/07/museo-de-colchagua.html.

Chile: Wine Route – Colchagua Museum. Accessed 15 May 2023, https://www.ladatco.com/CHI-WR-Colchagua%20Museum.htm.

Colchagua Wine Tours. Museums. Accessed 15 May 2023, https://www.colchaguawinetours.com/museums/.

Concha y Toro. Our history. Accessed 15 May 2023, https://conchaytoro.com/en/about-us/our-history/#:~:text=Concha%20y%20Toro's%20history%20begins,wine%20brand%20in%20the%20world.

Course Hero. The voyage of the beagle. Accessed 5 July 2023, https://www.coursehero.com/lit/The-Voyage-of-the-Beagle/chapter-15-summary/.

Dare2Go. La Campana National Park. Accessed 15 May 2023, https://dare2go.com/practical-information-la-campana-national-park/.

Darwin, C. 1845. *Journal of Researches into the Natural History and Geology of Countries Visited during the Voyage of HMS Beagle Round the World, under the Command of Capt. FitzRoy, R.N.* Second Edition, corrected, with additions. John Murray, London, 519 p.

Darwin, C. 1846. *Geological Observations on South America. Being the Third Part of the Geology of the Voyage of the Beagle, during the Years 1832 to 1836.* Smith, Elder and Co., London, 280 p.

Deckart, K., F. Hervé, C.M. Fanning, V. Ramírez, M. Calderón, and E. Godoy. 2014. U-Pb geochronology and Hf-O isotopes of zircons from the Pennsylvanian Coastal Batholith, South-Central Chile. *Andean Geology* v. 41 no. 1, p. 49–82.

Dunnell, T. 2017. A guide to the main wine regions near Santiago, Chile. Accessed 16 August 2020, https://www.savacations.com/guide-main-wine-regions-near-santiago-chile/.

Farias, M. 2007. Tectonique, Erosion et Evolution du Relief dans les Andes du Chili Central au Cours du Neogene. PhD dissertation, University Toulouse III, TC4001, 191 p.

Gana, P., R. Wall, and A. Gutiérrez. 1996. Mapa Geologico del Area de Valparaiso – Curacavi. Mapas Geologicas no. 1, Servicio Nacional de Chile, Geologia y MIneria, Subdireccion Nacional de Geologia, 1:100,000.

Giambiagi, L.B., P.P. Alvarez, E. Godoy, and V.A. Ramos. 2003. The control of pre-existing extensional structures on the evolution of the southern sector of the Aconcagua fold and thrust belt, southern Andes. *Teconophysics* v. 369, p. 1–19.

Giambiagi, L.B., M. Tunik, V.A. Ramos, and E. Godoy. 2009. The high Andean Cordillera of Central Argentina and Chile along the Piuquenes Pass-Cordon del Portillo Transect: Darwin's pioneering observations compared with modern geology. *Revista de la Asociación Geológica Argentina* v. 64 no. 1, p. 43–54.

Giambiagi, L.B., J. Mescua, N. Heredia, P. Farías, J. García Sansegundo, C. Fernández, S. Stier, D. Pérez, F. Bechis, S.M. Moreiras, and A. Lossada. 2014. Reactivation of Paleozoic structures during Cenozoic deformation in the Cordón del Plata and Southern Precordillera ranges (Mendoza, Argentina). *Journal of Iberian Geology* v. 40 no. 2, p. 309–320.

Giambiagi, L.B., S. Moreiras, and M. Strecker. 2017. The Fold-and-Thrust belts of the Southern Central Andes: A field excursion to the Mendoza & San Juan Provinces in NW Argentina. International Research Training Group – IGK 2018, Consejo Nacional de Investigaciones Cientificas y Tecnicas, Buenos Aires. 39 p. plus maps.

Giambiagi, L., A. Tassarab, A. Echaurren, J. Julveb, R. Quiroga, M. Barrionuevo, S. Liu, I. Echeverría, D. Mardónez, J. Suriano, J. Mescua, A.C. Lossada, S. Spagnotto, M. Bertoa, and L. Lothari. 2023. Crustal anatomy and evolution of a subduction-related orogenic system: Insights from the Southern Central Andes (22–35°S). Earth-Science Reviews, 44 p. Accessed 27 June 2023, https://www.sciencedirect.com/science/article/abs/pii/S0012825222002227?via%3Dihub.

Great Wine Capitals. 2023. Valparaiso – Casablanca Valley. Accessed 15 May 2023, https://www.greatwinecapitals.com/capitals/valparaiso-casablanca-valley-chile/.

Heredia, N., P. Farias, J. Garcia-Sansegundo, and L. Giambiagi. 2012. The basement of the Andean Frontal Cordillera in the Cordón del Plata (Mendoza, Argentina): Geodynamic Evolution. *Andean Geology* v. 39 no. 2, p. 242–257.

Hermanns, R.L., L. Fauque, and C.G.J. Wilson. 2014. ^{36}Cl terrestrial cosmogenic nuclide dating suggests Late Pleistocene to Early Holocene mass movements on the south face of Aconcagua mountain and in the Las Cuevas–Horcones valleys, Central Andes, Argentina. *In* Sepúlveda, S.A., L.B. Giambiagi, S.M. Moreiras, L. Pinto, M. Tunik, G.D. Hoke, and M. Farías (Eds.), *Geodynamic Processes in the Andes of Central Chile and Argentina.* Geological Society, London, Special Publications v. 399, p. 345–368.

KLS Geological Tours. 2019. Maipo Valley, Santiago, Chile. Accessed 12 October 2019, https://klsgeo .com/SC44.htm.

London Wine Competition. 2023. Casablanca Valley. Accessed 15 May 2023, https://londonwinec ompetition.com/en/resources/regions-3375/chile-3398/casablanca-valley-87.htm.

Lopez-Escobar, L., H. Moreno, M. Tagiri, K. Notsu, and N. Onuma. 1985. Geochemistry and petrology of lavas from San Jose volcano, Southern Andes (33°45'S). *Geochemical Journal* v. 19, p. 209–222.

Mackaman-Lofland, C., B.K. Horton, F. Fuentes, K.N.Constenius, and D.F. Stockli. 2018. Mesozoic to Cenozoic retroarc basin evolution during changes in tectonic regime, southern Central Andes (31–33°S): Insights from zircon U-Pb geochronology. *Journal of South American Earth Sciences.* Accessed 15 May 2023, https://www.sciencedirect.com/science/article/abs/pii/ S0895981118301251.

Martos, F.E., M. Naipauer, L.M. Fennell , E. Acevedo , N. Hauser, and A. Folguera. 2022. Neogene evolution of the Aconcagua fold-and-thrust belt: Linking structural, sedimentary analyses and provenance U-Pb detrital zircon data for the Penitentes basin. *Tectonophysics* v. 825, 9 p.

Moreiras, S. 2010. Geomorphological Evolution of the Mendoza River Valley. *In* del Papa, C. and R. Astini (Eds.), Field Excursion Guidebook, 18th International Sedimentological Congress, Mendoza, Argentina, FE-B2, p. 1–21.

Moreiras, S.M., and S.A. Sepúlveda. 2015. Megalandslides in the Andes of central Chile and Argentina (32°–34°S) and potential hazards. *In* Sepúlveda, S.A., L.B. Giambiagi, S.M. Moreiras, L. Pinto, M. Tunik, G.D. Hoke, and M. Farías (Eds.), *Geodynamic Processes in the Andes of Central Chile and Argentina*. Geological Society, London, Special Publications, v. 399, p. 329–344.

Mpodozis, C., and P. Conejo. 2012. Chapter 14, Cenozoic tectonics and porphyry copper systems of the Chilean Andes. Society of Economic Geologists Special Publication 16, p. 329–360.

Muñoz, M., F. Fuentes, M. Vergara, L. Aguirre, J.O. Nyström, G. Féraud, and A. Demant. 2006. Abanico East Formation: Petrology and geochemistry of volcanic rocks behind the Cenozoic arc front in the Andean Cordillera, central Chile (33°50'S). *Revista Geologica de Chile*, January 2006, 31 p.

Muñoz, M., F. Tapia, M. Persico, M. Benoit, R. Charrier, M. Farías, and A. Rojas. 2018. Extensional tectonics during late Cretaceous evolution of the Southern Central Andes: Evidence from the Chilean main range at ~35°S. *Tectonophysics* v. 744, p. 93–117.

PeakVisor.com. Campana National Park. Accessed 21 January 2023, https://peakvisor.com/park/ campana-national-park.html.

Ramos, V.A. 2008. Field trip guide: Evolution of the Pampean flat-slab region over the shallowly sub-ducting Nazca plate. *In* Kay, S.M., and V.A. Ramos (Eds.), *Field Trip Guides to the Backbone of the Americas in the Southern and Central Andes: Ridge Collision, Shallow Subduction, and Plateau Uplift*. Geological Society of America Field Guide 13, p. 77–116.

Ramos, V.A., and S.M. Kay. 1991. Triassic rifting and associated basalts in the Cuyo Basin, central Argentina. *In* Harmon, R.S., and C.W. Rapela (Eds.), *Andean Magmatism and its Tectonic Setting*. Geological Society of America Special Paper 265, Boulder, p. 79–91.

Ramos, V.A., R. Zapata, E. Cristallini, and A. Introcaso. 2004. The Andean thrust system— Latitudinal variations in structural styles and orogenic shortening. *In* McClay, K.R. (ed.), *Thrust Tectonics and Hydrocarbon Systems: AAPG Memoir* v. 82, p. 30–50.

Rivano, S., P. Sepulveda, R. Boric, and D. Espiñeira. 1993. Hojas Quillota y Portillo, Carta Geologica de Chile No. 73, Servicio Nacional de Geologia y Mineria, Santiago. 1:250,000.

Rossel, P., V. Oliveros, J. Mescua, F. Tapia, M.N. Ducea, S. Calderón, R. Charrier, and D. Hoffman. 2014. The Upper Jurassic volcanism of the Río Damas-Tordillo Formation (33°–35.5°S): Insights on petrogenesis, chronology, provenance and tectonic implications. *Andean Geology* v. 41 no. 3, p. 529–557.

Rundel, P.W., and P.J. Weisser. 1975. La Campana, a new National Park in Central Chile. *Biological Conservation* v. 8 (July), p. 35–46.

Salazar, C., and W. Stinnesbeck. 2015. Redefinition, stratigraphy, and facies of the Lo Valdés Formation (Upper Jurassic-Lower Cretaceous) in Central Chile. *Boletín del Museo Nacional de Historia Natural, Chile* v. 64, p. 41–68.

Schachner, M. 2022. Chile's Colchagua Valley combines deep-rooted history with Viticultural variety. *Wine Enthusiast Magazine.* Accessed 17 May 2023, https://www.winemag.com/2022/04/13/colchagua-valley-chile-wine-guide/.

Seia, M.G., P. Jara, M. Bertoa del Llano, A. Richard, L. Lothari, and L.B. Giambiagi. 2023. Pre-Andean deformation and its influence on the shortening ofthe Southern Precordillera, Mendoza, Argentina. *Journal of South American Earth Sciences* v. 126, 8 p.

Sempere, T., L.G. Marshall, S. Rivano, and E. Godoy. 1994. Late Oligocene-Early Miocene compressional tectosedimentary episode and associated land-mammal faunas in the Andes of central Chile and adjacent Argentina (32–37's). *Tectonophysics* v. 229, p. 251–264.

Sernageomin.cl. Online geologic map of Chile, 1:1,000,000. https://portalgeomin.sernageomin.cl/.

Thiel, R. 1980. Hoja Santiago, Region Metropolitana, Carta Geologica de Chile no. 39, Instituto de Investigaciones Geologicas, Santiago, 1:250,000.

Van Wyhe, J. 2010. What 1835 Chile quake taught Darwin. CNN, March 1, 2010. Accessed 17 May 2023, www.cnn.com/2010/OPINION/03/01/vanwyhe.quake.chile.darwin/index.html#.

Wikipedia. Coastal Batholith of Central Chile. Accessed 17 May 2023, https://en.wikipedia.org/wiki/Coastal_Batholith_of_central_Chile.

Wikipedia. Concha y Toro. Accessed 17 May 2023, https://en.wikipedia.org/wiki/Concha_y_Toro.

Wikipedia. La Campana National Park. Accessed 17 May 2023, https://en.wikipedia.org/wiki/La_Campana_National_Park.

Wikipedia. San Cristóbal Hill. Accessed 17 May 2023, https://en.wikipedia.org/wiki/San_Crist%C3%B3bal_Hill.

Wikipedia. Valparaiso. Accessed 17 May 2023, https://en.wikipedia.org/wiki/Valpara%C3%ADso.

Wikipedia. Viña Viu Manent. Accessed 17 May 2023, https://en.wikipedia.org/wiki/Vi%C3%B1a_Viu_Manent.

Winesearcher. 2018. Casablanca Valley Wine. Accessed 17 May 2023, https://www.wine-searcher.com/regions-casablanca+valley.

Winesearcher. 2022. Colchagua Valley Wine. Accessed 17 May 2023, https://www.wine-searcher.com/regions-colchagua+valley.

WineTourism. 2022. Colchagua Valley Wine Region. Accessed 17 May 2023, https://www.winetourism.com/wine-appellation/colchagua-valley/.

WineTourism.com. Viña Viu Manent. Accessed 17 May 2023, https://www.winetourism.com/winery/vina-viu-manent/.

World's Best Vineyards. 2021. Viu Manent. Accessed 17 May 2023, https://www.worldsbestvineyards.com/the-list/viu-manent.html.

Yáñez, G., M. Muñoz, V. Flores-Aqueveque, and A. Bosch. 2015. Gravity derived depth to basement in Santiago Basin, Chile: Implications for its geological evolution, hydrogeology, low enthalpy geothermal, soil characterization and geo-hazards. *Andean Geology* v. 42 no. 2, p. 147–172.

Index

For Product Safety Concerns and Information please contact our EU
representative GPSR@taylorandfrancis.com
Taylor & Francis Verlag GmbH, Kaufingerstraße 24, 80331 München, Germany